Chemical Risk Assessment

Chemical Risk Assessment: A Manual for REACH

Peter Fisk Associates Ltd

WILEY

Library of Congress Cataloging-in-Publication Data

Chemical risk assessment : a manual for REACH / Peter Fisk Associates Ltd.
    pages cm
  Includes index.
  ISBN 978-1-119-95368-5 (cloth)
  1. Chemicals–Law and legislation–European Union countries. 2. Chemicals–Safety regulations–European Union countries.
  3. Hazardous substances–Law and legislation–European Union countries. I. Peter Fisk Associates.
  KJE6011.C44 2014
  363.17′91 – dc23

                                                                                              2013028620

A catalogue record for this book is available from the British Library.

ISBN: 9781119953685

Typeset in 10/12 Times by Laserwords Private Limited, Chennai, India.
Printed and bound in Malaysia by Vivar Printing Sdn Bhd

1 2014

Contents

List of Figures

List of Tables

List of Contributors

Peter Fisk Associates includes the following staff members who have contributed to this book:

Peter Fisk – Managing Director

Oliver Warwick – Project Manager

Louise McLaughlin – Director

Rosalind Wildey – Director

Principal Consultants:

Helen Disley

Stephen Summerfield

Andrew Girling

Laura Robinson

Helen Barnes

Senior Consultants:

David Akinosho

Lucy Wilmot

Ola Dosunmu

Elina Kansikas

Sam Fisk

Consultants:

Gillian Federici

Additional Contributors:

Michel De Poortere by kind permission of Siletz sprl

Rohit Mistry by kind permission of eftec

Preface

How to Use This Book

This book about REACH is a handbook, providing practical advice aimed at the level of consideration of strategy, technical insights, and commercial realities. It does not intend to reproduce or summarise the official detailed guidance – of which there are many thousands of pages! Although that level of detail and amount of guidance is necessary, the abundance of guidance can in itself create problems. It is not possible for a single individual to absorb and understand such a huge amount of information, especially considering the diversity of issues concerning chemical properties, hazards, and risks. This book can help you understand REACH, whatever your responsibilities or interest in it.

What this handbook does do is provide a broad overview, but one that includes the details that the authors have found necessary in their experience in practice of all aspects of the REACH Regulation. The purpose of the book is to explain how REACH works, with an emphasis on an overall understanding and explanation of responsibilities, in order to help you to set strategy and priorities.

What is REACH Trying to Achieve?

REACH has many objectives, contained within what is one of the most comprehensive pieces of legislation ever enacted in the European Union. Its major objectives include:

- bringing all REACH-relevant substances, regardless of whether they have been previously assessed or not, into a common registration and evaluation process;
- moving the burden of responsibility more clearly into the hands of industry;
- placing priority of effort onto substances of very high concern;
- improving communication in the supply chain;
- encouraging the sharing of data between data owners;
- widening the availability of data on substances;
- removing unacceptably hazardous substances from the market.

How Can This Book Help You Understand REACH and Comply With It?

This book will help you to understand your part in REACH:

- **Policy maker** – business strategy in the chemicals industry in the European Union requires an awareness of the costs and impacts of REACH. This involves understanding the current and potential status of substances under REACH, and also how other related legislation may be affected.

- **Business manager** – if you are developing a substance or group of substances you will have to understand REACH, to ensure that appropriate and timely actions on substance are taken.
- **Scientist in R&D** – it is not possible to maintain the view that R&D can be seen as separate from acceptability in commercial and regulatory terms – you need to be informed about REACH and its impacts on substances with potentially hazardous properties.
- **Regulator** – if you are not working on REACH, then this book can help you understand how your work interfaces with REACH.
- **Specialist or Consultant** – you need to know where your speciality fits into the bigger picture of REACH.

How This Book is Set Out

Chapter Summary

Chapter	Title	Summary of contents
1	Introduction – policy and scientific context of chemicals risk and risk management.	An overview of REACH and other regulations, describing the purposes of hazard and risk assessment.
2	Roles and responsibilities in REACH.	What the supply chain and the various regulatory bodies must do.
3	Control of chemicals – legislative and policy context.	What REACH requires and where it fits in a global context.
4	Identification of substances for REACH – practicalities.	An introduction to a key step in establishing what needs to be done in REACH compliance.
5	Physico-chemical properties for REACH – purpose and practicalities.	These properties need to be understood before moving on to toxicology and environmental effects.
6	Assessing and documenting the intrinsic properties of substances in REACH.	Overview of REACH requirements in respect of the basics of how hazards and risks are assessed.
7	Overview of the assessment of the risks to the environment from chemicals.	Detail of REACH requirements in respect of the basics of how environmental hazards are assessed.
8	Environmental exposure.	Detail of REACH requirements in respect of the basics of how environmental exposures and risks are assessed.

Chapter	Title	Summary of contents
9	Assessing the hazards to human health from chemicals.	Detail of REACH requirements in respect of the basics of how hazards are assessed.
10	Human exposure to chemicals.	Detail of REACH requirements in respect of the basics of how environmental exposures and risks are assessed.
11	Managing hazard and risk.	How to deal with management of hazard in respect of classification and labelling, and controlling exposure to substances to the level necessary to establish safety.
12	Avoiding the use of hazardous substances: substitution and alternatives.	Overview of how to find viable alternatives to hazardous substances, in terms of new substances or technologies.
13	Hazards, risks and impacts – the development and application of frameworks for the assessment of risk.	Risk management options: how can risk be managed, appraisal of methods to control risk and the time and costs needed for implementation of control measures.
14	Socio-economic analysis in REACH.	The need for assessing the costs and benefits of control measures using socio-economic analysis as a tool in the context of the authorisation, restriction and risk management option appraisal processes of REACH.
15	REACH: how it is working and may develop.	What has worked well, what is failing. Future developments – neither science nor regulations are static.
16	Resources, official guidance, further reading and centres of expertise.	Reference sources for further information focussing on the technical resources.
Appendix A	Substance Classification and labelling under REACH.	The new CPL Regulations.
Appendix B	Further discussion of substance identification and sameness.	The importance of substance identification as a first technical step.
Appendix C	Tools for REACH compliance: IUCLID, Chesar and in-house databases.	Some practical advice for registrants.
Appendix D	Glossary	–

How Different Types of Reader Might Use This Book

The contents cover a wide range of needs. The following are suggestions for how to make the most of the content.

	Chapters
Strategy for REACH	1, 2, 3, 4, 11, 13–15
Substance assessment and registration	1, 4–10
Authorisation concerns	1, 13, 14
Substance substitution	12, 13

1

Introduction: Policy and Scientific Context of Chemicals Risk and Risk Management

This chapter acts as a foundation of understanding for the rest of the book. It introduces the regulatory systems that demand the evaluation of risk for chemical substances that are intended to be used and placed on the market. It sets out the development of risk assessment in the European, global and national contexts. This chapter also explains the key concepts of hazard and exposure. Hazard is defined as the inherent properties of a substance that may make it harmful – flammability, toxicity and so on. Exposure refers to the ways in which humans and the environment come into contact with substances. The reasons for bringing together hazard and exposure in order to understand risk are explained.

The focus of this book is the REACH Regulation (most often referred to just as 'REACH'), as this is the main regulatory driver for the risk assessment of chemical substances in the European Union. REACH (**R**egistration, **E**valuation, **A**uthorisation & restriction of **Ch**emicals), however, should be viewed in the context of other legislation that is either directly or indirectly connected to the REACH Regulation. With this in mind, the later sections of this chapter include consideration of United Kingdom legislation on chemicals, including worker and environmental protection. These sections are intended to serve as examples of how REACH is connected to prior legislation and how compliance with REACH works with such legislative regimes at national level.

The purpose of this book is to set out in a simple and concise way how to assess and manage the risks of chemicals to humans and the environment. This is done within the context of the main legislation that applies to the safety of the manufacture and use of chemicals in the European Union (EU) – the REACH Regulation. It is not the intention to give detailed guidance on each aspect of risk assessment or in depth assessment of specific aspects of REACH, but rather to explain the main aspects of chemical risk assessment and the processes that are applied, so that each aspect can be understood

Chemical Risk Assessment: A Manual for REACH, First Edition. Peter Fisk Associates Ltd.
© 2014 John Wiley & Sons, Ltd. Published 2014 by John Wiley & Sons, Ltd.

within the context of REACH. This book should act as a handbook, so the reader/user can find out about specific aspects of the process and technical elements in sufficient detail to understand where and how they fit in the risk assessment of chemicals, and where to look for more detailed information.

Legislation on chemicals has specific purposes and is aimed at control of particular processes or aspects of the manufacture, use, reuse and disposal of chemicals. In addition, some legislation is aimed at chemicals that are used in a particular way (for example pharmaceuticals or pesticides), or because they have specific dangerous qualities (carcinogens, explosives and highly flammable substances), and some legislation is aimed at protection of specific sections of the population (e.g. workers, consumers, pregnant workers). Other legislation is aimed at environmental protection by specific control of releases to the environment (e.g. integrated pollution potential and control – IPPC) or monitoring specific parts of the environment (e.g. the Water Framework Directive – WFD for water). Inevitably there is overlap between all this legislation on chemicals, and today companies manufacturing and using chemicals have to be aware of a wide range of legislation to ensure that they are complying with all the relevant laws to operate legally and safely.

REACH is concerned with the safety of chemical substances for placement on the European market. REACH is a 'Regulation' (as compared to a Directive[1]) meaning that REACH is a law that applies equally and with the same text in all EU Member States. REACH requires that industry supplies specific information on individual chemicals in order to demonstrate that its manufacture and specific uses in the EU market are safe[2] for humans and the environment.

Key features of REACH are:

- Those who make chemicals in the EU or import them into the EU (Manufacturers/Importers, i.e. 'industry') for placing on the EU and EEA (European Economic Area) market[3] are responsible for supplying information to demonstrate safe manufacture and use.
- The safety assessments done by industry are based on a risk assessment that examines the properties of the substance that may make it dangerous to humans and/or the environment, and the way the chemical is used that causes humans and/or the environment to come into contact with it.
- Information supplied is assessed by a central regulator: the European Chemicals Agency (ECHA).
- Each substance is assessed for safety on its own merit: that is, the potential risks or impact of each chemical and its uses are assessed on their own and not in combination with other substances.[4]
- The safety of chemicals is assessed only for the uses that the manufacturer puts forward; thus the assessment is valid for these uses only.
- The safety of all parts of the chemical's life cycle are relevant – from manufacture to final disposal (including recycling/reuse, if relevant).

---

[1] A Directive is applied (transcribed) by each Member State of the Union in its own law.

[2] What constitutes and is designated as 'safe' with the context of REACH is addressed in later chapters of this book.

[3] There are specific rules for substances that are used for the purposes of research – these are identified later (Section 4.1 and Appendix C).

[4] However, the breakdown products of the substance are relevant to the risk assessment.

- REACH is applied to the manufacture/import and use of chemicals substances, not chemicals that are used specifically as pharmaceuticals, biocides, plant protection products (pesticides), veterinary medicines and cosmetics. However, it does apply to the chemicals that are used to make these products.

The concepts that underpin the REACH Regulation are not new; what is new is the application of a single system for assessing the safety of chemicals being placed on the European market. To understand why REACH was created as a Regulation it is necessary to briefly look at what was in place prior to REACH coming into force in 2007.

The pre-REACH legislative framework comprised three main pieces of legislation, namely:

- Existing Substances Regulation (ESR).
- Dangerous Substances Directive (DSD) Seventh Amendment – concerning the placing on the market of 'new' substances (in the UK this was the Notification of New Substances Regulation or NONS). The legal basis was laid out in Directive 67/548/EEC.
- Marketing and Use (or 'Limitations') Directive.

In addition, the DSD set out the rules for the classification and labelling of substances. This is of key importance for hazard communication and also because the classification of substances leads to how the substance is dealt with in other legislation (Appendix A). This includes, importantly, how the substance should be handled and treated by users of the substance.[5]

Under this legislation prior to REACH, substances defined as 'new' (i.e. placed onto the European market after 1981) were required to be tested and notified before marketing in volumes above 10 kg. For higher volumes more in-depth testing – focused on long-term and chronic effects – had to be provided. On the basis of that information, the substances were assessed for their risks to human health and the environment. There were, however, no corresponding requirements for 'existing chemicals': chemicals that were on the European Community market between 1 January 1971 and 18 September 1981. These 'existing chemicals' were listed in the EINECS (European INventory of Existing Commercial chemical Substances), which consists of about 100 000 existing substances. This accounts for about 99% of the total volume of chemicals on the European market.

Risk assessment of new substances coming onto the market under pre-REACH legislation formed the basis of REACH and is the core of the registration of chemicals within REACH. For a new chemical to be placed on the EU market, the manufacturer had to chemically describe the substance and provide basic information on its properties in terms of hazard and use, and assess the potential risks to humans and the environment from the manufacture and use of the substance. The amount of information to be provided depended upon on the amount to be placed on the market. A manufacturer could present its information dossier to the 'Competent Authorities' of any of the Member States, who would assess the information and present a risk assessment of the substance and its uses for acceptance by all other Member States. The system was looked after

---

[5] Note that the regime for the classification and labelling of chemical substances is explained in Appendix A.

at EU level by the European Chemicals Bureau (part of the European Commission's Directorate General Joint Research Centre – DG JRC). The assessment could reach one of four possible conclusions:

1. **No immediate concern**: no need to consider again before next tonnage trigger.
2. **Concern**: define further information needs and requests at next tonnage trigger.
3. **Concern**: define further information needs and seek immediately.
4. **Concern**: immediately make recommendations for risk reduction.

For the existing EINECS substances, the risk assessment of these was the responsibility of the regulators at European and Member State level. Substances were placed on priority lists, four of which were established with about 50 substances on each. The prioritisation by the European Community (EC) and Member States was based on hazard, uses and high tonnage use. For this limited subset of substances, each substance was appointed a Member State 'Rapporteur' with the responsibility of conducting the risk assessment, retrieving and assessing all relevant information, and presenting the risk assessment to a Member State and EC expert group with representation from relevant industry sector groups for discussion and agreement. The assessments often required further information from industry (at the industry's expense) but the assessments were done by the Rapporteur and by their own government scientists. The assessments could conclude with one of three possible options for each of the different uses of the chemical and the risks they present to humans working with or using the substances as consumers, and each part ('compartment') of the environment (i.e. freshwater, marine, soil, air, and sewage treatment works). The three available conclusions were:

1. Need for further information and/or testing.
2. At present no need for further information and/or testing and no need for risk reduction measures.
3. Need for limiting the risk.

While the system for assessing new substances was generally regarded to work well and efficiently, the system for existing substances was slow (albeit thorough) and came under increasing criticism. This was both from industry, who wanted to show that their substances were risk assessed as safe, and pressure groups, who wanted to see the existing substances assessed and risky uses banned.

The solution was REACH, in which all substances new and existing are treated the same way, and the burden of information provision is on those who are placing the products on the market. Existing substances are brought into REACH as 'phase-in' substances; the timing for registration of these substances is based on particular hazard and the volume placed on the market. Former 'new' substances are considered to be registered within REACH, but the registrations must be updated before the next tonnage band is reached.

1.1   Overview of the Risk Assessment of Chemical Substances

This section describes the main concepts that underpin risk assessment of chemicals (although these concepts also underpin many other assessments of risk). The assessment of risk is based on the likelihood of something (usually undesirable) happening; this is based on assessing the quality of the 'thing' (in this case a chemical) that might have an effect and the likelihood that the effect will take place. Thus, how and how often the chemical comes into contact with the systems ('receptors') it can impact on (i.e. humans or the environment) forms part of the risk assessment.

The two sides of risk assessment of chemicals are:

1. The inherent properties of a substance that can cause harm (adverse effects).
2. The likelihood of contact with those hazards.

In terms of chemical risk assessment it is useful to think of 1 and 2 above as 'properties' and 'exposure', respectively.

It is the inherent properties of a substance that both define the hazard and influence how it comes into contact with humans and the environment. The likelihood of humans and the environment coming into contact with these hazards is determined by how the substances are used and how much of them are used. In this context, 'inherent properties' are those that cannot be altered since they are a consequence of the molecular structures of the constituents of the substance itself. In risk assessment of chemicals, the main hazards that relate to the ability of substances to poison humans or wildlife are referred to as toxicity. Toxicity can, of course, vary in severity or nature of effect and some substances are very toxic (i.e. small amounts can be very harmful). Other hazards, such as flammability, also vary in severity and the degree of process control needed to make the risk acceptable.

Risk is the likelihood that a hazard will actually cause its adverse effects, together with a measure of the effect. Likelihoods can be expressed as probabilities (e.g. '1 in a 1000'), frequencies (e.g. '1000 cases per year') or in a qualitative way (e.g. 'negligible', 'significant' etc.). The effect can be described in many different ways (HSE, nd) and this depends upon what effect is happening, that is what harm.

The 'risk' in chemical risk assessment, and in particular in REACH, is determined by establishing a national safe level, below which effects will not happen to a particular receptor (e.g. human or part specific part of a human such as skin or particular organ or system or the specific part of the environment), expressed as a concentration. This is then compared to the concentration that the receptor of concern in exposed to. If the exposure concentration is higher than the safe level, then there is a risk and that needs to be controlled to get back to a level (concentration) that is safe.

1.2 Chemical Hazard and Risk Programmes

1.2.1 REACH Overview

REACH brings all EU chemical regulation into a standardised approach, apart from the exemptions. It provides a system of hazard and risk assessment and sets out how these must be communicated. It does not deal with the overlap with other types of regulation but does in reality share many technical objectives with national and EU regulation. It is also at the centre of generic worldwide legislation, affecting decisions from research through to continuing commercial viability (Figure 1.1).

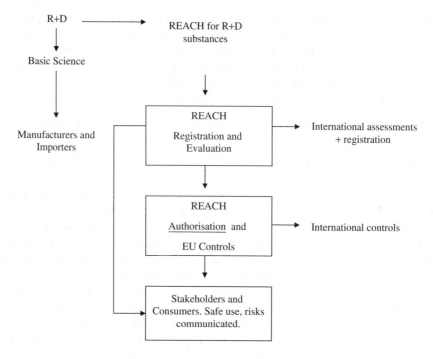

Figure 1.1   *REACH Overview*

1.2.2   Registration

The registration of chemicals under REACH applies to each importer or manufacturer of a substance (either on its own or present in a mixture) that is intentionally released in the EU in quantities ≥ 1 tonne/year.

1.2.2.1   *Registration Strategy: 'Existing' and 'New' Substances*

REACH applies different registration strategies depending on the tonnage level and whether a substance is considered as an 'existing substance' or a 'new substance'. Substances manufactured in or imported into the EU before December 2008, that is 'existing substances', were entitled to be pre-registered as phase-in during the pre-registration period (1 June to 1 December 2008). Substances eligible for 'phase-in' are:

• Those listed in the EINECS.
• Those that have been manufactured in the EU (including accession countries) but have not been placed on the EU market after 1 June 1992.
• Those that qualify as a so-called 'no-longer polymer' (ECHA, 2008).

These pre-registered substances are in the process of being phased-in according to their REACH requirements.

If a phase-in substance has not been pre-registered, it must be registered immediately (applicable from December 2008) in order for manufacturing and import in the EU to be legal, otherwise all activities must cease until registration is complete.

First-time manufacturers or importers[6] of a substance in quantities ≥ 1 tonne/year from December 2007, that is 'new substances', are required to register within six months of trading reaching the one tonne threshold and no later than twelve months before the relevant registration deadline.

New substances registered under the previous chemicals regulation process, that is the NONS, are considered to be registered under REACH. However, should the production volume increase or new information become available, REACH requires that the registration must be updated.

A process exists for substances to be used for R&D purposes, without full registration.

1.2.2.2   *Product and Process Oriented Research and Development (PPORD)*

REACH defines this as:

> *any scientific development related to product development or the further development of a substance, on its own, in preparations or in articles in the course of which pilot plant or production trials are used to develop the production process and/or to test the fields of application of the substance.*

(Article 3 (22))

The PPORD exemption of five years from the obligation of registration is for substances intended to be used for product and process orientated research and development (PPORD) (ECHA, nd-a). ECHA imposes the following constriction on PPORD Substances:

- The substance must be handled in a reasonably controlled conditions for the protection of workers and the environment.
- It is only made available to selected customers.
- The substance will be handled only by staff of a number of listed customers.
- The substance will not be made available to the general public at any time, either in the form of the substance on its own, in a preparation or in an article.
- Remaining quantities of the substance will be re-collected for disposal after the exemption period.

PPORD exemptions for less than 1 tonne/year can be for an indefinite length.

PPORD exemption for more than 1 tonne/year can be exempted for a maximum of five years. This exemption applies to the manufacturer, importer or producer of the articles and listed customers. The Regulation does not limit the quantities of the substance manufactured, imported, incorporated in articles or imported in articles, provided the quantities are limited to the purpose of PPORD.

A further five years needs to be justified (or 10 years for substances for human or veterinary use and substance not placed on the market). The justification must include the improvements and achievements obtained during the first five years of exemption, the reason for the previous research programme not being completed over the five-year exemption period and the expected achievement during the duration of the extension requested.

---

[6] First-time importers refers to companies which import or export a substance in the EU for the first time after REACH came into force, that is 1 June 2007.

1.2.2.3 Registration Process

The registration process consists of the submission to the ECHA (ECHA, nd-b) of a dossier containing hazard information and, where relevant, a risk assessment of the uses of the substance. The technical dossier should include a summary of the substance's properties and provide guidance on its safe handling. If a substance is produced or imported at > 10 tonnes/year, then a chemical safety report (CSR), which illustrates the safe use of the substance through the hazard and risk assessments, should also be provided.

REACH registration has been divided into three phases in order to process a high number of submissions:

Phase 1: Substances with a production or import volume in the EU of ≥ 1000 tonnes/year OR classified as Carcinogenic, Reprotoxic or Mutagenic 1 and 2 (CMR 1 and 2) OR substances which are classified for the environment as Aquatic Acute 1 or Aquatic Chronic 1, corresponding to the old DSD classification criteria R51–53, and manufactured or imported at ≥ 100 tonnes/year were required to be registered first, that is 1 December 2010.

Phase 2: Substances with a production or import volume 100–1000 tonnes/year required registration by 1 June 2013.

Phase 3: Substances with an import or production volume 1–100 tonnes/year require a registration by 1 June 2018.

The information requirements that need to be presented in the technical dossier depend on the tonnage and phase requirements. The information requirements can be found in Annexes VI to IX of the REACH regulation (ECHA, nd-c).

1.2.2.4 Substances Exempt from REACH

Not all substances imported or manufactured in the EU at ≥ 1 tonne/year require registration. REACH regulation Annex IV lists specific substances exempt from REACH. These include well-understood substances such as water, hydrogen, oxygen and the noble gases, as well as some naturally occurring substances such as ores and minerals. Other classes of substances that currently do not require registration under REACH include: polymers, monomers bound into polymers at < 2%, cosmetics, food additives, by-products, and products from reaction with additives or waste.

1.2.3 Evaluation

After the registration is submitted, one or more forms of evaluation are carried out by the authorities.

1.2.3.1 Technical Completeness Check (TCC)

The evaluation process initially takes the form of an automated electronic check that all the required technical contents of the dossier are included at a basic level. This is referred to as the technical completeness check (TCC). If the TCC is failed, it will lead to immediate rejection of the dossier, and it will then be necessary for the registrant to make the necessary corrections and re-submit. A tool is available for the registrant to ascertain in advance if the TCC will be passed. This tool works as a plug-in or application within the IUCLID (International Uniform ChemicaL Information Database) software (which most registrants use to compile their technical dossier). The tool has been regularly updated and registrants should ensure that they use the latest version.

The TCC does not evaluate the science or approach, only that entries are present and/or specific fields completed, including:

- substance composition
- business information
- volume and use pattern
- chemical property data (or waiver) for all of the endpoints associated with the appropriate Annex level (VII–X)
- guidance on safe use
- attachments.

A TCC pass does not necessarily indicate a successful registration.

1.2.3.2   Compliance Checks

ECHA may make a more in-depth review on a selective basis. These compliance checks are made by ECHA technical staff. The review is likely to cover such elements as:

- Adequacy and completeness of the data in technical terms
- Grounds for any data waiving
- Full compliance with the regulatory requirements
- Exposure
- Suitability of scientific approaches used in the chemical safety assessment (CSA).

During the phase-in period, the compliance check is the most commonly used in-depth review of scientific approach. Due to the volume of dossiers received, ECHA expects to conduct compliance checks for only approximately 5% of submissions received in that period. Dossiers are prioritised for compliance checking based on specific criteria: for example, if the substance is hazardous or used in widely-dispersed applications, or contains numerous data waivers. However, a proportion of dossiers are compliance checked on the basis of randomised selection.

The ECHA communicates its findings to the registrant through Decision Letters, which alert the registrant to non-compliance with the regulatory requirements, and/or Quality Observation Letters, which recommend adjustments to the methods used in the submission. The changes called for are required rather than optional. The ECHA gives feedback in a practical way, making clear reference to specific guidelines, and has been willing to participate in discussion meetings. The ECHA may request additional further testing as a result of the Compliance Check.

The ECHA is obliged to undertake consultation with the registrant, the member states and any other interested parties.

1.2.3.3   Testing Proposal Examination (TPE)

In accordance with the REACH Regulation, certain types of new experimental studies should be proposed by the registrant rather than being conducted before registration. ECHA reviewers make specific checks on dossiers which contain such test proposals, to assess whether the proposed tests are appropriate. Whilst scientific factors from the chemical safety assessment and chemical data from elsewhere in the dossier are taken into account, the testing proposal examination checks are not comparable with the compliance check.

Test proposals for vertebrate animal studies are published for consultation and calling in of any existing data held within Europe for sharing. The ECHA communicates its findings to the registrant through Decision Letters. All test proposals must be checked by the ECHA. Fixed deadlines have been set out for completion of these checks associated with each tranche of phase-in registrations.

1.2.3.4   *Substance Evaluation – Community Rolling Action Plan (CoRAP) Programme*

Substances which are identified as posing a particularly serious concern are prioritised for full substance evaluation, undertaken by member state competent authorities. Through this form of evaluation, under the community rolling action plan (CoRAP) programme, information from all individual registrations of the same substance is brought together by the member state reviewers.

The priority lists for this form of evaluation are developed on the basis of hazards, tonnage, and exposure. This being a rolling action plan, the priority list will be regularly updated to reflect current issues and concerns.

The purpose of the CoRAP evaluation is to ascertain, using risk-based methods, whether the substance is adequately controlled and to identify whether any courses of action, such as EU-wide risk management, are necessary. In the course of this evaluation, additional information may be required from registrants (e.g. monitoring data; property data outside the normal regulatory requirements outlined in the relevant REACH Annex).

Experience from the member state-led assessments under the previous legislation (Existing Substances Regulation) and the early work suggest that CoRAP will involve a process of detailed assessment made on a substance-by-substance basis, led by a nominated member state; with regular in-depth technical discussion meetings Some Member States appear to view the CoRAP process as a risk management option (RMO) process within which an evaluation of the risk is made followed by an assessment of the most appropriate measures for control of the risk/s. The risk management option process is somewhat similar in essence to the risk reduction strategy process under the Existing Substances Regulation and should involve the consideration of the practicability, effectiveness, ease of monitoring and proportionality of the proposed measures. This involves some consideration and comparison of the costs and benefits of the possible measures. The measures considered may be processes within REACH (such as Authorisation or Restriction) as well outside REACH, for example specific legislation at EU level (e.g. the Carcinogens and Mutagens Directive, Water Framework Directive or Industrial Emissions Directive) or specific legislation at national level.

1.2.3.5   *Annual Reporting*

Every February, the ECHA issues an annual Evaluation Report via its web site. This presents some statistical details of progress, useful reference information on common issues and findings from the evaluation processes conducted in the year to date.

1.2.4 Authorisation and Restriction

This section explains these processes within REACH; both authorisation and restriction are revisited and discussed in more depth in relation to additional supporting analysis including socio-economic analysis (SEA) in Chapters 13 and 14 of this book.

The main objective of REACH is the systematic collection of data on the properties and uses of substances that are intended to be placed on the EU market and the assessment of those data to show safe use. The main process driving that in REACH is the registration of individual substances, providing information in dossiers as described in Chapters 2–10 of this book. The concept of safe use in REACH is expressed as adequate control, in which the levels of exposure of humans and the environment due to use are compared to notional safe levels that are deemed protective specific receptors, that is for humans in the workplace, for the general public and for specific parts of the environment. The data presented in registration dossiers must demonstrate adequate control for each use pattern and for each relevant receptor in order for the use to fulfil the requirements of REACH, and thus to be placed on the market for those uses.

For some substances, due to their intrinsic hazardous properties and associated uncertainties on the type of harmful effects they may have, it is not possible to indicate a safe level and, therefore, in theory adequate control cannot be demonstrated (i.e. substance of very high concern (SVHC) – Chapters 6 and 13 of this book). Alternatively, there may be substances that are not SVHCs and for which is possible to demonstrate adequate control, but there is still the possibility of a risk from cumulative uses (not accounted for by individual and separate chemical safety assessments from a number of different registrants). The process within REACH for additional controls on SVHCs is *Authorisation* and the process by which additional controls on substances that may pose risks not accounted for individual registrations is *Restriction*.

Authorisation and Restriction, therefore, represent an additional layer of precaution and assessment that is applied to hazardous and risky substances. The aim of Authorisation is to progressively replace SVHCs with safer substances and the aim of Restriction is to provide a 'safety net' for the imposition of risk reduction measures that would not be otherwise put in place though the registration process. The processes of Authorisation and Restriction in REACH are concerned with placing limits on the use of dangerous substances. They are similar processes with complimentary objectives, but have key differences.

The process of Authorisation effectively places 'ban' of all uses of a SVHC unless an authorisation is granted to allow a specific use or uses to continue. It is for industry (manufacturers and users) to make the case for continued use in the form of an application for authorisation and for ECHA committees and, ultimately, the European Commission and the European Parliament to decide on the validity of the case.

The authorisation process is driven by hazard (i.e. the intrinsic properties of the substances that cause it to be dangerous) and starts with identification of a substance as an SVHC (Chapters 6 and 13 of this book). Substances identified as a SVHC (by way of a dossier submitted by a Member State to the ECHA or ECHA on behalf of

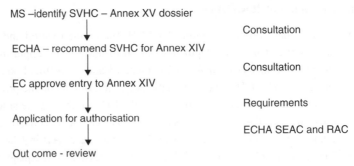

MS –identify SVHC – Annex XV dossier

↓   Consultation

ECHA – recommend SVHC for Annex XIV

↓   Consultation

EC approve entry to Annex XIV

↓   Requirements

Application for authorisation

↓   ECHA SEAC and RAC

Out come - review

Figure 1.2   *Flow Diagram of SVHC, Placing on ANNEX XIV, Authorisation and Review Process*

the Commission) are placed on a 'Candidate List'.[7,8] From this list substances may be selected for assessment against criteria for recommendation to require authorisation (placing on Annex XIV). There are specific exemptions to authorisation; it applies only to the marketing and use of substances; it does not apply to manufacture alone, that is substances manufactured solely for export out of the EU, or use as an intermediate only.

The assessment for placing a substance on Annex XIV is done by the ECHA and takes account of the volume placed on the market, the uses (in particular if there is a large number of different uses, distributed widely across the EU, so-called wide and dispersive use) and alternatives (particularly if there appear to be viable alternatives, since authorisation seeks to replace the most hazardous substances, this depends upon alternatives being available). On the basis of this assessment a recommendation (with a supporting report) is made to the Commission for inclusion on Annex XIV. The recommendation also includes proposals for when the substance should be no longer used from unless an authorisation is granted for specific uses – this is called the sun-set date as well as possibly a review date (i.e. a maximum time that the authorisation can be granted for before it is re-assessed by the ECHA and the Commission). Note that all authorisations are time limited; the length of time that the authorisation is granted for is dependent on the production cycle of the substance and the information submitted in the authorisation application.

Figure 1.2 illustrates the 'route' for a substance from identification as a SVHC though possible selection for the need for authorisation – placing on Annex XIV of REACH.

There are two possibilities for an authorisation to be granted:

1. The substance does not have a safe level (i.e. non-threshold CMR (carcinogen, mutagen, reproductive toxin) or PBT/vPvB (persistent, bioaccumulative and toxic/very persistent and very bioaccumulative)); therefore, 'adequate control' cannot be demonstrated. For these substances an authorisation can only be granted if it is demonstrated in the authorisation application that there are no technically or economically feasible

---

[7] Designation as a SVHC and listing on the Candidate List have consequences within REACH, even if the substance is not then selected for Annex XIV. Obligations under Article 33 of REACH mean that producers of articles (products that are objects) that contain 0.1% of a SVHC or more by weight must inform their customers about that and supply information on how to use the article safely with respect to the SVHC content.

[8] Once a substance is placed on the Candidate List, it will not be removed, unless there is new information to demonstrate that is no longer fulfils the criteria for SVHC – this is, of course, unlikely.

or available alternatives for the substance and also that the socio-economic benefits outweigh the risks. This is called the socio-economic analysis route.
2. Substances for which a safe level can be derived, that is adequate control can be demonstrated. For these substances, an application can be granted so long as adequate control is demonstrated. If suitable alternatives are available, then a substitution plan must be presented.

The application for authorisation must include a chemical safety report (unless one has already been submitted to ECHA in a registration dossier) and an analysis of alternatives, in addition it *may* include a socio-economic analysis (although in practice this *must* be included because it will be very difficult to present the socio-economic arguments to support the benefits outweighing the risk without such an analysis). The theory and practice of socio-economic analysis in support of authorisation applications and restriction proposals are explained in Chapters 13 and 14 of this book.

Restriction is a ban on a specific use or uses, with all other uses being permitted (so long as they have been registered appropriately). The cases for restrictions are made by Member State Competent Authorities or by the ECHA (at the request of the Commission). As with authorisation, the case for restriction is assessed by the ECHA committees, with the ultimate decision made by the Commission and the Parliament.

There may also be potential risks from substances for which it is possible to demonstrate safe use. Because registration in REACH is done per substance and per legal entity, it means that it is possible that a number of chemical safety assessments for the same substance (submitted by different registrants) can demonstrate safe use, but there may be a cumulative risk when all the safety assessments are considered together. For example, it is only possible for a registrant that is a manufacturer to consider releases from its plants and processes and those of its customers (whose uses are being supporting in the registration). The registrant can assess these uses as safe when it considers the releases and control of releases of the substance. There are also other manufacturers with customers with similar or different uses, which are also assessed as safe. However, when the cumulative releases are considered, there may be the need for additional measures to control the risks – that is risks that are only apparent when the releases from all uses are considered together. Since it can only be the ECHA that can be in possession of all the dossiers (and therefore chemical safety reports) for a substance, it can only be the ECHA that can identify such risks. There is particular concern when there is additional risk at a European-wide scale.

The objective of Authorisation is the progressive removal and replacement of a SVHC from the EU market, whilst the objective of Restriction is to act as a 'safety net' to impose limitations (restrictions) on the uses of a substance that present a European-wide risk. Restriction applies to all substances, whereas authorisation applies only to a SVHC (and substances which have properties of equivalent concern to a SVHC).

Authorisation and restriction are possibly the most contentious, controversial and misunderstood parts of the REACH process. This is largely because, at the time of writing, these processes within REACH are untested. Certainly for authorisation, no substances have yet been fully through the process, as no application has yet been assessed by the ECHA. However, a few restriction proposals have been made and decisions on restriction for specific uses of specific substances made.

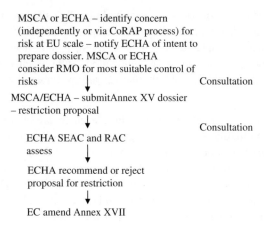

MSCA or ECHA – identify concern
(independently or via CoRAP process) for
risk at EU scale – notify ECHA of intent to
prepare dossier. MSCA or ECHA
consider RMO for most suitable control of
risks   Consultation

MSCA/ECHA – submitAnnex XV dossier
– restriction proposal

                                         Consultation

ECHA SEAC and RAC
assess

ECHA recommend or reject
proposal for restriction

EC amend Annex XVII

Figure 1.3 *Flow Diagram of CoRAP, ANNEX XV Dossier and Placing on ANNEX XVII, Restriction Process*

Figure 1.3 illustrates the 'route' for restriction from registration, the CoRAP process to a restriction being placed on Annex XVII of REACH.

1.2.5   Hazard and Risk Communication

In this section, the chemical supply chain as defined in REACH is discussed. This is an area which caused many stakeholders concerns before any substance was registered. With the experiences gained since then, it is possible to refine and, in part, eliminate those fears, but many areas remain difficult for all.

The composition of what constitutes the supply chain is well-defined in the REACH guidance. Although taken from the environmental guidance 'Guidance on information requirements and chemical safety assessment, Chapter R.16: Environmental Exposure Estimation', page 12 (ECHA, 2012), Figure 1.4 is a useful starting point, since it deals with life cycle rather than legal matters:

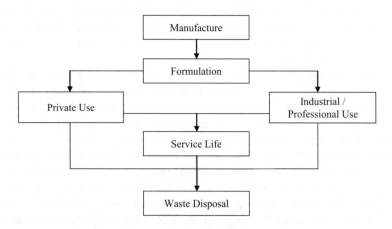

Figure 1.4 *Overview of Substance Use Patterns*

Within that structure, the supply chain can be seen as consisting of the following actors:

- **Manufacturers or importers**, the import of substances, preparations or articles are all included here.
- **Downstream users**, someone who uses a substance, either on its own or in a preparation, in the course of their industrial or professional activities. Many different types of companies can be downstream users (DUs), including formulators of preparations, producers of articles, craftsmen, workshops and service providers or re-fillers.
- **Retailers**, may be involved and do have responsibilities in respect of provision of information.
- **Private consumers**, where relevant in the life cycle.
- **Waste processors and recyclers**.

Some key elements of supply chain actions can be identified:

- Formulators have a key place in this chain, because the process of mixing substances to make a preparation adds a whole new dimension to the assessment of hazard and risk. Formulators need to prepare safety data sheets to cover the preparation.
- REACH requires a considerable amount of information to be communicated to downstream users by suppliers, to enable them to use chemicals safely. In addition, REACH requires downstream users to communicate new information on hazards and also use information when the supplier's advice seems to be incomplete.
- Downstream users need to communicate upstream and downstream, for example when identifying uses to a supplier or collecting information on customers' uses.

In such a context, there may be occasions when a DU simply does not want to take the commercial risk of giving information to a supplier about what exactly it is doing. Also, on occasion it has been found that suppliers are asking DUs questions that are far more detailed than is actually needed to fulfil their responsibilities. Such scenarios do not need to give rise to a tense relationship – that is not in anyone's interests. However, a well-prepared supplier could perhaps gain advantage over another supplier by handling DUs well. The DU needs sufficient knowledge of REACH (and good advice) to be able to judge these and other matters. Indeed, it could be seen as one of the consequences of REACH that DUs and their support associations have had to learn many more 'new things' than suppliers have!

The general fears around REACH caused many DUs to make pre-registrations of substances just in case their suppliers did not! It is to be hoped that such examples will not recur. Since 2010 the focus of concern has shifted to DUs looking at eSDS (extended Safety Data Sheet) supplied to them, and finding them to be inconsistent and very hard to understand. Section 11.8 of this book helps with that, but some general statements can be made. How has it arisen that DUs have concerns about an eSDS from a manufacturer/importer? Whilst REACH has presented severe difficulties due to a highly compressed time line, typical problems include:

- Suppliers failing to communicate about hazards and exposure within the SIEF (Substance Information Exchange Forum).
- Suppliers failing to communicate with their customers.

- Suppliers and DUs have assumed that establishment of what the use pattern is (and assignment of descriptor codes) is enough information to characterise risk; in reality it often is not, because the default exposure values driven by the codes are very high. Agreement of codes must be accompanied by real understanding of what is actually happening!
- DUs not checking with suppliers that their use will be covered.
- Poorly-written eSDS.
- Insufficient technical expertise at both levels, including basic knowledge of admittedly very complex Guidance.
- Inadequate awareness of the implementation of the globally-harmonised system of Classification, Labelling and Packaging (CLP) alongside the implantation of REACH (Appendix A). This has had a positive benefit in gaining more consistency with transport regulations.

Another area of challenge for formulators is the communication of hazard and risk of preparations. They will need to collate information about all the ingredients and make an assessment, then report it in the SDS.

Whilst the above problems have been identified, the efforts of industry sector groups to establish clear and agreed descriptions of use pattern and exposure models must also be acknowledged. As with all things in REACH, findings must be supported by evidence, and this takes time to assemble.

The responsibilities in the supply chain are covered in more detail in Section 11.5 of this book.

1.2.6   Hazards

1.2.6.1   Physico-Chemical Hazards

Physico-chemical hazards such as flammability, explosivity, and auto-flammability may be manifested in standard laboratory tests when a sufficient quantity of the substance is present. There are established worldwide classification criteria for these properties.

1.2.6.2   Toxicological Hazards

Potential effects on humans are assessed firstly on the basis of studies with non-animal laboratory tests (termed '*in vitro* assays') or with animals (if required by legislation or there is no alternative to provide the data needed). Studies with animals are only models for possible effects on humans and, as such, are not perfect predictors.

Study schemes are set out to assess hazard step-by-step, with the following targets:

- Genetic toxicology
- Short-term toxicity
- Long-term toxicity
- Effects on reproduction and development
- Carcinogenicity.

Animal testing can be minimised through the use of read-across strategies and other alternatives, such as (Q)SAR modelling. However, it is widely thought that there is much research to be done before use of *in vivo* animal models of human hazard can be eliminated.

For certain carcinogenic, mutagenic or reprotoxic substances, authorisation of each use must be obtained. This is independent of any risk characterisation.

1.2.6.3   Environmental Hazards

A substance's potential environmental hazard is investigated through testing for toxicity to environmentally-relevant organisms (ecotoxicity). There are standardised tests for ecotoxicity to aquatic organisms (including sediment) and terrestrial organisms. Effects on microorganisms present in biological waste-water treatment plant are also examined.

REACH sets out precautionary criteria as regards environmental fate. If a substance has a high potential for bioaccumulation in the absence of degradability, it is regarded as a hazard, even if there is no evidence of toxicity.

As part of this precautionary approach, substances decided to be PBT or vPvB require authorisation for specific uses (Section 1.2.4 in this chapter).

1.2.7   Overview of Types of Exposure

The inherent hazard of a substance may trigger various risk and safety concerns for humans and the environment during its use. The potential for and consequences of such risk are determined, to a large extent, by the nature and level of exposure during the use of the substance. The term 'exposure' in risk assessment of chemical substances means the form and/or route of contact, and potential interaction of different substances with humans and the environment. The exposure may be short term or long term, once or repeatedly, by different pathways, in low or possibly in high concentrations (CEFIC/VCI, 2009).

The exposure of a chemical substance can be direct and intentional, as in the case of consumer use of personal care products or washing-up liquid. Exposure can also be indirect and unintentional, as in the case of the loss of dyestuff from a dyeing process. The dyestuff which was intended to be taken up by the fibre or fabric is then lost to waste water, leading to exposure of the environment (rivers: fresh water and/or sea: marine water) and possible subsequent exposure of humans through intake of fish and/or drinking water.

1.2.7.1   Human Health Exposure

The underlining principle/concept for human health exposure is very well captured and explained by the definition of exposure by the International Programme on Chemical Safety (IPCS, 2001): the 'contact of an organism with a chemical or physical agent, quantified as the amount of chemical available at the exchange boundaries of the organism and available for absorption'.

In the case of human health exposure, the contact with a physical or chemical agent occurs with the visible exterior of a person (i.e. target), such as the skin, and openings, such as the mouth, nostrils and lesions (EPA, 2011). Depending on the use and life cycle stages of the physical or chemical agent, the target individual may cut across different population categories: workers, professionals and consumer users of the substance or products containing the substance.

The way a substance is used and the population that would be affected (e.g. workers, adult consumers or children) can also define the significance of the point of contact and route of exposure. For example, the oral route of exposure is typically generally not considered relevant for workers. Mouth contact and oral exposure are, however, significant for children, as they are exposed during the normal oral exploration of their environment (i.e. hand-to-mouth behaviour) and by touching floors, surfaces and objects such as toys (EPA, 2011).

The individual's population category and activity patterns, as well as the nature and concentration of the chemical, will determine the frequency, duration, route and magnitude of the exposure.

1.2.7.2   Environmental Exposure

The process of environmental exposure begins with a chemical substance released into the environment from a point source (e.g. emissions from an industrial stack), or wide dispersive sources such as multiple emissions from cars. Point source releases are to the 'local environment', while releases into a larger area from multiple point sources or wide dispersive uses are to the 'regional environment' (ECHA, 2012; EPA, 2011). A chemical substance's contact with the environment will be through one or more of the various 'environmental compartments': air, water and soil.

Once in the environment, the chemical substance follows an exposure pathway, along which it can be transformed and transported through the environment via air, water, soil, dust and diet. The physico-chemical properties of the substance determine the overriding fate and transport mechanisms of the substance in the environment.

1.2.7.3   Exposure of Humans via the Environment

As well as the direct exposure from use of a chemical substance or its products, indirect exposure of humans can also occur through releases to the environment, exposure to humans via the environment may result from releases to the air and water, and solid and hazardous waste disposal.

Chemical releases to rivers, lakes and streams may result in a substance's accumulation in fish and other marine life. These may be subsequently used as a source of food, or ingested by persons using the downstream reaches of rivers as a water supply. Those living downwind of a chemical manufacturing facility may be exposed to fugitive and point source releases of chemical toxins to the atmosphere. Disposal of solid and hazardous wastes on the land, either in repositories such as landfills, or into subterranean strata by injection into wells, may result in contamination of potable groundwater if the waste is not isolated from the surrounding environment.

The significance of the exposure of humans via the environment is dependent on the tonnage of the substance produced, the tonnage used and the inherent chemical hazard of the substance.

Risk assessment of a chemical substance involves producing an 'exposure scenario'. This means the set of conditions, including operational conditions and risk management measures, which describe how the substance is manufactured or used during its life cycle. An exposure scenario also includes how the manufacturer or importer controls, or recommends downstream users to control, exposures of humans and the environment.

These exposure scenarios may cover one specific process or use, or several processes or uses, as appropriate (ECHA, 2012).

1.2.8 Overview of Risk Characterisation

The purpose of risk characterisation is to identify uses of a substance that present an unacceptable risk to the environment or to human health, and to reduce these risks to an acceptable level. Under REACH, risk characterisation must be carried out and documented for all substances that are either:

1. Classified according to the CLP Regulation (European Regulation (EC) No 1272/2008 on Classification, Labelling and Packaging of Substances and Mixtures) as having health, environmental or physical hazards or
2. Meet the criteria set out under REACH for PBT substances, and are manufactured or imported in the EU at a volume of > 10 tonnes/year.

As discussed in the introduction to this section, risk characterisation brings together two different types of information:

1. The properties of the substance and an assessment of any hazards, where possible quantified to give a maximum safe level of exposure (discussed further in Chapters 7 and 9 of this book);
2. The use patterns of the substance and quantification of predicted exposure levels for the environment and humans (this is discussed further in Chapters 8 and 10 of this book).

Whenever possible, risk is quantified by calculation of a risk characterisation ratio (RCR). This is the ratio of the predicted exposure level to the safe level of exposure. A RCR greater than one (where the predicted exposure is greater than the safe level) indicates an unacceptable risk, whereas RCRs less than one indicate that the use of the substance may be considered safe.

A separate risk characterisation is performed for each identified use of the substance (e.g. manufacturing, use as a chemical intermediate or use in household cleaning products). For human health, RCRs are calculated for the different routes of exposure (oral, dermal or inhalation), short- and long-term exposure and the different types of people who may be exposed (workers at industrial sites, professionals, adult consumers and children and humans exposed via the environment). If exposure via more than one route or from more than one source is possible, a combined RCR is calculated. For the environment, RCRs are calculated for the different types of environment that may be exposed to the substance (marine and fresh water, marine and fresh water sediment, soil, microorganisms in the sewage treatment plant and predators) at both regional and local scales.

A substance may possess hazardous properties for which it is not possible to define a safe level, either because there is no threshold below which exposure can be considered safe (e.g. genotoxicity or carcinogenicity) or because the standard tests for that property do not provide this information (e.g. flammability or irritation). For the former group, semi-quantitative risk characterisation may be appropriate. This involves calculating a level corresponding to a low, possibly theoretical, risk. RCRs are then calculated in the same way as for a quantitative risk characterisation. For the latter group of substances, qualitative risk assessment must be carried out. This involves consideration of the severity of the hazard and how risks associated with it are controlled. A qualitative assessment is then made of the likelihood that these controls are adequate to prevent harm to relevant groups. For health hazards, the general principle is to limit or avoid contact with the substance. For physical hazards, the aim is to eliminate or reduce the likelihood of accidental events occurring. The degree of control required is proportional to the severity of the hazard. If a substance possesses both hazards that can be quantified and those that cannot, both quantitative and semi-quantitative or qualitative risk characterisation is required.

All uses to be covered in a submission under REACH must be shown to be safe by the risk characterisation. If an initial risk characterisation gives one or more RCRs higher than one (or the qualitative risk characterisation indicates that a hazard is inadequately controlled), there are several options:

- Refine the estimates of exposure levels. Initial estimates of exposure levels often make use of various default values (e.g. the percentage of a substance unintentionally lost to the environment during mixing of substances, or the concentration available for inhalation when spraying a substance). These defaults are intended to cover the worst case of a very general situation and, therefore, are often unrealistically high. If use of the defaults still results in RCRs less than one, there is no need for refinement. However, consideration of the specific use often means that it is possible to justify reducing these estimates if necessary. This can involve making measurements of exposure levels.
- Refine the estimates of the safe level of exposure. In some cases, gaining greater knowledge of the properties of the substance (for example by carrying out longer-term animal studies) can result is a more accurate estimate of a safe level of exposure (although this may be higher or lower than the original estimate).
- Put in place additional measures to manage the identified risks. For example: wearing gloves whenever a substance is handled can reduce worker exposure; reducing the concentration of the substance in a cleaning product can reduce consumer exposure; and fitting technology at a manufacturing site to limit emissions can reduce environmental exposure.
- Do not support the affected use(s). If a risk is still present after all possible refinement of the estimates, and it is not possible (or not commercially viable) to introduce further controls, it may be necessary to advise against the use.

The choice of approach can depend on many factors, both technical and commercial. An iterative approach of refinement followed by review of the risk characterisation is usually appropriate. During this process, uncertainty in the calculated RCRs should be considered. Uncertainty in the RCRs results from uncertainties in the estimates of both hazards and exposures, and is relevant to deciding whether risks are adequately controlled. If there is too much uncertainty in the outcome of the risk characterisation, further iterations may be required.

The iteration ends when the qualitative and quantitative risk characterisation indicates that risks are controlled to a level of very low concern, or it is concluded that it is not possible to demonstrate control of the risks. The conditions which allow safe use, or the information that the use is advised against, must then be communicated down the supply chain.

1.2.9   Successful Interaction with REACH: Registration, Evaluation and Authorisation

1.2.9.1   Introduction: How EU Chemical Legislation Evolved

REACH is, of course, the culmination of a series of directives and regulations enacted by the European Parliament since the end of the 1960s (Council Directive 67/548/EEC of 27 June 1967 on the approximation of laws, regulations and administrative provisions relating to the CLP of dangerous substances). It is difficult to understand how REACH relates to other chemical legislations without taking into consideration how they came by in the first place. Historically, the primary consideration of the European Commission and of the Member States was initially worker and consumer protection. Later on, especially after the accession of Sweden to the European Union (in 1995), more emphasis has been placed on environmental protection. In particular, many of the non-enforceable environmental goals of OSPAR (Oslo–Paris Commission, historically piloted by Sweden and composed by several EU and non-EU states), which predates the European Union, were gradually included in European legislation, such as the Water Framework Directive (2000/60/EC). While, retrospectively, the gradual development of chemical legislation may have been seen as haphazard, the underlying reason was that both regulators and industry (through various associations) were struggling to find middle ground between command and control and hazard-based legislation, potentially affecting the competitiveness of the European chemical industry, and a more constructive approach based on industry's experience in the safe handling of hazardous substances and self-regulation.

While REACH was intended to provide a EU-wide, directly applicable (as opposed to nationally enacted) legislation ensuring a harmonised market within its borders, the conflicting positions of non-governmental organisations (relayed by media and by public opinion) and industry, with the EU authorities caught in the middle, continue. The former advocate rapid bans of potentially problematic substances, based solely on their hazard profiles, while the latter accepts the need to adapt the EU's chemical policy but at the same time enhance or at least maintain its competitiveness, protecting and strengthening the internal market and reaching decisions based on risk.

Until the early 1970s the first task of the chemical industry and of the regulators (EU Commission and the Member States represented by their various health and environment protection agencies) was to inventory the thousands of chemicals in commerce and their known properties. Simultaneously, internationally agreed test protocols to generate (eco)toxicological data under Good Laboratory Practices were developed under the auspices of organisations such as the OECD (Organisation for Economic Co-operation and Development).

It should be noted that the Classification and Labelling Directive 65/548/EEC did not impose an obligation for manufacturers to develop new (eco)toxicological data. Manufacturers had a tendency to focus on measuring physical property data (critical for transportation and storage safety) and short-term toxicity data. With the exception of Germany, which had developed its own system of water endangering classification (WGK – Wassergefährdungsklassen), and with its manufacturers agreeing to voluntarily sponsor basic environmental testing of its products, generally little of the more expensive long-term toxicity data were developed. The only exceptions were for certain chemicals that belonged to families suspected of repeated-exposure hazard properties, such as sensitisation, carcinogenicity, reprotoxicity and mutagenicity, or for other substances in wide dispersive use or with a high potential for human exposure, as in cosmetics. More recently the potential effects of endocrine-mimicking substances on the unborn child are being investigated but there is still some controversy whether existing methods to determine reproductive effects are sufficient. Similarly, the concepts of synergistic health effects between exposure to low doses of chemicals and the long-term environmental effects or secondary poisoning due to a combination of persistency, bioaccumulation and toxic properties have gained increased traction, with the latter now fully part of REACH.

The chemical industry had accumulated a huge amount of practical experience in handling safely several hazardous substances without necessarily undergoing extensive testing. This experience was based on the effects observed during the time where the methodology to assess them in a systematic way was not yet fully developed. Some notorious examples are the use of some metals, inorganic or organic compounds in cooking (lead utensils and lead acetate as a condiment by the Romans), medicine (arsenic for the treatment of leukaemia, psoriasis, mercury for syphilis etc.), cosmetics (lead in eyeliners), jewellery (nickel plating), marine coatings (organotin antifouling additives), pesticides (DDT), herbicides (2,4-D or dichlorophenoxy acetic acid), refrigerants (chlorofluorocarbons), felt hats (mercury) and even transformer cooling oils (polychlorinated biphenyls). The remaining uses of all of these substances are now strictly controlled or banned. New technologies have made possible complete substitution of some of these substances in essential applications, such as mercury in thermometers or in the production of chlorine by separation cells.

Margot Wallström, the EU Environment Commissioner when REACH was enacted, had called for decisive action by claiming that had REACH been in place, asbestos would not have caused and still be causing 100 000 industrial deaths, although one can hardly call asbestos a man-made chemical, the control of which is the primary purpose of REACH. While that remark was widely criticised or applauded, depending on which side the comments were coming from, it was effective in instituting the principle of substitution of CMR Cat. 1a or 1b substances. However, for the reasons explained above, such as greater awareness of the effects of certain chemicals, exposure to them has dropped significantly between the 1930s and the 1970s, showing that the early chemicals legislation has achieved its goals.

Nevertheless, the growing awareness that exposure to certain substances can have some long-term effects that were originally unsuspected has resulted in adopting a more precautionary attitude towards innovation. In the USA, for example, there is a requirement for industry to report to the EPA (Environmental Protection Agency) significant new uses rule (SNUR), which would be authorised once they are demonstrated to be safe. In the EU, the Commission published in 2000 its interpretation of the Precautionary Principle (EC, 2000),[9] which is referred to in all chemical legislations enacted since then.

At the same time it was realised that the undesirable effects of these substances were also a function of exposure. By restricting the use of these substances to those

---

[9] EC, 2000: 'The precautionary principle enables rapid response in the face of a possible danger to human, animal or plant health, or to protect the environment. In particular, where scientific data do not permit a complete evaluation of the risk, recourse to this principle may, for example, be used to stop distribution or order withdrawal from the market of products likely to be hazardous'.

applications where the benefits can be demonstrated versus the absence of risk, these negative effects can be avoided. For example, DDT is still used to treat mosquito nets, nickel is in every euro coin, a mercury compound is an essential preservative in certain vaccines and without lead metal and chemicals in batteries consumers would certainly have problems running automobiles.

The result of this approach was the so-called Marketing and Use Directive (Limitations Directive, i.e. Directive 76/769/EEC) restricting or banning the use of certain hazardous chemicals and the DSD 67/548/EEC setting up an in-depth review of existing priority chemicals selected for their hazard properties (CMR, PBT or vPvB, in other words: Carcinogenic, Mutagenic, Reprotoxic, Persistent, Bioaccumulative, Toxic or very Persistent and very Bioaccumulative) and all new chemicals. The experience with the implementation of both Directives was retrospectively viewed as too slow by the European Parliament and some member states (200 new and existing substances were assessed over a period of 30 years). The solution, ironically, was to turn over to industry the responsibility of preparing dossiers on all 30 000 commercial substances with strict deadlines.

The World Health Organization (WHO) had initiated a programme called IPCS (International Programme for Chemical Safety) which published 241 EHC (environmental health criteria) critical reviews on the effects of chemicals and physical and biological agents on human health and the environment. The first environmental health criterion concerned mercury and was published in 1976. One of the last for chemicals (2005) was on clay minerals but more recently the focus has been on methodology and on physical agents, such as extremely low frequency fields (2007).

The next step was to define a 'base set' of data that would cover the complete hazard profile of a chemical substance and from there to classify the substance, therefore ensuring that appropriate measures would be taken when managing the risk related to exposure. About the same time the OECD and industry through its trade associations (CEFIC in Europe, the ACC (American Chemistry Council) in the United States and the JCIA (Japan Chemical Industry Association) for Japan) reached an agreement to submit a defined data set (and to fill gaps if any) to an international review panel. The programme, called HPV (high production volume, prioritising substances manufactured at a rate greater than 1000 tonnes/year) is still underway but in a sense has been superseded in the EU by REACH. In the USA, the EPA has gone further, firstly by making this programme obligatory for US manufacturers and importers and, secondly, by lowering the reporting threshold to 1 million pounds/year or 450 tonnes/year. Also, more recently, the EPA has set up an Extended HPV programme by including substances that have reached the volume threshold since HPV was started.

As a result of all of this, knowledge about the hazards posed by chemical substances has increased considerably since the 1970s. The process of self-classification based on the new data has already resulted considerable changes in the production and the use of certain chemicals, which is the reason why industry is confident that the majority of existing chemical uses will be shown to be safe.

The following chapters look in more detail at the impacts and the links with REACH on three other important pieces of chemical legislation and their interactions with some of existing national implementations.

How in practice the potential risks associated with chemicals are identified and managed, and their impact on costs and liabilities, are reviewed in more detail. The specific risks are identified by scope and their relationship with REACH is noted.

As seen in Table 1.1, the regulations concerning chemicals fall into several categories:

- Regulations that overlap REACH in some respects:
  - Air
  - Water Framework Directive
  - Carcinogens at Work Directive
  - The cosmetics regulations
  - Biocidal products
  - Plant protection products.
- Regulations that may apply resulting from compliance with REACH, if new data or an evaluation generated under REACH trigger a change in the hazard classification of a substance:
  - Biocides
  - Construction products
  - Cosmetics
  - Ozone depleting substances (ODS)
  - 'Seveso' directives
  - Toys
  - Prior informed consent (PIC)
  - Transport
  - Waste.

The six regulations that have potential for overlapping REACH are now be reviewed in more detail.

1.2.10   Regulation and Assessment of Hazardous Chemicals Outside of the European Union

The REACH programme exists within a global context of regulatory regimes, many of which have similar objectives and methods, although the scope and focus varies considerably from one sphere of regulation to another.

Multinational companies are present or dominant in almost all areas of the modern chemicals industry. Therefore, it is not uncommon for the same company to face similar or equivalent legislative requirements for the same substance in different parts of the world. It is notable that the ECHA has entered into mutual memoranda of understanding with several other regulators around the world. Whilst not being legally binding, this implies recognition that acceptance of registration under other regulatory schemes suggests that certain technical standards have been met, hence this is taken into account when a registration is necessary under REACH (or vice versa). The ECHA website has further information and should be consulted for the latest information of MoU (Memorandum of Understanding) in place; those existing at the time of writing are shown in Table 1.2.

There are in place some voluntary programmes involving assessment of chemical hazard and risk. For example, HPV programmes, in progress internationally and in various global sectors, share some similar approaches with the assessments made under regulatory systems. In most cases, these programmes focus on hazard assessment and

Table 1.1   *Relation to REACH of chemical risk management.*

Chemical risk scope	EU legislation	Relation to REACH

Air
  Urban
  Indoor
	Particulates, industrial emissions, solvent emissions 1999/19/EC and 2004/42/EC, IPPC 2008/1/EC	Each solvent use must be risk assessed and demonstrated to be safe

Biocidal products	Directive 98/8/EC and Regulation (EU) No 528/2012	Annex I listed substances are considered registered under REACH and are therefore exempted
Classification and labelling	CLP-Regulation (EC) No 1272/2008	Basis for identification as a substance of very high concern (SVHC)
Construction products	Construction Products Regulation (305/2011/EU – CPR)	Declaration of content of hazardous substances and identification of risks posed by construction materials
Cosmetics	Cosmetics Directive 76/768/EEC and recast as Regulation (EC) No 1223/2009	The environmental impact of cosmetics ingredients must be assessed. CMR ingredients are regulated
Food	Additives (Directive 89/107/EEC)	Excluded from REACH
Fresh and coastal waters	Water Framework Directive 2000/60/EC, IPPC 2008/1/EC	The Water Framework Directive (WFD) provides a framework to set Environmental Quality Standards for chemical substances found in surface waters. The methodologies may differ from REACH
Ozone depleting substances (ODS)	ODS (Office of Dietary Supplements) legislation, USA; Regulation (EC) No 2037/2000, Regulation (EC) 1005/2009, 2010/372/EU and (EU) 744/2010	By 2012 the consumption of ozone-depleting substances has dropped by 98%. REACH provides for an assessment of the ozone depleting potential of volatile substances
Greenhouse gases	EU implementation of the Kyoto Protocol	Monitoring of greenhouse gases emissions (especially certain fluorinated gases)

Health
  Consumers
  Workers
	Carcinogens at Work Directive 2004/37/EC	REACH provides a mechanism for authorisation of Cat. 1a and 1b carcinogens

(continued overleaf)

Table 1.1   (continued)

Chemical risk scope	EU legislation	Relation to REACH
Laboratory animals	The Cosmetics Regulation (EC) No 1223/2009 contains provisions restricting animal testing of cosmetic ingredients	REACH contains provisions for adapting testing requirements and for reading across to minimise animal suffering and use
Major accident prevention	'Seveso' Directives: 96/82/EC, 2003/105/EC	Applies to storage of hazardous chemicals. Not in the scope of REACH
Medicinal products	Directive 2001/83/EC and Regulation (EC) No 726/2004	Applies to safety of active pharmaceutical ingredients and components of medicinal products. Not in the scope of REACH
Plant protection products	Directive 2009/128/EC	Normally exempted from REACH. PBT/vPvB assessment may differ from ECHA REACH guidance
Radioactive substances	Directive 96/29/Euratom	Not in the scope of REACH
Toys	Directive 2009/48/EC	CMR ingredients, allergenic substances are regulated
Trade (international)	Rotterdam convention and prior informed consent (PIC) – Council Decision 2006/730/EC and Regulation (EC) n° 689/2008	Applies to banned or extremely restricted and to extremely hazardous pesticides. Not in scope of REACH
Transportation	United Nations Orange Book, International Air Transport Association (IATA) , EU directive 2008/68 – inland transport of dangerous goods	Excluded from REACH
Waste	WEEE (Waste Electrical and Electronic Equipment) Directive 2012/19/EU, RoHS (Restrictions of Hazardous Substances) Directive 2002/95/EC, Landfill Directive	Waste is excluded from REACH

Table 1.2 *Examples of regulatory systems for hazardous chemicals outside EU.*

Region/ country	Scheme and/ or associated legislation	Regulator	Objectives	Scope	Comments
Europe (non-EU)					
Norway	Regulations relating to restrictions on the manufacture, import, export, sale and use of chemicals and other products hazardous to health and the environment (Product Regulations) (2004)	Klima-OG Forurensnings-Direktoratet	Human health and environment throughout life cycle	Restriction of manufacturing, supply and use of specified substances, substance types and chemical families	Same Regulator manages the Norwegian Product Register
Switzerland	813.11 Ordinance of 18 May 2005 on Protection against Dangerous Substances and Preparations (Chemicals Ordinance, ChemO)	FOPH; FOEN; SOCA		Notification of new substances; classification and labelling; SDS	
Middle East					
Turkey	Regulation on the inventory and control of chemicals (2008)		Notification of new substances and reporting data to an Inventory	All substances > 1 tonne	

(continued overleaf)

Table 1.2 *(continued)*

Region/country	Scheme and/or associated legislation	Regulator	Objectives	Scope	Comments
African nations	Little in the way of focussed legislation on supply		Many nations use, or are in the process of adopting, GHS or similar evaluation and/or apply occupational health and safety regulations		
Australasia Australia	National Industrial Chemicals Notification and Assessment Scheme (NICNAS); Industrial Chemicals (Notification and Assessment) Act	Australian Government Department of Health and Ageing		Regulation of industrial chemicals	2011 major review to make the scope more comprehensive Memorandum of Understanding established with the ECHA
New Zealand	Hazardous Substances and New Organisms (HSNO) Act (1996)	Environmental Protection Authority		Hazardous substances (physico-chemical, tox or ecotox)	

	Legislation	Authority	Coverage	Substances	Notes
Asia					
China		Chinese Ministry of Environment Protection			
Japan	Chemical Substances Control Law (2009)	MITI	New chemicals notification; classification and labelling	All substances > 1 tonne (same as REACH)	Statement of Intent established with the ECHA
	Industrial Safety and Health Law (ISHL)		Workplace H&S		
South Korea	Toxic Chemicals Control Act (1991)	NIER			New legislation similar to REACH is expected soon
Taiwan	Occupational Safety and Health Act	Institute of occupational safety and health under council of labour affairs (CLA)			
India	No centralised scheme. Various pieces of specific legislation focussing on individual requirements		Manufacture and supply; use in industry and by consumers; covering health and environment		

(continued overleaf)

Table 1.2 (continued)

Region/country	Scheme and/or associated legislation	Regulator	Objectives	Scope	Comments
North America United States of America	Toxic Substances Control Act (TSCA)	EPA			A more comprehensive programme is expected to be brought in the near future, with similarities to REACH Statement of Intent established with the ECHA
	Chemicals 'Right to know' HPV programme (voluntary)	EPA	Collecting of hazard and use information for release in the public domain	Chemicals at high volume in USA	
Canada	Canadian Environmental Protection Act (1999)	Environment Canada	Management of new and existing chemicals and control of pollution		Memorandum of Understanding established with ECHA
	Government of Canada Challenge programme	Environment Canada; Health Canada	Risk assessment and management (in depth or screening)	Priority list (at time of writing, 200 substances) for in-depth assessment	

	Pest Control Products Act, the Canada Consumer Product Safety Act and the Food and Drugs Act	Health Canada	
South America			
Uruguay	No centralised scheme. Various different specific items of legislation		GHS and SDS; restriction of specific dangerous chemicals
Venezuela	No centralised scheme. Various different specific items of legislation		CLP and SDS; restriction of specific dangerous chemicals
Peru	No centralised scheme. Various different specific items of legislation		CLP; restriction of specific dangerous chemicals
Colombia	No centralised scheme. Various different specific items of legislation		CLP and SDS; restriction of specific dangerous chemicals
Paraguay	No centralised scheme. Various different specific items of legislation		CLP and SDS; restriction of specific dangerous chemicals

(continued overleaf)

Table 1.2 (*continued*)

Region/ country	Scheme and/ or associated legislation	Regulator	Objectives	Scope	Comments
International programmes					
	UNECE (United Nations Economic Commission for Europe) Protocol; Stockholm Convention		Identification, monitoring and strategic long term control of persistent organic pollutants	Operates through a candidates list process with separate final lists of substances banned; restricted; or for which unintentional production must be avoided	
	ICCA (International Council of Chemical Associations)/OECD HPV programme (voluntary)	OECD with substance specific 'sponsorship' by a national authority	Collecting of hazard and use information for release in the public domain	Chemicals at high volume in any global region	
	Globally Harmonised system (GHS) for classification and labelling. Enacted into law (fully or partially), through various locally applying regulations, for example CLP 2008 in EU	UN (UNECE), administrated where applicable by local regulators, for example the ECHA	Identification and clear communication of chemical hazards in the supply chain		

International Programme on Chemical Safety

IOMC (Inter-Organization Programme for the Sound Management of Chemicals) (international cooperation between WHO ILO OECD UNEP FAO UNIDO UNITAR and Canadian COHS)

Communication of data by inter-governmental organisations

OECD MAD (mutual acceptance of data)

Basis of international acceptability for non-clinical data sharing gathering information on properties and use pattern of substances manufactured and supplied in large volumes.

This section briefly summarises these regimes as they stand at the time of writing, though it does not attempt to be comprehensive, especially with regard to laws and programmes at Member State (e.g. the UK) or national (e.g. England/Wales, Scotland and Northern Ireland) levels. Such legislation continues to develop, and global regulation and safety standards for chemicals in supply can be expected to harmonise increasingly in future. Many commentators have noted that existing regulatory systems are beginning to move towards approaches similar to those used in REACH, as countries seek to avoid becoming the default market for certain chemicals that are deemed unsafe in other areas where more stringent standards are applied.

1.2.10.1   Risk

Under REACH, in many circumstances, risk must be assessed.

In this context, risk can be summarised as the likelihood of the undesirable property being expressed under foreseeable circumstances. Under REACH it is necessary to give consideration to the normal life cycle of industrial chemicals carried out in EU, from manufacturing to processing of end-of-life wastes. Major industrial accidents and misuse of substances are outside the scope.

Assessment of risk is not always required as part of the REACH submission. There are often misunderstandings surrounding whether or not an assessment of risk is necessary. The following questions are pertinent:

- Does the substance have any identifiable hazard(s)?
- What is the tonnage?
- What is the pattern of use? Could the foreseeable life cycle of the substance lead to exposure of humans or the environment?
- Are any reasons associated with limited exposure used as part of waiving for any of the property data endpoints?

Assessment of exposure and risk is necessary when a substance has any identifiable hazard (see above), has a tonnage of at least 10 tonnes/year, and is used in any application not operated under 'strictly controlled conditions' (an exceptional level of control which, if demonstrated to specified standards, can mean a reduced registration package is possible) and in the event that exposure-based adaptation is used in data waiving anywhere in the technical dossier.

1.2.10.2   Qualitative versus Quantitative Approaches

In the case of many hazards, a quantitative approach to assessing risk is possible. This is normally the case for human health short-term and long-term exposure and for environmental effects.

In the case of human health, the effects data are used to derive an estimated safe dose (derived no-effect level or DNEL – for various exposure pathways and types of user). The exposure modelling leads to an estimated dose to humans associated with workplace or consumer use, or exposure via the environment. The risk is then characterised by taking the ratio of estimated exposure/DNEL, leading to a RCR. A value of RCR $<$ 1 indicates that the risk is acceptable.

In the case of the environment, the estimated safe dose for an ecosystem is called the predicted no-effect level (PNEC – for various environmental compartments, e.g. aquatic organisms, soil organisms). Exposure modelling leads to predicted environmental concentrations (PECs) for the equivalent compartments. The risk is then characterised by taking the ratio of PEC/PNEC, leading to a RCR (occasionally referred to as risk quotient). A value of RCR $<$ 1 indicates that the risk is acceptable.

1.2.10.3 Management of Risks if RCR Is Equal to or Greater Than One

It is important for registrants to be aware that if any RCR $\geq$ 1 is found, indicating an unacceptable risk, then further work is needed on the part of the registrant. The RCR must be refined to give a value less than one, before the registration can be made. This would usually be done by adapting the exposure assessment to take account of additional measures to restrict the relevant exposure. Some tips are presented in this book in Chapters 8 and 10, discussing exposure assessment and refinements, and in Chapter 11 on managing risks.

1.2.10.4 Test Proposal Rule

For chemical properties that are required under Annexes IX–X of REACH, new testing must not be conducted without the permission of the ECHA. If the endpoint applies at the tonnage band for the substance, and no existing data (including prediction) are available, the registrant must include a testing proposal in the registration. The ECHA will review the testing proposals as part of its evaluation process (Section 1.2.3 in this chapter). For any proposed test in vertebrates, a public consultation procedure is additionally undertaken as part of the evaluation, to establish if any studies already exist and invite comments on the proposal. See Example 1.1 and Case Study 1.1.

Example 1.1

The registrant has no information regarding the 90-day repeated dose toxicity.
The registrant proposes a study.
The ECHA conducts a public consultation, stating that the registrant proposal appears correct in the light of the available data.
Other industry bodies and commentators review the proposal.
A non-governmental organisation is aware that there is relevant data held in the database of a non-EU country.
The registrant gains access to that data and therefore proposes to not test.
The ECHA considers that the study is not compliant and insists on a test.
The registrant appeals.
The Member States and the ECHA review the information, including further representations from all stakeholders, and a final decision is reached. No appeal is possible.

Case Study 1.1   Test proposal procedure

The Lead Registrant of a substance (requiring an Annex IX-compliant data set) has evaluated the available data. The SIEF members own several studies and two of the three substances are well described in the published chemical literature; also the substances are within the applicability domain of established (Q)SARs. The registrant concludes that the physico-chemical, environmental and parts of the mammalian toxicity data sets are adequately covered but that the substance lacks subchronic repeated dose toxicity data and sensitisation and eye irritation data. The registrant concludes that testing of these three endpoints is necessary to complete the hazard assessment and risk characterisation.

1. The registrant has already established that no such data exist already within the SIEF or in the public domain.
2. The sensitisation study is required at Annex VIII, so the registrant proceeds and commissions a suitable laboratory to undertake the test using the method recommended in Guidance part R7a, following the Guideline in Regulation EC 440/2008.
3. Eye irritation data are required at Annex VII. For eye irritation a stepwise approach is necessary, because it is required to initially assess the effect in an *in vitro* test. An *in vivo* test may need to be conducted afterwards, depending on the outcomes. Since the *in vivo* eye irritation test is required at Annex VIII, the registrant may also proceed and commission the test providing this stepwise approach is followed.
4. The 90-day repeated dose test is required at Annex IX, so the registrant must not proceed to undertake this test. The registrant includes a Testing Proposal in the IUCLID dossier in the relevant endpoint section, including details of the substance identity to be tested and the guideline method to be used.
5. Following registration, the ECHA evaluates the Test Proposal. The ECHA publishes on its web site the substance identity with the study that has been proposed, and a data holder for a structurally similar substance comes forward with an existing study which can be read-across.

1.2.10.5   *Availability of Existing Data and Rights of Access*

In the course of preparing a dossier, it is often discovered that there are a lot of useful data already in existence. While 'completeness' is highly desirable, the rights and investments of each data owning organisation must be respected. Copyright must also be held for published data. It is very important to ensure appropriate rights of access are put in place between the registrant and the data owner before the registration is made. A fee is normal, and is generally in proportion to the typical cost of the test in question. This also applies to data originating from regulatory authorities and published data. Some publishers require a fee or impose terms and conditions.

Table 1.3 *Definitions of the Klimisch reliability codes.*

Klimisch	Definition	Comment
1	Reliable	Compliant with the required test guideline, test conducted with no significant shortcomings and in compliance with Good Laboratory Practice (GLP).
2	Reliable with restrictions	Close to Klimisch 1 but with a shortcoming, such as a departure from the guideline, incomplete reporting, or the result is a prediction rather than a measurement. Scientific papers in journals can meet these requirements, if exceptionally well documented.
3	Not reliable (also Invalid)	Definitely unreliable due to a deficiency in acceptable scientific practice; such data are usually of no worth.
4	Reliability not assignable	Uncertain reliability, but insufficient information available to resolve the uncertainty; this can include reports of uncertain origin or poorly-reported scientific papers. If the origin of the data is unknown, then Klimisch 3 might be more appropriate. These results could be useful as part of weight of evidence, but are usually of no worth.

Example 1.2

For example, if a measurement of water solubility is made, but for reasons of analytical difficulties a definitive result cannot be calculated, the study could still be Klimisch 1 or 2. The uncertainty in this case is due to a technical limitation rather than the study being poorly executed or reported. However, this should not become an excuse for work falling short of widely-accepted quality criteria.

1.2.10.6   Data Reliability

When a study report is reviewed for inclusion in a registration data set, a Klimisch code should be assigned. These codes are summarised in Table 1.3.

Some degree of expertise should be applied.

References

CEFIC/VCI (The European Chemical Industry Council/Verband der Chemischen Industrie e.V.) (2009) REACH Practical Guide on Exposure Assessment and Communication in the Supply Chains.

EC (2000) Commission of the European Communities Brussels, 2.2.2000 COM(2000) 1 Final Communication from the Commission on the Precautionary Principle.

ECHA (European Chemicals Agency) (nd-a) PPORD Exemption, http://echa.europa.eu/web/guest/regulations/reach/substance-registration/ppord-exemption (last accessed 11 July 2013).

ECHA (European Chemicals Agency) (nd-b) REACH-IT Login, http://reach-it.echa.europa.eu/ (last accessed 11 July 2013).

ECHA (European Chemicals Agency) (nd-c) REACH Legislation, http://echa.europa.eu/web/guest/regulations/reach/legislation (last accessed 11 July 2013).

ECHA (European Chemicals Agency) (2008) Pre-registration, Data Submission to ECHA and REACH, http://echa.europa.eu/documents/10162/17096/press_memo2_20080603_en.pdf (last accessed 30 July 2013).

ECHA (European Chemicals Agency) (2012) Guidance on Information Requirements and Chemical Safety Assessment, Chapter R.16: Environmental Exposure Estimation.

EPA (US Environmental Protection Agency) (2011) Exposure Factors Handbook (http://www.epa.gov/ncea/efh/report.html) (last accessed 11 July 2013).

HSE (UK Health and Safety Executive) (nd) ALARP "at a glance". http://www.hse.gov.uk/risk/theory/alarpglance.htm (last accessed 30 July 2013).

IPCS (International Programme on Chemical Safety) (2001) Glossary of Exposure Assessment-Related Terms: A Compilation, IPCS Exposure Terminology Subcommittee, International Programme on Chemical Safety, World Health Organization, http://www.who.int/ipcs/publications/methods/harmonization/en/compilation_nov2001.pdf (last accessed 11 July 2013).

2

Roles and Responsibilities in REACH

2.1 The Structure and Responsibilities of the Authorities

2.1.1 Role of the ECHA

The ECHA (European Chemical Authority) is a regulatory authority operating within the European Union, based in Helsinki, which manages the following aspects of REACH:

- Technical
- Scientific
- Administrative.

The ECHA not only assists companies with information regarding the use of chemicals, but also the legislation around these chemicals, and safety information.

To make it easier to comprehend how the European Chemical Authority operates, firstly the structure of the ECHA needs to be developed. Like all organisations, there is a Management Board, as well as an Executive Director; which are respectively responsible for the financial planning and the annual reporting of the agency, as well as the legal representative of the agency. In addition to this, there is a Member State Committee, comprising of the 27 European Union Member States and the European Economic Area (EEA) countries[1] Moreover, the Risk Assessment Committee and the Committee for Socio-economic Analysis prepare opinions on evaluation, proposals for restrictions on classification and labelling and applications for authorisation. Lastly, there is a Forum, which coordinates the network between Member State Authorities responsible for the enforcement of REACH.

[1] Currently: Norway, Iceland and Lichtenstein. Although these countries participate in the bodies as well as the network, they do not have the right to vote.

Chemical Risk Assessment: A Manual for REACH, First Edition. Peter Fisk Associates Ltd.
© 2014 John Wiley & Sons, Ltd. Published 2014 by John Wiley & Sons, Ltd.

2.1.2   The Role of the Member State Committee (MSC)

The Member State Committee (MSC) participates primarily in evaluation and authorisation regarding REACH. It is up to the MSC to resolve differences of opinions between Member States and the identification of substances of very high concern. The committee then provides opinions and thoughts concerning the ECHA's drafts regarding the substance evaluation process. If an agreement cannot be met within the MSC, the matter is referred to the European Commission. To avoid this referral to the European Commission, the MSC attempts to obtain unanimous agreement on Member State draft decisions regarding substance evaluation. Additionally, the MSC seeks to resolve cases when two or more Member States have expressed an interest in evaluating the same substance. Furthermore, the MSC also provides thoughts on the CoRAP (draft community rolling action plan), which is prepared by the ECHA regarding substances that need to be evaluated by Member States due to the fact that they could pose a risk to not only the environment but also human health. Secondly, the MSC also deals with any other proposals from Member States, seeking to add additional substances onto the CoRAP outside the annual update.

2.1.3   The Role of the Member State Competent Authorities (MSCA)

The role of MSCAs is defined in Article 123 of REACH.[2] However, this really does need expanding, so that the responsibilities of the authorities can be better understood.

- **The general public** does not merely mean people who are exposed to the substance, in so far that they are exposed to a release of chemicals, but also the final consumer. Despite this, it is also important to recognise the role which other organisations (Governmental or non-Governmental) have in engaging with the public about the risks chemicals pose and how to use them safely.
- **What are the risks arising from substances?** As mentioned previously, there is a Risk Assessment Committee and plenty of internal communication; this question is posed regarding the safety of the general public, in so far as informing them of particular risks of certain chemicals, for example household bleach.
- **When is communication necessary?** By communicating via the Risk Assessment Network, Member States are able to take a coordinated approach, but the final decision remains with the MSCAs.

The competent authorities for REACH and CLP (Classification, Labelling and Packaging) are the Member States. A number of these are agencies in the environmental sector. The designation of the competent authorities was completed by the Member States' governments in 2010, allowing for closer cooperation; regarding not only the exchange of information but also providing support, as well as training, to other Member States.

---

[2] The competent authorities of the Member States shall inform the general public about the risks arising from substances where this is considered necessary for the protection of human health or the environment. The Agency, in consultation with competent authorities and stakeholders and drawing as appropriate on relevant best practice, shall provide guidance for the communication of information on the risks and safe use of chemical substances, on their own, in preparations or in articles, with a view to coordinating Member States in these activities.

2.2   Forum Enforcement Project – REACH-EN-FORCE-1

REACH-EN-FORCE-1 was the first of a number of EU-wide REACH enforcement projects. Although, the term enforcement often co-notates prosecution, recently it involves less forceful action. The project was designed to enforce the principle: 'no data, no market', which is at the core of what REACH seeks to combat. Although this example has already ended, it is a good demonstration of what the authorities are trying to accomplish with REACH, and it has led to future projects, including the pre-registration enforcement campaign, launched in 2009.

2.3   Future Aims of the HSE (an Example of a ECHA-Related Authority Acting in the UK)

Although Phase 1 has passed, the initial slogan of 'no data, no market' is still at the core of REACH. For example, restriction no. 50 (Annex XVII of REACH) (EU, 2006) limits the amount of polycyclic-aromatic hydrocarbons (PAHs) that can be found in tyres placed on the market after 1 January 2010. Polycyclic-aromatic hydrocarbons are a group in excess of 100 chemical substances, eight of which are carcinogenic, damaging not only human beings but also the environment.

Other responsibilities of the Competent Authority under REACH are to:

- Provide advice to interested parties on their respective responsibilities under REACH through a 'helpdesk'.
- Enforce compliance with REACH registration.
- Conduct in depth evaluation of selected prioritised substances and prepare draft decisions.
- Propose harmonised classification and labelling.
- Identify substances of very high concern for authorisation.
- Propose restrictions.
- Nominate candidates to membership of ECHA committees on risk assessment and socio-economic analysis.
- Appoint members for the Member State committee.
- Appoint a member to the Forum for Information Exchange and meet to discuss enforcement matters.
- Liaise as appropriate with relevant enforcing organisations in relation to downstream responsibilities under REACH; for example, the implementation by users of the risk management measures designated by suppliers, the adherence to restrictions or to the refusal of authorisation.

2.4   What Does REACH Require as Regards Enforcement?

Titles XIII and XIV of the REACH Regulations require each Member State to select a competent authority (as mentioned above) (Table 2.1). This competent authority must maintain a control system for enforcement (required date 1 December 2008) that needs

Table 2.1 *Regulatory authorities in the United Kingdom.*

	Environmental protection	Health and safety
England and Wales	Environmental Agency	HSE and district councils
Scotland	SEPA	HSE and district councils
Northern Ireland	NIEA	HSENI and district councils
Offshore	Department of Energy and Climate Change	HSE

to provide dissuasive penalties for those who choose not to comply with the regulations set forward in REACH. The other aspect of enforcement comes through the Forum, which will coordinate enforcement projects and inspections – furthering the levels of cooperation and coordination.

2.5   What Powers Do Enforcing Authorities Have?

The REACH Enforcement Regulations (2008) provide each authority the powers required to carry out their responsibilities; these are not only regarding inspection but also for enforcement of the legislation – the proviso of enforcement notices as well as being able to prosecute those who breach the regulations.

2.6   The Responsibilities of Industry

The industry actors for the purpose of REACH are:

1. The Manufacturer
2. The Importer
3. 'Only representative'
4. The downstream user (DU)

   Although these actors in the chemical supply chain have different responsibilities, one of the key aims of REACH is sharing of data and communication within an industry sector and among the various industry actors. REACH puts primacy of responsibility on industry.

2.6.1   Responsibilities of the Manufacturer

For the purpose of REACH, the manufacturer is any 'natural or legal person'[3] established within the European Community and who produces or extracts (in its natural state)

---

[3] The phrase 'natural person' or 'legal person' is one used in legal systems and simply means a person or entity recognised as having capability, right and obligations under the law. Refer to the REACH guidance on registration (ECHA, 2012) for more details and clarification of this phrase.

a chemical substance in the European Community. The manufacturer has the first and main responsibility of assessing the hazards and risks of chemical substances it produces, whether on its own, in a mixture or in an article. The assessment should cover the whole life cycle of the substance, that is manufacturing, use and waste disposal.

The manufacturer then has the obligation to register the substance with ECHA (see above).

The manufacturer also has the responsibility of:

1. Implementing risk management measures to prevent identified risks.
2. Covering an identified use provided by the downstream user in its registration, or informing the downstream user of uses not covered.
3. Communicating hazardous properties and relevant risk management measures down the supply chain by using an extended Safety Data Sheet (eSDS).

In addition, producers should supply information on the safe use of articles to industrial and professional users, and consumers on request.

2.6.2   Responsibilities of the Importer

The importer is any 'natural or legal person' established within the European Community who physically introduces a chemical substance on its own or in a mixture into the customs territory of the European Community. Such 'physical introduction' is regarded under REACH as placing on the market,[4] and gives the importer the main responsibility of assessing the risks and hazards and of chemical substances s/he imports.

All the responsibilities of the manufacturer (Section 2.6.1 of this chapter) also apply to the importer.

2.6.3   The Only Representative

A manufacturer, formulator or article producer outside the European Community cannot make a registration under REACH. It can, however, appoint an 'only representative', established in the European Community, to make a REACH registration for substances it supplies to the EU. The 'only representative' then has the obligation to make a REACH registration and has the responsibilities of the importer (ECHA, 2012).

2.6.4   Responsibilities of the Downstream User

The downstream user is any 'natural or legal person' established within the European Community who uses a substance, either on its own or in a mixture by way of processing, formulation, in the course of its industrial or professional activities but is not the manufacturer or the importer of the substance. The downstream user is one that is not the first in line in the supply chain of the substance used. Thus, it is possible for an industry actor which is a manufacturer of a substance to be a downstream user of another substance.

---

[4] Placing on the market means supplying or making available, whether in return for payment or free of charge, to a third party.

Unlike the manufacturer or importer, the downstream user does not have the obligation to register substances under REACH. The downstream user, however, has the responsibility of:

1. informing their supplier (the manufacturer or importer) of transfer from one container to another, mixing, production of an article or any other utilisation (Article 3(24));
2. checking that their use is within the boundary of use(s) covered by the supplier in the eSDS;
3. implementing risk management measures for safe use of substances;
4. performing a chemical safety assessment (CSA) and preparing a chemical safety report (CSR) for any use(s) not covered by the supplier in the eSDS.

2.7   Communication in the Supply Chain and with Regulators

2.7.1   Use Descriptor System

The assessment of the full life cycle of all substances in terms of human exposure is required for compliance with current REACH requirements. The Use Descriptor System, discussed in detail in ECHA Guidance R12 (ECHA, 2010), has been established to facilitate communication up and down the supply chain. The purpose of the Use Descriptor System is to standardise how different exposure scenarios are described. The five descriptor lists defined by the ECHA are shown in Tables 2.2–2.6; they serve as the starting point for exposure estimation. These are divided into:

1. Sector of Use (SU)
2. Chemical Product Category (PC)
3. Process Category (PROC)
4. Environmental Release Category (ERC)
5. Article Category (AC)

The environment descriptors are discussed further in Environmental Exposure Modelling (Section 8.7 of this book). Various exposure estimation tools are available; these make use of the defined use descriptors as input parameters as the starting point for human health exposure assessment. The approach to human exposure modelling is tiered, as recommended by REACH guidance. The underlying purpose of a tiered approach is to prioritise chemicals where the potential risk to human health and the environment requires a more in-depth level of risk assessment. Three phases of the tiered approach exist: Tier 0, Tier 1 and Tier 2. Tier 0 is aimed at screening out substances and associated exposure scenarios that are considered low risk. Chemicals and conditions that are not excluded in the Tier 0 screen are then evaluated in Tier 1. Tier 1 is considered a more conservative system which discriminates between scenarios of some concern and no concern without much refinement. The risk characterisation ratio (RCR), which compares actual exposure levels to derived no-effect levels (DNELs), defines the level of concern as the basis of the discrimination between exposure scenarios. Tier 1 models also use the use descriptor codes as their starting point. In turn, a Tier 2 approach makes use of more refined exposure models and/or existing exposure measurements. Tier 2 models are usually employed where a Tier 1 model has highlighted a possible risk and where, based

Table 2.2   *Descriptor list for sectors of use (SU).*

Key descriptor: main user groups	
SU 3	Industrial uses: uses of substances as such or in preparations[a] at industrial sites
SU 21	Consumer uses: private households (= general public = consumers)
SU 22	Professional uses: public domain (administration, education, entertainment, services, craftsmen)

Supplementary descriptor: sectors of end-use		NACE21 codes
SU1	Agriculture, forestry, fishery	A
SU2a	Mining, (without offshore industries)	B
SU2b	Offshore industries	B6
SU4	Manufacture of food products	C10,11
SU5	Manufacture of textiles, leather, fur	C13–15
SU6a	Manufacture of wood and wood products	C16
SU6b	Manufacture of pulp, paper and paper products	C17
SU7	Printing and reproduction of recorded media	C18
SU8	Manufacture of bulk, large scale chemicals (including petroleum products)	C19.2 + 20.1
SU9	Manufacture of fine chemicals	C20.2–20.6
SU10	Formulation (mixing) of preparations and/or re-packaging (excluding alloys)	C20.3–20.5
SU11	Manufacture of rubber products	C22.1
SU12	Manufacture of plastics products, including compounding and conversion	C22.2
SU13	Manufacture of other non-metallic mineral products, for example plasters, cement	C23
SU14	Manufacture of basic metals, including alloys	C24
SU15	Manufacture of fabricated metal products, except machinery and equipment	C25
SU16	Manufacture of computer, electronic and optical products, electrical equipment	C26 to 27
SU17	General manufacturing, for example machinery, equipment, vehicles, other transport equipment	C28–30,33
SU18	Manufacture of furniture	C31
SU19	Building and construction work	F
SU20	Health services	Q86
SU23	Electricity, steam, gas water supply and sewage treatment	C35–37
SU24	Scientific research and development	C72
SU0	Other	

[a]For the sake of consistency with the descriptor system in IUCLID 5.2, in these lists the term "preparation" has not been replaced by "mixture".

Table 2.3   *Descriptor list for chemical product categories (PC).*

Chemical product category (PC)

	Category for describing market sectors (at supply level) regarding all uses (workers and consumers)	Examples and explanations
PC1	Adhesives, sealants	–
PC2	Adsorbents	–
PC3	Air care products	–
PC4	Antifreeze and de-icing products	–
PC7	Base metals and alloys	–
PC8	Biocidal products (e.g. Disinfectants, pest control)	PC 35 should be assigned to disinfectants being used as a component in a cleaning product
PC9a	Coatings and paints, thinners, paint removers	–
PC9b	Fillers, putties, plasters, modelling clay	–
PC9c	Finger paints	–
PC11	Explosives	–
PC12	Fertilisers	–
PC13	Fuels	–
PC14	Metal surface treatment products, including galvanic and electroplating products	This covers substances permanently binding with the metal surface
PC15	Non-metal-surface treatment products	Such as, for example, treatment of walls before painting
PC16	Heat transfer fluids	–
PC17	Hydraulic fluids	–
PC18	Ink and toners	–
PC19	Intermediate	–
PC20	Products such as pH regulators, flocculants, precipitants, neutralisation agents	This category covers processing aids used in the chemical industry
PC21	Laboratory chemicals	–
PC23	Leather tanning, dye, finishing, impregnation and care products	–
PC24	Lubricants, greases, release products	–
PC25	Metal working fluids	–
PC26	Paper and board dye, finishing and impregnation products: including bleaches and other processing aids	–
PC27	Plant protection products	–
PC28	Perfumes, fragrances	–
PC29	Pharmaceuticals	–
PC30	Photochemicals	–

Table 2.3   (continued)

Chemical product category (PC)		
	Category for describing market sectors (at supply level) regarding all uses (workers and consumers)	Examples and explanations
PC31	Polishes and wax blends	–
PC32	Polymer preparations and compounds	–
PC33	Semiconductors	–
PC34	Textile dyes, finishing and impregnating products; including bleaches and other processing aids	–
PC35	Washing and cleaning products (including solvent based products)	–
PC36	Water softeners	–
PC37	Water treatment chemicals	–
PC38	Welding and soldering products (with flux coatings or flux cores.), flux products	–
PC39	Cosmetics, personal care products	–
PC40	Extraction agents	–
PC0	Other (use UCN codes: see last row)	–

Table 2.4   Descriptor list for process categories (PROC).

Process categories (PROC)		
	Process categories	Examples and explanations
PROC1	Use in closed process, no likelihood of exposure	Use of the substances in high integrity contained system where little potential exists for exposures, for example any sampling via closed loop systems.
PROC2	Use in closed, continuous process with occasional controlled exposure	Continuous process but where the design philosophy is not specifically aimed at minimising emissions. It is not high integrity and occasional expose will arise. for example through maintenance, sampling and equipment breakages.
PROC3	Use in closed batch process (synthesis or formulation)	Batch manufacture of a chemical or formulation where the predominant handling is in a contained manner, for example through enclosed transfers, but where some opportunity for contact with chemicals occurs, for example through sampling.
PROC4	Use in batch and other process (synthesis) where opportunity for exposure arises	Use in batch manufacture of a chemical where significant opportunity for exposure arises, for example during charging, sampling or discharge of material, and when the nature of the design is likely to result in exposure.

(continued overleaf)

Table 2.4 *(continued)*

Process categories (PROC)

	Process categories	Examples and explanations
PROC5	Mixing or blending in batch processes for formulation of preparations* and articles (multistage and/or significant contact)	Manufacture or formulation of chemical products or articles using technologies related to mixing and blending of solid or liquid materials, and where the process is in stages and provides the opportunity for significant contact at any stage.
PROC6	Calendering operations	Processing of product matrix calendering at elevated temperature an large exposed surface.
PROC7	Industrial spraying	Air dispersive techniques. Spraying for surface coating, adhesives, polishes/cleaners, air care products, sandblasting. Substances can be inhaled as aerosols. The energy of the aerosol particles may require advanced exposure controls; in the case of coating, overspray may lead to waste water and waste.
PROC8a	Transfer of substance or preparation (charging/discharging) from/to vessels/large containers at non-dedicated facilities	Sampling, loading, filling, transfer, dumping, bagging in non-dedicated facilities. Exposure related to dust, vapour, aerosols or spillage, and cleaning of equipment to be expected.
PROC8b	Transfer of substance or preparation (charging/discharging) from/to vessels/large containers at dedicated facilities	Sampling, loading, filling, transfer, dumping, bagging in dedicated facilities. Exposure related to dust, vapour, aerosols or spillage, and cleaning of equipment to be expected.
PROC9	Transfer of substance or preparation into small containers (dedicated filling line, including weighing)	Filling lines specifically designed to both capture vapour and aerosol emissions and minimise spillage.
PROC10	Roller application or brushing	Low energy spreading of, for example, coatings. Including cleaning of surfaces. Substance can be inhaled as vapours, skin contact can occur through droplets, splashes, working with wipes and handling of treated surfaces.
PROC11	Non-industrial spraying	Air dispersive techniques. Spraying for surface coating, adhesives, polishes/cleaners, air care products, sandblasting. Substances can be inhaled as aerosols. The energy of the aerosol particles may require advanced exposure controls.

Table 2.4   *(continued)*

Process categories (PROC)

	Process categories	Examples and explanations
PROC12	Use of blowing agents in manufacture of foam	–
PROC13	Treatment of articles by dipping and pouring	Immersion operations. Treatment of articles by dipping, pouring, immersing, soaking, washing out or washing in substances; including cold formation or resin type matrix. Includes handling of treated objects (e.g. after dying, plating). Substance is applied to a surface by low energy techniques such as dipping the article into a bath or pouring a preparation onto a surface.
PROC14	Production of preparations* or articles by tabletting, compression, extrusion, pelletisation	Processing of preparations and/or substances (liquid and solid) into preparations or articles. Substances in the chemical matrix may be exposed to elevated mechanical and/or thermal energy conditions. Exposure is predominantly related to volatiles and/or generated fumes, dust may be formed as well.
PROC15	Use as laboratory reagent	Use of substances at small scale laboratory (< 1 l or 1 kg present at workplace). Larger laboratories and R+D installations should be treated as industrial processes.
PROC16	Using material as fuel sources, limited exposure to unburned product to be expected	Covers the use of material as fuel sources (including additives) where limited exposure to the product in its unburned form is expected. Does not cover exposure as a consequence of spillage or combustion.
PROC17	Lubrication at high energy conditions and in partly open process	Lubrication at high energy conditions (temperature, friction) between moving parts and substance; significant part of process is open to workers. The metal working fluid may form aerosols or fumes due to rapidly moving metal parts.
PROC18	Greasing at high energy conditions	Use as lubricant where significant energy or temperature is applied between the substance and the moving parts.
PROC19	Hand mixing with intimate contact and only personal protective equipment (PPE) available	Addresses occupations where intimate and intentional contact with substances occurs without any specific exposure controls other than PPE.
PROC20	Heat and pressure transfer fluids in dispersive, professional use but closed systems	Motor and engine oils, brake fluids. Also in these applications, the lubricant may be exposed to high energy conditions and chemical reactions may take place during use. Exhausted fluids need to be disposed of as waste. Repair and maintenance may lead to skin contact.

(continued overleaf)

Table 2.4   *(continued)*

Process categories (PROC)

	Process categories	Examples and explanations
PROC21	Low energy manipulation of substances bound in materials and/or articles	Manual cutting, cold rolling or assembly/disassembly of material/article (including metals in massive form), possibly resulting in the release of fibres, metal fumes or dust
PROC22	Potentially closed processing operations with minerals/metals at elevated temperature. Industrial setting	Activities at smelters, furnaces, refineries, coke ovens. Exposure related to dust and fumes to be expected. Emission from direct cooling may be relevant.
PROC23	Open processing and transfer operations with minerals/metals at elevated temperature	Sand and die casting, tapping and casting melted solids, drossing of melted solids, hot dip galvanising, raking of melted solids in paving. Exposure related to dust and fumes to be expected.
PROC24	High (mechanical) energy workup of substances bound in materials and/or articles	Substantial thermal or kinetic energy applied to substance (including metals in massive form) by hot rolling/forming, grinding, mechanical cutting, drilling or sanding. Exposure is predominantly expected to be to dust. Dust or aerosol emission as result of direct cooling may be expected.
PROC25	Other hot work operations with metals	Welding, soldering, gouging, brazing, flame cutting. Exposure is predominantly expected to fumes and gases.
PROC26	Handling of solid inorganic substances at ambient temperature	Transfer and handling of ores, concentrates, raw metal oxides and scrap; packaging, unpackaging, mixing/blending and weighing of metal powders or other minerals 23
PROC27a	Production of metal powders (hot processes)	Production of metal powders by hot metallurgical processes (atomisation, dry dispersion).
PROC27b	Production of metal powders (wet processes)	Production of metal powders by wet metallurgical processes (electrolysis, wet dispersion).

[a]For the sake of consistency with the descriptor system in IUCLID 5.2, in these lists the term "preparation" has not been replaced by "mixture".

Table 2.5 *Descriptor list for environmental release categories (ERC).*

Environmental release categories

	Name	Description
ERC 1	Manufacture of substances	Manufacture of organic and inorganic substances in chemical, petrochemical, primary metals and minerals industry including intermediates, monomers using continuous processes or batch processes applying dedicated or multipurpose equipment, either technically controlled or operated by manual interventions.
ERC 2	Formulation of preparations	Mixing and blending of substances into (chemical) preparations in all types of formulating industries, such as paints and do-it-yourself products, pigments paste, fuels, household products (cleaning products), lubricants, and so on.
ERC 3	Formulation in materials	Mixing or blending of substances which will be physically or chemically bound into or onto a matrix (material), such as plastics additives in master batches or plastic compounds. For instance plasticisers or stabilisers in polyvinyl chloride (PVC) master batches or products, crystal growth regulator in photographic films and so on.
ERC 4	Industrial use of processing aids in processes and products, not becoming part of articles	Industrial use of processing aids in continuous processes or batch processes applying dedicated or multipurpose equipment, either technically controlled or operated by manual interventions. For example, solvents used in chemical reactions or the 'use' of solvents during the application of paints, lubricants in metal working fluids, anti-set-off agents in polymer moulding/casting.
ERC 5	Industrial use resulting in inclusion into or onto a matrix	Industrial use of substances as such or in preparations (non-processing aids), which will be physically or chemically bound into or onto a matrix (material) such as binding agent in paints and coatings or adhesives, dyes in textiles fabrics and leather products, metals in coatings applied through plating and galvanising processes. The category covers substances in articles with particular function and also substances remaining in the article after having been used as a processing aid in an earlier life cycle stage (e.g. heat stabilisers in plastic processing).
ERC 6a	Industrial use resulting in manufacture of another substance (use of intermediates)	Use of intermediates in, primarily, the chemical industry using continuous processes or batch processes applying dedicated or multipurpose equipment, either technically controlled or operated by manual interventions, for the synthesis (manufacture) of other substances. For instance, the use of chemical building blocks (feedstock) in the synthesis of agrochemicals, pharmaceuticals, monomers and so on.

(continued overleaf)

Table 2.5   *(continued)*

Environmental release categories

	Name	Description
ERC 6b	Industrial use of reactive processing aids	Industrial use of reactive processing aids in continuous processes or batch processes applying dedicated or multipurpose equipment, either technically controlled or operated by manual interventions. For example, the use of bleaching agents in the paper industry.
ERC 6c	Industrial use of monomers for polymerisation	Industrial use of monomers in the production of polymers, plastics (thermoplastics), polymerisation processes. For example the use of vinyl chloride monomer in the production of PVC.
ERC 6d	Industrial use of auxiliaries for polymerisation processes in production of resins, rubbers, polymers	Industrial use of chemicals (cross-linking agents, curing agents) in the production of thermosets and rubbers, polymer processing. For instance, the use of styrene in polyester production or vulcanisation agents in the production of rubbers.
ERC 7	Industrial use of substances in closed systems	Industrial use of substances in close systems. Use in closed equipment, such as the use of liquids in hydraulic systems, cooling liquids in refrigerators and lubricants in engines and dielectric fluids in electric transformers and oil in heat exchangers. No intended contact between functional fluids and products foreseen and thus low emissions via waste water and waste air to be expected.
ERC 8a	Wide dispersive indoor use of processing aids in open systems	Indoor use of processing aids by the public at large or professional use. Use (usually) results in direct release into the environment/sewage system, for example detergents in fabric washing, machine wash liquids and lavatory cleaners, automotive and bicycle care products (polishes, lubricants, de-icers), solvents in paints and adhesives or fragrances and aerosol propellants in air fresheners.
ERC 8b	Wide dispersive indoor use of reactive substances in open systems	Indoor use of reactive substances by the public at large or professional use. Use (usually) results in direct release into the environment, for example sodium hypochlorite in lavatory cleaners, bleaching agents in fabric washing products, hydrogen peroxide in dental care products.
ERC 8c	Wide dispersive indoor use resulting in inclusion into or onto a matrix	Indoor use of substances (non-processing aids) by the public at large or professional use, which will be physically or chemically bound into or onto a matrix (material) such as binding agent in paints or coatings or adhesives, dyeing of textile fabrics.

Table 2.5 *(continued)*

Environmental release categories

	Name	Description
ERC 8d	Wide dispersive outdoor use of processing aids in open systems	Outdoor use of processing aids by the public at large or professional use. Use (usually) results in direct release into the environment, for example automotive and bicycle care products (polishes, lubricants, de-icers, detergents), solvents in paints and adhesives.
ERC 8e	Wide dispersive outdoor use of reactive substances in open systems	Outdoor use of processing aids by the public at large or professional use. Use (usually) results in direct release into the environment, for example the use of sodium hypochlorite or hydrogen peroxide for surface cleaning (building materials).
ERC 8f	Wide dispersive outdoor use resulting in inclusion into or onto a matrix	Outdoor use of substances (non-processing aids) by the public at large or professional use, which will be physically or chemically bound into or onto a matrix (material) such as binding agent in paints and coatings or adhesives.
ERC 9a	Wide dispersive indoor use of substances in closed systems	Indoor use of substances by the public at large or professional (small scale) use in closed systems. Use in closed equipment, such as the use of cooling liquids in refrigerators, oil-based electric heaters.
ERC 9b	Wide dispersive outdoor use of substances in closed systems	Outdoor use of substances by the public at large or professional (small scale) use in closed systems. Use in closed equipment, such as the use of hydraulic liquids in automotive suspension, lubricants in motor oil and break fluids in automotive break systems.
ERC 10a	Wide dispersive outdoor use of long-life articles and materials with low release	Low release of substances included into or onto articles and materials during their service life in outdoor use, such as metal, wooden and plastic construction and building materials (gutters, drains, frames etc.)
ERC 10b	Wide dispersive outdoor use of long-life articles and materials with high or intended release	Substances included into or onto articles and materials with high intended release during their service life from outdoor use. Such as tyres, treated wooden products, treated textile and fabric like sun blinds and parasols and turniture, zinc anodes in commercial shipping and pleasure craft, and brake pads in trucks or cars. This also includes releases from the article matrix as a result of processing by workers. These are processes typically related to PROC 21, 24, 25, for example: sanding of buildings (bridges, facades) or vehicles (ships).

(continued overleaf)

Table 2.5   (continued)

Environmental release categories

	Name	Description
ERC 11a	Wide dispersive indoor use of long-life articles and materials with low release	Low release of substances included into or onto articles and materials during their service life from indoor use. For example, flooring, furniture, toys, construction materials, curtains, footwear, leather products, paper and cardboard products (magazines, books, news paper and packaging paper), electronic equipment (casing).
ERC 11b	Wide dispersive indoor use of long-life articles and materials with high or intended release	Substances included into or onto articles and materials with high or intended release during their service life from indoor use. For example: release from fabrics, textiles (clothing, floor rugs) during washing. This also includes releases from the article matrix as a result of processing by workers. These are processes typically related to PROC 21, 24, 25. For example removal of indoor paints.
ERC 12a	Industrial processing of articles with abrasive techniques (low release)	Substances included into or onto articles and materials are released (intended or not) from the article matrix as a result of processing by workers. These processes are typically related to PROC 21, 24, 25. Processes where the removal of material is intended but the expected release remains low, including for example: cutting of textile, cutting, machining or grinding of metal or polymers in engineering industries.
ERC 12b	Industrial processing of articles with abrasive techniques (high release)	Substances included into or onto articles and materials are released (intended or not) from/with the article matrix as a result of processing by workers. These processes are typically related to PROC 21, 24, 25. Processes where the removal of material is intended, and high amounts of dust may be expected, includes for example: sanding operations or paint stripping by shot blasting.

Table 2.6   *Descriptor list for substances in articles (AC).*

Article categories, no release intended

Article categories (and non-exhaustive examples) for describing the type of article in which the substance is contained during service life and waste life	Suitable TARIC (TARiff Integre Communautaire) chapters
Categories of complex articles	
AC1 Vehicles Examples: Trucks, passenger cars and motor cycles, bicycles, tricycles and associated transport equipment; other vehicles: railway, aircraft, vessels, boats	86–89

Table 2.6 *(continued)*

Article categories, no release intended

	Article categories (and non-exhaustive examples) for describing the type of article in which the substance is contained during service life and waste life	Suitable TARIC (TARiff Integre Communautaire) chapters
AC2	Machinery, mechanical appliances, electrical/electronic articles Examples: Machinery and mechanical appliances; electrical and electronic articles, for example computers, video and audio recording, communication equipment; lamps and lightening; cameras; refrigerator, dish washer, washing machines	84/85
AC3	Electrical batteries and accumulators	8506/07

Categories of material-based articles

AC4	Stone, plaster, cement, glass and ceramic articles Examples: Glass and ceramic article: for example dinner ware, drinking glasses, pots, pans, food storage containers; construction and isolation articles; natural or artificial abrasive powder or grain, on a base of textile material, of paper, of paperboard or of other materials	68/69/70
AC5	Fabrics, textiles and apparel Examples: Clothing, bedding, mattresses, curtains, upholstery, carpeting/flooring, car seats, textile toys	50–63, 94/95
AC6	Leather articles Examples: Gloves, purses, wallets, footwear, furniture	41–42, 64, 94
AC7	Metal articles Examples: Cutlery, cooking utensils, pots, pans, jewellery, toys, furniture, construction articles	71, 73–83, 95
AC8	Paper articles Examples: Paper articles: tissues, towels, disposable dinnerware, nappies, feminine hygiene products, adult incontinence products; paper articles for writing, office paper; printed paper articles: for example newspapers, books, magazines, printed photographs; wallpaper	48–49
AC10	Rubber articles Examples: Tyres, flooring, gloves, footwear, toys	40, 64, 95
AC11	Wood articles Examples: Flooring, walls, furniture, toys, construction supplies	44, 94–95
AC13	Plastic articles Examples: Tyres, flooring, gloves, footwear, toys	39,94/95, 85/86

(continued overleaf)

Table 2.6   *(continued)*

Use descriptor for articles with intended release of substances
Descriptor based on an indicative list of examples

AC30	Other articles with intended release of substances, please specify
AC31	Scented clothes
AC32	Scented eraser
AC33	*Entry removed after the REACH CA meeting in March 2008*
AC34	Scented toys
AC35	Scented paper articles
AC36	Scented CD
AC38	Packaging material for metal parts, releasing grease/corrosion inhibitors

upon observations or physical properties of a substance, the result of a Tier 1 model is deemed too conservative. Exposure estimation tools which may be applied for Tier 1 and Tier 2 exposure estimation are discussed in more detail in Chapter 10 of this book, including ECETOC (European Centre for Ecotoxicology and Toxicology of Chemicals) Targeted Risk Assessment (TRA), the Advanced REACH Tool (ART), ConsExpo and Stoffenmanager.

References

ECHA (2010) Guidance on Information Requirements and Chemical Safety Assessment Chapter R.12: Use Descriptor System, Version 2, March 2010.

ECHA (2012) Guidance on Registration, Guidance for the Implementation of REACH, Version 2, May 2012.

EU (2006) Annex XVII. Restrictions on the manufacture, placing on the market and use of certain dangerous substances, preparations and articles. *Official Journal of the European Union*, **L 396**, 395, http://ec.europa.eu/enterprise/sectors/chemicals/files/markrestr/annex_xvii_301206_en.pdf (last accessed 12 July 2013).

3

Control of Chemicals – Legislative and Policy Context

The focus of this book is REACH. However, REACH exists alongside other legislation that is also aimed at the control of chemicals. Putting REACH in context, therefore, merits some understanding of why and how REACH was put on the EU (European Union) statute book and what other EU and national level legislation interacts with REACH. The aim of this chapter, therefore, is to set REACH in context and to indicate other legislation that interacts with REACH in either a direct or indirect way. The purpose is not to give a deeper legal understanding, there are numerous legal texts that can do that, but rather to give an overview of the legislative 'landscape' in which REACH operates.

3.1 How EU Chemical Legislation Evolved

REACH is, of course, the culmination of a series of directives and regulations enacted by the European Parliament since the end of the 1960s (Council Directive 67/548/EEC of 27 June 1967 on the approximation of laws, regulations and administrative provisions relating to the classification, packaging and labelling of dangerous substances). It is difficult to understand how REACH relates to other chemical legislations without taking into consideration how they came by in the first place. Historically, the primary consideration of the European Commission and of the Member States was initially worker and consumer protection. Later on, especially after the accession of Sweden to the EU (in 1995), more emphasis was placed on environmental protection. In particular, many of the non-enforceable environmental goals of OSPAR (Oslo–Paris Commission, historically piloted by Sweden and composed by several EU and non-EU states), which predates the EU, were gradually included in European legislation such as the Water Framework Directive (WFD) (2000/60/EC). While, retrospectively, the gradual development of chemical legislation may have been seen as haphazard, the underlying reason was

Chemical Risk Assessment: A Manual for REACH, First Edition. Peter Fisk Associates Ltd.
© 2014 John Wiley & Sons, Ltd. Published 2014 by John Wiley & Sons, Ltd.

that both regulators and industry (through various associations) were struggling to find middle ground between command and control and hazard-based legislation, potentially affecting the competitiveness of the European chemical industry, and a more constructive approach based on industry's experience in the safe handling of hazardous substances and self-regulation.

While REACH was intended to provide an EU-wide, directly applicable (as opposed to nationally enacted) legislation ensuring a harmonised market within its borders, the conflicting positions of non-governmental organisations (relayed by media and by public opinion) and industry, with the EU authorities caught in the middle, continue. The former advocates rapid bans of potentially problematic substances, based solely on their hazard profiles, while the latter accepts the need to adapt the EU's chemical policy but at the same time enhance or at least maintain its competitiveness, protecting and strengthening the internal market and reaching decisions based on risk.

Until the early 1970s the first task of the chemical industry and of the regulators (EU Commission and the Member States represented by their various health and environment protection agencies) was to inventory the thousands of chemicals in commerce and their known properties. Simultaneously, internationally agreed test protocols to generate (eco)toxicological data under Good Laboratory Practices were developed under the auspices of organisations such as the OECD (Organisation for Economic Co-operation and Development).

It should be noted that the Classification and Labelling Directive 65/548/EEC did not impose an obligation for manufacturers to develop new (eco)toxicological data. Manufacturers had a tendency to focus on measuring physical property data (critical for transport and storage safety) and short-term toxicity data. With the exception of Germany which had developed its own system of water endangering classification (WGK – Wassergefährdungsklassen) and with its manufacturers agreeing to voluntarily sponsor basic environmental testing of its products, generally little of the more expensive long-term toxicity data was developed. The only exceptions were for certain chemicals that belonged to families suspected of repeated-exposure hazard properties such as sensitisation, carcinogenicity, reprotoxicity and mutagenicity, or for other substances in wide dispersive use or with a high potential or human exposure, as in cosmetics. More recently, the potential effects of endocrine-mimicking substances on the unborn child are being investigated but there is still some controversy whether existing methods to determine reproductive effects are sufficient. Similarly, the concepts of synergistic health effects between exposure to low doses of chemicals and the long-term environmental effects or secondary poisoning due to a combination of persistency, bioaccumulation and toxic properties have gained increased traction, with the latter now fully part of REACH.

The chemical industry had accumulated a huge amount of practical experience in handling safely several hazardous substances without necessarily undergoing extensive testing. This experience was based on the effects observed during the time where the methodology to assess them in a systematic way was not yet fully developed. Some notorious examples are the use of some metals, inorganic or organic compounds in cooking (lead utensils and lead acetate as a condiment by the Romans), medicine (arsenic for the treatment of leukaemia, psoriasis, mercury for syphilis etc.), cosmetics (lead in eyeliners), jewellery (nickel plating), marine coatings (organotin anti-fouling additives), pesticides (DDT), herbicides (2,4-D or dichlorophenoxy acetic acid), refrigerants (chlorofluorocarbons), felt hats (mercury) and even transformer cooling oils (polychlori-nated biphenyls). The remaining uses of all of these substances are now strictly controlled or banned. New technologies have made possible complete substitution of some of these substances in essential applications, such as mercury in thermometers or in the production of chlorine by separation cells.

Margot Wallström, the EU Environment Commissioner when REACH was enacted, had called for decisive action by claiming that had REACH been in place, asbestos would not have caused and still be causing 100 000 industrial deaths,[1] although one can hardly call asbestos a man-made chemical, the control of which is the primary purpose of REACH. While that remark was widely criticised or applauded, depending on which side the comments were coming from, it was effective in instituting the principle of substitution of CMR (Carcinogenic, Mutagenic, Reprotoxic) Cat. 1a or 1b substances. However, for the reasons explained above, such as greater awareness of the effects of certain chemicals, exposure to them has dropped significantly between the 1930s and the 1970s, showing that the early chemicals legislation has achieved its goals.

Nevertheless, the growing awareness that exposure to certain substances can have some long-term effects that were originally unsuspected has resulted in adopting a more precautionary attitude towards innovation. In the USA, for example, there is a require-ment for industry to report to the EPA (Environmental Protection Agency) significant new uses (SNUR – significant new uses rule), which would be authorised once they are demonstrated to be safe. In the EU, the Commission published in 2000 its interpretation of the Precautionary Principle (EC, 2000) which is referred to in all chemical legislations enacted since then.

At the same time it was realised that the undesirable effects of these substances were also a function of exposure. By restricting the use of these substances to those applications where the benefits can be demonstrated versus the absence of risk, these negative effects can be avoided. For example, DDT is still used to treat mosquito nets, nickel is in every euro coin, a mercury compound is an essential preservative in certain vaccines and without lead metal and chemicals in batteries consumers would certainly have problems running automobiles.

The result of this approach was the so-called Marketing and Use Directive (Limitations Directive, i.e. Directive 76/769/EEC) restricting or banning the use of certain hazardous chemicals and the Dangerous Substances Directive (DSD) 67/548/EEC setting up an in-depth review of existing priority chemicals selected for their hazard properties (CMR, PBT or vPvB, in other words: Carcinogenic, Mutagenic, Reprotoxic, Persistent, Bioac-cumulative, Toxic or very Persistent and very Bioaccumulative) and all new chemicals.

---

[1] Margot Wallström European Commissioner for the Environment Action on Environment and Health, Harvard Medical School, USA 27 April 2004: 'For example, asbestos was once seen as a valuable, versatile material and was used extensively in buildings. Every year people are now dying from exposure to asbestos. It is estimated that, in developed countries alone, 100 000 more people will die. The costs of removing asbestos from buildings and contaminated sites have been enormous. Man-made chemicals accumulate in our bodies. Many workers are exposed to chemicals that can cause allergies, respiratory diseases, cancers and problems with reproduction. In Europe, occupational skin diseases result in the loss of 3 million working days each year. These eczemas and other skin problems can often be directly related to chemicals. They occur in most industries, forcing many affected workers to change jobs. The cost of 3 million working days lost has been estimated at 600 million Euro per year. Globally, the ILO estimates that over 440 000 deaths a year worldwide are related to chemicals at work. Scientists have also spotted a worrying decline in sperm counts and fertility rates in Europe and the US, which may be caused by chemicals. In the sea, chemicals used in antifouling paint lead to sex changes in molluscs, which has exterminated certain species and seriously affected the productivity of aquacultures'.

The experience with the implementation of both Directives was retrospectively viewed as too slow by the European Parliament and some member states (200 new and existing substances were assessed over a period of 30 years). The solution, ironically, was to turn over to industry the responsibility of preparing dossiers on all 30 000 commercial substances with strict deadlines.

The WHO (World Health Organization) had initiated a programme called IPCS (International Programme for Chemical Safety) which published 241 EHC (environmental health criteria) critical reviews on the effects of chemicals and physical and biological agents on human health and the environment. The first EHC concerned mercury and was published in 1976. One of the last for chemicals (2005) was on clay minerals but more recently the focus has been on methodology and on physical agents such as extremely low frequency fields (2007).

The next step was to define a 'base set' of data that would cover the complete hazard profile of a chemical substance and from there to classify the substance, therefore ensuring that appropriate measures would be taken when managing the risk related to exposure. About the same time the OECD and industry through its trade associations (CEFIC in Europe, ACC in the United States and JCIA for Japan) reached an agreement to submit a defined data set (and to fill gaps if any) to an international review panel. The programme, called HPV (high production volume, prioritising substances manufactured at a rate greater than 1000 tonnes/year), is still underway but in a sense has been superseded in the EU by REACH. In the USA, the EPA has taken further steps, firstly by making this programme obligatory for US manufacturers and importers and, secondly, by lowering the reporting threshold to 1 million pounds/year or 450 tonnes/year. Also, more recently, the EPA has set up an Extended HPV programme by including substances that have reached the volume threshold since HPV was started.

As a result of all of this, knowledge about the hazards posed by chemical substances has increased considerably since the 1970s. The process of self-classification based on the new data has already resulted in considerable changes in the production and the use of certain chemicals, which is the reason why industry is confident that the majority of existing chemical uses will be shown to be safe.

The following subsections look in more detail the impacts and the links with REACH on three other important pieces of chemical legislation and their interactions with some of existing national implementations.

Reviewed in more detail are how, in practice, the potential risks associated with chemicals are identified and managed, and their impact on costs and liabilities. The specific risks are identified by scope and their relationship with REACH is noted. How the risks are managed in practice is then discussed.

As seen in Table 3.1, the regulations concerning chemicals fall into several categories:

- Regulations that overlap REACH in some respects:
  - Air
  - Water Framework Directive
  - Carcinogens at Work Directive
  - The cosmetics regulations
  - Biocidal products
  - Plant protection products (PPPs).

Table 3.1 *Relation to REACH of chemical risk management.*

Chemical risk scope	EU legislation	Relation to REACH

Air
  Urban
  Indoor
	Particulates, industrial emissions, solvent emissions 1999/19/EC and 2004/42/EC, IPPC 2008/1/EC	Each solvent use must be risk assessed and demonstrated to be safe.

Biocidal products	Directive 98/8/EC and Regulation (EU) No. 528/2012	Annex I listed substances are considered registered under REACH and are therefore exempted.
Classification and labelling	CLP-Regulation (EC) No. 1272/2008	Basis for identification as a substance of very high concern (SVHC).
Construction products	Construction Products Regulation (305/2011/EU – CPR)	Declaration of content of hazardous substances and identification of risks posed by construction materials.
Cosmetics	Cosmetics Directive 76/768/EEC and recast as Regulation (EC) No. 1223/2009	The environmental impact of cosmetics ingredients must be assessed. CMR ingredients are regulated.
Food	Additives (Directive 89/107/EEC)	Excluded from REACH.
Fresh and coastal waters	Water Framework Directive (WFD) 2000/60/EC, IPPC 2008/1/EC	The WFD provides a framework to set Environmental Quality Standards for chemical substances found in surface waters. The methodologies may differ from REACH.
Ozone depleting substances (ODS)	ODS legislation: Regulation (EC) No. 2037/2000, Regulation (EC) 1005/2009, 2010/372/EU and (EU) 744/2010	By 2012 the consumption of ozone-depleting substances has dropped by 98%. REACH provides for an assessment of the ozone depleting potential of volatile substances.
Greenhouse gases	EU implementation of the Kyoto protocol	Monitoring of greenhouse gases emissions (especially certain fluorinated gases).

Health
  Consumers
  Workers
	Carcinogens at Work Directive 2004/37/EC	REACH provides a mechanism for authorisation of Cat. 1a and 1b carcinogens.

(*continued overleaf*)

Table 3.1 *(continued)*

Chemical risk scope	EU legislation	Relation to REACH
Laboratory animals	The Cosmetics Regulation (EC) No. 1223/2009 contains provisions restricting animal testing of cosmetic ingredients	REACH contains provisions for adapting testing requirements and for reading across to minimise animal suffering and use.
Major accident prevention	'Seveso' Directives: 96/82/EC, 2003/105/EC	Applies to storage of hazardous chemicals. Not in the scope of REACH.
Medicinal products	Directive 2001/83/EC and Regulation (EC) No. 726/2004	Applies to safety of active pharmaceutical ingredients and components of medicinal products. Not in the scope of REACH.
Plant protection products	Directive 2009/128/EC	Normally exempted from REACH. PBT/vPvB assessment may differ from ECHA REACH guidance.
Radioactive substances	Directive 96/29/Euratom	Not in the scope of REACH.
Toys	Directive 2009/48/EC	CMR ingredients, allergenic substances are regulated.
Trade (international)	Rotterdam convention and prior informed consent (PIC) – Council Decision 2006/730/EC and Regulation (EC) n° 689/2008	Applies to banned or extremely restricted and to extremely hazardous pesticides. Not in scope of REACH.
Transportation	United Nations Orange Book, IATA, EU directive 2008/68 – inland transport of dangerous goods	Excluded from REACH.
Waste	WEEE 2012/19/EU, RoHS 2002/95/EC, Landfill Directive	Waste is excluded from REACH.

- Regulations that may apply resulting from compliance with REACH, if new data or an evaluation generated under REACH trigger a change in the hazard classification of a substance:
  - Biocides
  - Construction products
  - Cosmetics
  - Ozone depleting substances (ODS)
  - 'Seveso' directives
  - Toys
  - Prior informed consent (PIC)
  - Transportation
  - Waste.

The six regulations that have potential for overlapping REACH are now reviewed in more detail.

3.2 Air Quality Regulations

Air quality can in principle be managed in several ways:

- By controlling releases to air from production and use of certain substances.
- By setting exposure limits, particularly in the workplace.

With respect to potential exposure to the environment (not only air, but also soil and surface waters) during manufacturing of chemical substances, the applicable legislation would be the Industrial Emissions Directive 2010/75/EU (IED). As this is a European Directive, each Member state enacts this in its own legislation and imposes limits on emissions installation by installation (however, technical standards are set out at EU level in BREF (best available techniques reference) notes as provided by the IPPC (Integrated Pollution Potential and Control) Directive). Compliance with emission limits may mean that releases from relevant processes are controlled to a level that means that they comply with REACH. Not all processes are covered bythe IED, nor all substances. In addition, the Directive calls for implementation of best available techniques (BAT). This requirement goes beyond the requirements of REACH, since the latter are satisfied as long as emissions do not result into an unacceptable risk to health or the environment.

What follows are examples of how the regulators have been dealing historically with certain solvents.

For solvents in coatings and aerosol formulations, the applicable legislations are 1999/19/EC and 2004/42/EC. The driver is the prevention of tropospheric ozone formation. Emissions are controlled by specifying their concentration in the formulation, based on their vapour pressure and without taking into account their contribution to tropospheric ozone formation. The exemption regime for certain solvents that have a lower potential of ozone formation which exists in the USA is not applicable here. However, in some cases where exposure to the solvent might create a high risk, more stringent measures may be taken, such as banning consumer use of paint strippers containing dichloromethane. The applicable regulation controlling the use of certain substances used to be the Marketing and Use Directive 89/677/EEC but this has now been superseded by REACH. The scope of REACH is limited to the hazard characterisation and to the evaluation and control of risks posed by uses of solvents.

For several years the dry-cleaning industry has been under pressure to substitute perchlorethylene ('PERC') with other solvents due its carcinogenic properties (Cat. 2 GHS (globally harmonised system) classification). In the meantime the Scientific Committee for Occupational Exposure Limits (SCOELs) and national authorities have set exposure limits for occupational settings. Some countries such as France have specific legislation covering all safety aspects of dry-cleaning operations. In the meantime industry has taken steps to reduce PERC losses by volatilisation during processing.

REACH guidance allows the setting of a derived no-effect level (DNEL) based on agreed occupational exposure limits (OELs). As explained elsewhere in this book,

DNELs are important in that they help define safe use conditions. However, historically OELs have been set on an existing database of toxicological data. It is probable that as more data are generated under REACH, some DNELs and, consequently, the OELs will have to be updated. In theory at least, the greater certainty provided by a broader database may result in a less severe classification, while data showing previously unsuspected effects may alternatively result in a more severe classification. Evidence of CMR Cat. 1a/1b or PBT/vPvB properties will trigger an evaluation as potential SVHCs (substance of very high concerns) under REACH, as well as possible Art. XIV listing.

Formaldehyde is extensively used in the manufacturing of binder resins for wood panels used in construction. Formaldehyde is classified as a Cat. 2 (GHS) carcinogen. Emission limits from wood panels have been set according to European Standard EN13986 to 0.1 ppm (according to an emission chamber test EN 717-1) to ensure consumer safety. There are also OELs. Again, formal evaluation of industry's formaldehyde chemical safety report (CSR) by the ECHA (European Chemicals Agency) might result in changes to these levels.

Potential consumer exposure to substances emitted from construction materials has led to increasing interest by the European and national legislators in regulating indoor air quality. However, the realisation that the priority contaminating substances determining indoor air quality are, for the most part, dependent on human behaviour (smoking, cooking, candle burning, heating, ventilation) and other natural environmental factors (radon from granitic soils) that are not always easy to regulate, raises other issues such as personal freedom. The Scientific Committee for Health and Environment Risks (SCHER) considers that formaldehyde, carbon monoxide, nitrogen dioxide, benzene, naphthalene, environmental tobacco smoke (ETS), radon, lead and organophosphate pesticides are compounds of concern in indoor environment in addition to moulds and bacteria. Others consider that emissions may be tackled at source. However, this may not always be necessary in terms of public health protection, as REACH already provides a process to evaluate whether exposures to given substances are below levels of concern and if not to take action designed to reduce emissions.

3.3 Water Framework Directive

The aim of the WFD 2000/60/EC is to improve the ecological quality of Community surface waters and, consequently, of groundwater and drinking water. Through specific measures, it provides 'for the progressive reduction of discharges, emissions and losses of priority substances and the cessation or phasing-out of discharges, emissions and losses of the priority hazardous substances (PHS)'.

As the Community and Member States are parties to international agreements aiming at the protection of the marine environment, the WFD also serves as an instrument contributing to compliance with these agreements. For instance, the Convention for the Protection of the Marine Environment of the North-East Atlantic (OSPAR), signed in Paris on 22 September 1992 and approved by Council Decision 98/249/EC, aims at the cessation of emissions of priority substances and the cessation or phasing out of discharges, emissions and losses of the priority hazardous substances. The ultimate aim is to achieve concentrations in the marine environment near background values for naturally occurring substances and close to zero for man-made synthetic substances.

The WFD provides for broad criteria for selection of priority hazardous substances, meaning that potentially many substances may be impacted by the Directive. In reality, any human activity is bound to be reflected in some emissions of chemical substances (synthetic or naturally derived) to the environment which will be detectable depending on the sensitivity of analytical methods. That is why EU regulations tend to steer away from zero environmental levels as an unreachable target. In 2011 about 30 priority substances (polyaromatic hydrocarbons, plant protection products, certain metals, pharmaceuticals and flame retardants) were identified for a programme of environmental monitoring, based on risk assessments. Member States and the European Commission have recognised the magnitude of the workload resulting from the obligation to monitor environmental levels and are further considering a process of prioritisation based on data from REACH. A proposal for a new directive amending the WFD has been tabled.[2] This proposal lists 48 priority hazardous substances and may grow in the future.

Guidance for the setting of environmental quality standards (EQS) was published in 2011.[3] While also risk-based, this guidance may not always be aligned with the methodology used in the implementation of REACH, since greater emphasis is placed on possible long-term effects.

In principle, evidence of compliance to EQS could be used to request exemption from Authorisation under REACH, on the basis that the WFD already provides for adequate environmental protection.

3.4   Carcinogens at Work

The Carcinogens at Work Directive 2004/37/EC creates an obligation for employers to evaluate the risks of exposure or of likely exposure to carcinogens and mutagens, and in particular to reduce their use 'in particular by replacing it, in so far as is technically possible, by a substance, preparation or process which, under its conditions of use, is not dangerous or is less dangerous to workers' health or safety, as the case may be'. If substitution by a less hazardous substance is not possible technically, exposure must be reduced, for example by manufacture and use in a closed system. Again, if a closed system is not technically possible, exposure levels should be reduced to as low a level as is technically possible and, in any case, shall not exceed the limit value set in Annex III of the Directive.

More recently,[4] a legislative initiative based on the lessons learnt by implementing the Carcinogens at Work Directive has noted that 'while exposure to most carcinogenic substances is regulated under the REACH provisions, there are some – process generated – substances which are not covered by REACH. A significant number of workers

---

[2] Proposal for a Directive of the European Parliament and of the Council amending Directives 2000/60/EC and 2008/105/EC as regards priority substances in the field of water policy.

[3] Technical Guidance for Deriving Environmental Quality Standards (TGD-EQS, 2011), Guidance Document No. 27.

[4] Legislative initiative to amend Directive 2004/37/EC on the protection of workers from the risks related to carcinogens and mutagens at work.

are exposed to these substances, including Diesel engine exhaust emissions, Respirable crystalline silica, Rubber process fume and dust and Mineral oils'.

Some stakeholders have concluded that the promotion of substitution provided by the Directive has not been very effective. In that respect, the Authorisation provisions of REACH should provide a strong legal instrument forcing substitution of some substances, although REACH has some built-in exemptions from Authorisation, for example of intermediates.

The Directive sets OELs in Annex III of the Directive. These OELs have been used as justification for setting DNELs in REACH. Several registrants have been using strict compliance to these OELs as a justification for exemption from REACH authorisation, because the proof of protection is set in legislation other than REACH. A legal argument can be developed on the basis that different regulations cannot pursue the same objective competing with each other.

Similarly, OELs have been based on other endpoints than carcinogenicity. Strong evidence of compliance to these levels down the supply chain can also justify exemptions from Authorisation, or demonstrating that an adequate level of human health protection is achievable by eventually implementing targeted restrictions while minimising the potential socio-economic impact.

3.5   Cosmetics

For obvious reasons, the health implications of exposure to substances contained in personal care products and cosmetics require a high level of scrutiny to ensure consumer protection. However, with the exception of CMR substances, which are banned from use in cosmetics except in certain cases where a risk assessment concludes that they can be used safely, an evaluation of substances is done by the Scientific Committee for Consumer Safety (SCCS) at the discretion of the authorities. After the Directive was revised, economic operators now have an obligation to maintain a product information file (PIF) and to make it available to authorities on request.

The methodology used by the Scientific Committee for Consumer Safety to evaluate the safety of cosmetics ingredients also differs in the detail from the methodology used by REACH, sometimes reaching different conclusions. Also, for obvious reasons, greater emphasis is put on dermal exposure.

Environmental properties of cosmetics ingredients remain fully within the scope of REACH and all cosmetic ingredients in use in the EU must be registered as long as they are above the trigger volumes. The Cosmetics Regulation has requirements concerning nanomaterials.

REACH and the Cosmetics Regulation also differ in the requirements to minimise animal testing. As with REACH, non-animal alternative testing methods are obligatory as they become scientifically validated and, furthermore, their use in cosmetics is forbidden if an animal-based test is undertaken when an alternative method has been validated. The only exception is if the substance has been tested for another purpose, such as to meet regulatory requirements in non-EU countries.

3.6   Biocidal and Plant Protection Products

Directive 98/8/EC, or the Biocidal Products Directive (BPD) was recently updated and became Regulation (EU) No. 528/2012 (BPR). The two principal changes were:

- The obligation for biocides contained in articles to be authorised (listed on Annex I of the Directive/Regulation), creating a level playing field between biocide-containing imported articles and articles manufactured in the EU.
- Labelling requirements on biocides contained in articles.

After much discussion, it was agreed that the primary criterion for differentiating a biocidal product, which can be an article containing a biocide which is intended to be released, and a treated article containing a biocide primarily designed to protect it or the user from microorganism development or to repel insects, is the primary function of the article.

For example, a carpet impregnated with an antimoth compound and a sleeping bag with insect repellent are not considered to be biocidal products, even if, in this case, the activity of the biocide extends beyond the treated article. Both carpets and sleeping bags are not intended to be used primarily as biocide delivering devices.

REACH has also specific requirements for articles that release substances at a rate greater than 1 tonne/year during normal use conditions.

REACH and the BPR/BPD also differ in their approaches in that only biocides listed in Annex I are authorised after a transition period, while REACH-assessed substances may require authorisation only if they meet certain CMR or PBT/vPvB criteria.

Substances listed on Annex I of the BPD/BPR are considered registered under REACH but only for that use. They are exempted from authorisation under REACH but not from evaluation or from restrictions.

The other overlap with REACH is data sharing. Since test data developed for the purpose of registration under the biocides regulations are relevant for REACH, it is necessary for data holders to join a SIEF (Substance Information Exchange Forum).

As with biocidal products, PPPs are exempt from REACH registration but still fall under the scope of the legislation insofar as there are data sharing requirements.

Non-hazardous ('classified') substances do not need to be subjected to a risk assessment in REACH. However, PPPs and biocides are always risk assessed under their respective regulations.

Several differences between exposure assessments for biocides, REACH and PPPs have been noted:

- For biocides and PPPs, the environmental risk assessments are primarily focused on application and service life; no quantitative estimate of exposure from waste is performed.
- For PPPs soil exposure is directly dependent on application rate.
- For biocides product-type specific emission scenario documents (ESDs) are used as a basis for emission estimation.
- There are differences in CMR and PBT/vPvB criteria between the respective legislations.

- There is a 0.1 μg/l threshold limit in groundwater for biocides and for PPPs which is not relevant in REACH, where secondary poisoning is the determining factor.
- Other differences were noted in the risk assessment methodologies.

3.7   Nationally (UK)-Implemented Legislative and Policy Frameworks

As described in Section 3.1 of this chapter, the REACH Regulation is one of a number of EU Directives and Regulations relating to the supply and use of chemicals. EU Regulations, including REACH, are direct-acting; in other words they must be adopted directly across all EU Member States (EC, 2012a). EU Directives set out objectives for obligations that must be fulfilled by a given date but is open with regard to how this is done (EC, 2012b). EU Directives are transposed onto the national legal framework of the Member States to which they apply. Separate national legislation that goes above and beyond EU required standards may also apply.

The purpose of this section is to focus on some key aspects of UK law concerning occupational and environmental exposure to chemicals, and how these relate to other EU Directives and the REACH Regulation. It is not intended to be comprehensive, nor should it be regarded as definitive guidance on how to comply with the legislation described. Similar provisions apply in other member states.

3.7.1   Workplace Exposure

3.7.1.1   Background

Occupational health legislation in the United Kingdom is a complex network of regulations, much of which originates from the 1974 Health and Safety at Work Act. The Health and Safety Executive (HSE), alongside local authorities, is responsible for enforcing UK law relating to many workplaces in England, Wales and Scotland.[5] The responsibilities of the HSE and the local authorities are not limited to exposure to hazardous chemicals but also include a number of other aspects of workplace safety, such as noise, vibrations, radiation and workplace transport. Likewise, the Health and Safety Executive for Northern Ireland (HSENI) fulfils a similar role in Northern Ireland.

As well as its role in workplace safety, the HSE acts as the **REACH Competent Authority** for the UK (Section 2.1.3 of this book) and also works on behalf of the UK government in negotiating and transposing EU legislation relating to occupational health and safety (HSE, nd-a). HSENI does not have any legal responsibilities relating to REACH.

Further information on the responsibilities of the HSE and HSENI may be found on their respective web sites (HSE, nd-b; HSENI, nd).

The principal national regulations that are relevant to non-accidental chemical exposure and management in England, Scotland and Wales are:

- Control of Substances Hazardous to Health (COSHH) Regulations 2002
- Dangerous Substances and Explosive Atmospheres Regulations 2002 (DSEAR).

---

[5] In many cases there is separate legislation for Scotland.

Separate legislation applies for Northern Ireland:

- COSHH Regulations (Northern Ireland) 2003
- DSEAR (Northern Ireland) 2003.

3.7.2   Control of Substances Hazardous to Health Regulations (COSHH) 2002

3.7.2.1   Summary of the Regulations

The COSHH Regulations 2002 (UK Gov, 2002) place a duty on all employers (including the self-employed) to ensure that substances hazardous to health are adequately controlled in the workplace. For the purposes of COSHH, hazardous substances are defined as:

- Substances or mixtures (preparations) that are classified according to the criteria of the CHIP Regulations[6] and carry any of the following indications of danger: very toxic; toxic; harmful; corrosive or irritant[7].
- Substances for which the Health and Safety Commission has set a maximum exposure limit (MEL) or occupational exposure standard.
- Biological agents.
- Nuisance dusts ($\geq 10\,\text{mg/m}^3$ inhalable particles or $\geq 4\,\text{mg/m}^3$ respirable particles, 8 hour time-weighted average).
- Substances which do not fall into the categories above but due to their properties and use pattern can pose a risk to humans.

Employers must fulfil their obligations under COSHH by completing a risk assessment for all processes that use any substance meeting the above criteria. For employers with five or more employees, the findings of this assessment must be recorded and made available to the employees. The risk assessment must demonstrate that exposure to the hazardous substance is either prevented or controlled to an extent that protects against risk to human health. This can be achieved by use of appropriate processes and equipment, engineering controls and personal protective equipment (including respiratory protection).

The COSHH Regulations also include provisions for occupational monitoring and health surveillance for certain substances.

3.7.2.2   COSHH and REACH

There are many similarities between COSHH and REACH, and in principle the two should work side-by-side. Both regulations require assessment of occupational exposure and risk, and demand that safe use can be demonstrated for all relevant processes and conditions under which hazardous substances are used. Information generated for the purposes of REACH can and should be used to complete the COSHH assessment and, similarly, existing COSHH assessments can be used as a starting point for downstream users to determine if their processes are adequately described by the exposure scenarios presented in their suppliers' extended Safety Data Sheets (eSDSs).

---

[6] Chemicals (Hazard Information for Packaging for Supply) Regulations 2009 (implementing Annex I of the Dangerous Substances Directive 67/548/EEC) in the United Kingdom, currently in transition to Regulation No. (EC) 1272/2008 on classification, labelling and packaging of mixtures (CLP Regulation). CLP will fully replace CHIP by June 2015.

[7] Substances which are sensitising to either skin or the respiratory tract carry the Xi: Irritant symbol.

Any risk management measures (RMMs) specified in the REACH exposure scenarios should correspond with good practice measures recommended in the COSHH assessment (HSE, nd-c), but it is also acceptable to specify additional measures in the COSHH assessment if local conditions require them.

However, there are also some fundamental differences between the two Regulations. These are summarised in the following sections.

Responsibility for Making the Assessment.   Responsibility for completion of a COSHH assessment lies with individual companies or employers (including the self-employed), whereas the REACH chemical safety assessment (CSA) must in the first instance be completed by the EU manufacturer or importer of the substance. Downstream users of registered substances also have certain obligations under REACH as discussed in Section 2.6.4 of this book.

Scope of the Assessment.   The first key difference in scope of the assessment is the definition of 'hazardous substances'. A COSHH assessment must take into consideration all hazardous substances used (including biological agents and nuisance dusts, and hazardous substances such as fumes that are produced in the process), irrespective of the quantities involved. For REACH, however, the CSA is required for chemical substances produced in or imported into the EU in quantities above 10 tonnes/year. REACH does not assess biological agents, dusts (unless these are considered dangerous due to (eco)toxicological properties) or process fumes.

Secondly, the COSHH assessment must be based on the specific processes and conditions used at the individual workplace. In contrast, the CSA should contain Exposure Scenarios for all the intended uses of the substance, including all production and formulation that take place within the EU, as well as the final end use of the substance or product(s) containing the substance. The CSA must also consider waste disposal and service life of articles that contain the substance. In many cases, the CSA will therefore be generic, describing processes and uses only in broad terms. These Exposure Scenarios are reported in the eSDS.

In addition to the COSHH assessment, the **downstream users** of a substance have an obligation to review the exposure scenarios in the eSDS to ensure that processes are adequately described.

Qualitative or Quantitative Assessment.   There are different ways in which exposure assessment and risk characterisation can be conducted. A fully quantitative assessment would require actual exposure to the substance to be quantified for all relevant routes of exposure (usually the inhalation and dermal (skin contact) routes for occupational situations). The exposure concentrations are then compared to a no observed adverse effect level (NOAEL), DNEL or another equivalent value (Sections 1.2.10.2 and 9.9 of this book) to obtain a risk characterisation ratio (Section 1.2.8 of this book). The DNEL or other value is usually determined based on the results of standard toxicity tests with laboratory animals, or occasionally may be based on experience with humans. This is not exactly the same as compliance with a statutory workplace exposure limit (WEL) established under COSHH (in the UK), or a European OEL. In the case of a workplace exposure limit experts from the authorities will have considered in detail a (usually) comprehensive and long-standing data set. DNEL values are assessed by registrants from smaller data sets and with a frequently less certain extrapolation from laboratory data to protection of workers.

Exposure could be assessed quantitatively using monitoring data (i.e. measurements of the concentration of the substance in the air, or use of patches to determine the amount of skin contact). More commonly for REACH purposes, worker exposure assessment is based on modelling, where exposure to the substance is predicted. A number of tools are available for this purpose; these are described in some detail in Section 10.2 of this book.

For the REACH CSA, quantitative assessment is preferred provided that it is possible and valid to do so, that is a DNEL value or equivalent can be established. Qualitative assessment is appropriate when no quantitative effect level has been established (for example skin or eye irritation).

In most circumstances, quantitative assessment would not typically be reported for COSHH. A useful tool from the HSE, COSHH Essentials (HSE, nd-d), assigns substances to hazard groups according to their classification. By inputting details of the substance properties (physical form, volatility) and process details, the tool provides recommendations for the types of operational conditions (OCs) or RMMs that are appropriate to control exposure, but the exposure estimation calculations are not available to the user. The tool can also be used to assess mixtures.

The COSHH Essentials scheme is referred to in REACH Technical Guidance and, with careful application, is a potentially useful 'Tier 1' assessment model (Section 10.2 of this book).

Duties as a Result of the Assessment. Under the UK COSHH regulations, employers have a duty to prevent or control exposure to all hazardous substances. Substitution of the hazardous substance(s) with a less hazardous alternative is, by default, the preferred option provided it is practical to do so. If prevention or substitution is not viable, then adequate control measures must be implemented according to the following order of priority:

1. Design and use of appropriate work processes, systems and engineering controls and the provision and use of suitable equipment and materials.
2. Control of exposure at source, including adequate ventilation systems and appropriate organisational measures.
3. Provision of suitable personal protective equipment in addition to (1) and (2).

These are similar to the operational conditions and RMMs specified in the REACH CSA.

The duty of the REACH registrant is to develop exposure scenarios and specify the conditions of safe use of the substance in those scenarios. Such conditions could include the same operational conditions or RMMs as are specified in the COSHH assessment. The registrant can also indicate if there are any uses of the substance which are advised against. However, at the registration stage the scope of the REACH CSA does not include proposals for substitution.

Possible outcomes of the COSHH assessment can include a need to either regularly monitor occupational exposure (particularly for specific substances named in the regulations), or to implement health surveillance programmes for workers. These would not normally be recommended for REACH purposes although the results of continuing programmes may of course be useful for the CSA.

3.7.2.3   Documentation

The REACH Regulation proscribes the documentation required for registration and CSR. The full CSA (including a technical dossier and a CSR, where required) would usually be confidential, although some parts are published on the ECHA web site (ECHA, 2013). However, the key elements of the assessment, including exposure scenarios, operational conditions, RMMs and DNELs are made publicly available via the eSDS, which replace previous Material Safety Data Sheets required under the previous legislation. All suppliers of hazardous substances are required to make the eSDS available to their customers.

Employers with five or more employees are required to document the findings of their COSHH assessments and to make those documents readily available to relevant parties, including employees, safety representatives or regulatory inspectors. There is no standard documentation because the nature of the assessments is so variable, but some typical example assessments are given on the HSE's web site (HSE, nd-e). It is not usual for these documents to be published outside the workplace.

3.7.3   Dangerous Substances and Explosive Atmospheres Regulations (DSEAR) 2002

3.7.3.1   Summary of the Regulations

The DSEAR 2002 (HSE, 2003) implement two EU Directives in the UK:

• Chemical Agents Directive (98/24/EC)
• Explosive Atmospheres Directive (99/92/EC).

The purpose of the DSEAR are to control risks due to fire, explosions or similar energetic events (for example thermal runaway during use of an extreme oxidiser). The scope of the regulations extends to dangerous substances that are classified as explosive, oxidising, highly flammable or flammable, explosive dust atmospheres and others (for example liquids with flashpoint above the cut-offs for classification as highly flammable or flammable but which are used at elevated temperature and, thus, have the potential to produce flammable vapours).

It is the responsibility of employers to protect both workers and members of the public who may be put at risk by activities involving the types of dangerous substance described above. Measures must be put in place to prevent or control risks, and if risks cannot be eliminated then suitable measures must be implemented to reduce the severity of the effects resulting from any fire or explosion. Suitable measures can include engineering controls, avoidance of ignition sources or other adverse conditions, segregation of incompatible substances, limiting the number of workers exposed and provision of personal protective equipment (this list is not exhaustive).

3.7.3.2   DSEAR and REACH

In addition to human health hazards due to the toxicological properties of a substances, the REACH CSA must also give due consideration to risks posed to humans as a result of hazardous physico-chemical properties. Most commonly these would be explosive properties, flammability and oxidising properties. Other relevant properties include violent reaction with water, release of flammable or toxic vapours due to reaction with water or acids and corrosion of metals.

Quantitative exposure assessment is not possible for most physico-chemical hazards in the same way as toxicological hazards; rather, a qualitative assessment must be conducted of the likelihood or frequency of an accident occurring under normal conditions of use, and the likely severity of such an accident in respect of loss of life or injury.

It should be noted that neither DSEAR nor REACH consider major accident situations, which are covered by separate legislation under the Seveso II Directive (98/82/EC), implemented in the United Kingdom under the COMAH (Control of Major Accident Hazards) Regulations (HSE, nd-f).

REACH Technical Guidance (ECHA, nd) is limited in respect of risk characterisation for physico-chemical hazards. The guidance suggests a simplified questionnaire approach for downstream users, based on a scheme developed by DG Employment (EC, nd) for the purposes of the Chemical Agents Directive (98/24/EC), which would typically be used for the purposes of compliance with DSEAR and other national equivalents. To date, there has been little real experience of following these recommendations in the context of REACH. In addition, it is unclear how practical such a questionnaire process would be for high volume substances or those with complex supply chains, not to mention the willingness of downstream users to provide detailed information on their processes to suppliers.

The approach is described in more detail in Section 10.6 of this book but it should be clearly understood that any simple assessment carried out for REACH purposes cannot be considered as adequate for individual employers to ensure compliance with DSEAR with respect to classified physicochemical hazards. Furthermore, DSEAR require other considerations outside the scope of REACH, to be taken into account, such as compressed gases, chemical reaction hazard assessment (which could lead to hazardous heat or pressure effects in a reaction vessel), incompatible storage materials or conditions and fumes, vapours or dusts given off during a process. Nevertheless, if available the findings of an assessment performed for the purposes of DSEAR can be used to inform the control measures risk characterisation reported in the REACH CSA and eSDS.

Similar to COSHH, DSEAR specify that substitution (of either the substance or the process) is the preferred option in complying with the employer's duty to reduce or eliminate risk.

3.7.3.3   Workplace Exposure Levels

Workplace exposure limits have been defined in the UK for around 500 substances (documented in EH40 Workplace exposure limits) (HSE, 2011). These values have been approved by the Health and Safety Commission and represent the maximum safe working levels (in most cases specified for 15 minute 'peak' exposure and 8 hour time-weighted average exposure). If the concentration of a substance in the workplace exceeds these limits, control measures must be implemented to reduce exposure (for example, provision of appropriate respiratory protective equipment).

EU-wide OELs (indicative limit values (ILVs) and indicative occupational exposure limit values (IOELVs) are also set for a number of substances in four separate Directives.[8] For these substances, national limits should be equivalent or else more conservative.

3.7.3.4   *Important Note Regarding WELs and DNELs*

DNEL values calculated for REACH should be regarded as a benchmark, not as an exposure limit. However, in some circumstances REACH allows provision for IOELVs to be used instead of DNELs for risk characterisation purposes (Section 9.9 of this book).

3.8   UK Environmental Regulation

3.8.1   Overview and IPPC

In England, the Department for Food, Environment and Rural Affairs (DEFRA) is responsible for most aspects of environmental policy in England, though some environmental policy falls to different Government departments.

The day to day implementation and enforcement of environmental law and policy lies with regulatory bodies.

The control of pollution lies with the Environment Agency (EA) in England and Wales. In Scotland, it is the Scottish Environment Protection Agency (SEPA), and in Northern Ireland, the Department of the Environment for Northern Ireland.

The EA must seek to achieve air and water quality objectives when exercising its functions in determining authorisations and consents.

REACH will, over time, produce a large amount of information on the hazardous properties, use and exposure of chemicals in the EU. Other environmental legislation, in particular legislation where EQS apply, will benefit from this information.

In England and Wales, a single regulatory framework for pollution control is in place under the Environmental Permitting (England and Wales) Regulations 2010. These Regulations have streamlined and integrated the following:

- Waste management licensing
- Pollution prevention and control permitting
- Water discharge consenting
- Groundwater permitting and
- Radioactive substances authorisations and registrations.

These Regulations are in line with the EU IPPC Directive (Directive 96/61/EC; codified Directive 2008/1/EC) concerning integrated pollution prevention and control. This EU Directive allows for emissions to air, water (including discharges to sewer) and land, plus a range of other environmental effects, to be considered together (DEFRA, 2010a).

Under the IPPC Directive, competent authorities in member states must set permit conditions so as to achieve a high level of protection for the environment as a whole.

The IPPC Directive sets out the main principles for the permitting and control of installations based on an integrated approach and the application of best available

---

[8] 91/322/EEC; 2000/39/EC; 2006/15/EC; 2009/161/EU.

techniques (BAT). These are the most effective techniques to achieve a high level of environmental protection, taking into account the costs and benefits of implementation.

The IPPC Directive specifies the types of installations that are covered, in terms of thresholds and industry sectors, and what the permits must contain. It does not, however, specify EQS. Coordination of standards is achieved via the exchange of information on BAT between the Commission and Industry, which are defined in best available techniques reference documents for each industrial sector covered. Member states must take these documents into account when setting permit conditions. emission limit values (ELVs) are set by reference to BAT. However, if an EQS set under European or National law would be breached, it is possible to set conditions stricter than the BAT or to refuse a permit.

The best available techniques reference documents are published online (EC JRC, nd).

The IPPC Regulations set out the facilities that need environmental permits or need to be registered as exempt.

The EA regulates: Part A(1) installations; Part A(1) mobile plant; waste mobile plant; waste operations; mining waste operations; radioactive substances activities; water discharge activities; groundwater activities.

The relevant local authority regulates: Part A(2) installations and Part A(2) mobile plant including any waste operations, water discharge activities or groundwater activities carried on as part of the installation or mobile plant; Part B installations and Part B mobile plant (DEFRA, 2010b).

Local authorities are responsible for regulating some 19 000 facilities, mostly to control air emissions, but in limited cases to enforce integrated pollution prevention and control. They also have limited responsibilities in relation to waste exemptions (DEFRA, 2013).

Facilities that have IPPC permits are likely to have site-specific information about their plants relating to total permitted releases of substances or substance types. Such site-specific information is useful for exposure and risk assessment under REACH.

The EA has powers of enforcement under the REACH Enforcement Regulations 2008 SI 2852. The EA has the right to enter and examine or investigate any premises.

3.8.2   Waste (England and Wales) Regulations 2011 SI 988

Requires businesses to apply the waste management hierarchy, introduces a two-tier system for waste carrier and broker registration and excludes some categories of waste from waste controls. These amend the Environmental Permitting Regulations 2010.

3.8.3   Water Legislation in the UK

The EU WFD (Directive 2000/60/EC of the European Parliament and of the Council establishing a framework for the Community action in the field of water policy) applies to all surface freshwater bodies, groundwater, estuaries and coastal water out to one mile from low-water.

The WFD is based on the concept of integrated pollution control. It integrates the protection of surface and groundwater, as well as the provisions relating to water quality and quantity. It introduces the concept of integrated river basin management, whereby river basin districts (RBDs) are established within which demanding environmental objectives must be met. The WFD rewrites existing water legislation into a new overarching programme.

The key objective of the WFD is to achieve 'good water status' for all bodies of surface water and groundwater by 2015, which will ensure satisfying human needs, ecosystem functioning and biodiversity protection.

A strategy for dealing with chemical pollution of water is set out in Article 16 of the WFD, which defines:

> *For those pollutants which present a significant risk to or via the aquatic environment, measures shall be aimed at the progressive reduction and, for priority hazardous substances, at the cessation or phasing-out of discharges, emissions and losses.*

The European Commission selects priority substances from those substances which present a significant risk to or via the aquatic environment, and proposes lists of priority substances at least every four years. The Priority Substances Directive (Directive 2008/105/EC) set EQS for 33 priority substances. The priority substances are identified in Annex II of the Directive, which replaces Annex X to the WFD. Member States shall take actions to meet those quality standards by 2015 as part of achieving good chemical status. The Directive establishes (in Part A of Annex I) the maximum available concentration and/or annual average concentration which, if met, allows the chemical status of the waterbody to be described as 'good'. Sediment EQS are not required, however, if a substance is unlikely to partition to, or accumulate in, sediment. Similarly, biota EQS are not required if a substance does not bioaccumulate. Water column EQS, based on direct toxicity to pelagic organisms, are always determined.

DEFRA has policy responsibility for the implementation of the WFD in England, however much of the implementation work is undertaken by the EA in England and Wales.

The Water Environment (WFD) Regulations 2003 make provision for the purpose of implementing in RBDs within England and Wales Directive 2000/60/EC. The Regulations identify 11 RBDs.

The EA is required to carry out detailed monitoring and analysis in relation to each RBD. Appropriate environmental objectives and a Programme of Measures to achieve these will then be proposed for each RBD. The WFD is flexible in its objective of achieving 'good water status' for all bodies of water by 2015. Member States may extend the achievement date by up to 12 years, or set objectives less stringent than 'good status', if obtaining proves to be infeasible. Where the environmental objective of a RBD is not to achieve 'good status' by 2015, reasons for this will be stated.

The environmental objectives and Programmes of Measures are brought together in a river basin management plan for each RBD. Management plans are detailed accounts of how the environmental objectives set for the river basin are to be reached. River basin management plans should have been prepared by December 2009, and they must be reviewed and updated every six years. DEFRA, the EA and other public bodies are required to have regard to river basin management plans, in exercising their functions in relation to RBDs.

River Basin Management Plans require a new strategic planning process to be established for the purposes of managing, protecting and improving the quality of water resources. The EA is the competent authority and is to prepare river basin management plans for Secretary of State/the Welsh Ministers approval. The plans are to set environmental objectives and to set out programmes of measures to fulfil the plans.

The first cycle of plans were produced, approved and published in December 2009, and the programmes of measures should have been made operational by December 2012, to meet the objectives identified in the Plan by December 2015. All Protected Areas have to meet their objectives by then. A second plan and programmes of measures are required for the following six years, and a third one for the six years thereafter. Along with other public bodies, the Agency is required to have regard to river basin management plans and to any supplementary plans in exercising their functions in relation to RBDs. Their preparation and execution may influence regulation of activities under environmental permitting although it is too early to determine those influences (DEFRA, 2010b).

3.8.4 Directive 2006/118/EC on the Protection of Groundwater against Pollution and Deterioration

This requires the setting of criteria for assessing the groundwater chemical status of groundwater bodies, including the establishment of threshold values; procedures for assessing chemical status of groundwater bodies; and identification of significant and sustained upward trends of pollutants. It is thus mainly concerned with the classification of groundwater bodies. However, it also requires measures to prevent or limit the input of pollutants to groundwater. The Environmental Permitting Regulations are one of the mechanisms for meeting the requirements for such measures.

3.8.5 Groundwater (England and Wales) Regulations 2009 (SI 2009 No. 2902)

These implement parts of the WFD (2000/60/EC) that applies to groundwater, and Article 6 of the 2006 Groundwater Directive (2006/118/EC). They supplement the Environmental Permitting (England and Wales) Regulations 2007 and existing water pollution legislation.

3.8.6 Air Legislation in the UK

The Environment Act 1995 requires the UK Government and the devolved administrations for Scotland and Wales to produce a national air quality strategy containing standards, objectives and measures for improving ambient air quality and to keep these policies under review. There is equivalent legislation in Northern Ireland.

The most recent Air Quality Strategy for England, Scotland, Wales and Northern Ireland was published by the UK Government and the devolved administrations in July 2007. The strategy contains policies for the assessment and management of United Kingdom air quality and the implementation of EU and International agreements.

The strategy sets air quality standards relating to the quality of air and objectives for the restriction of the levels at which particular substances are present in the air.

The Environment Act 1995 requires that the EA and the Scottish Environment Protection Agency have regard to the Air Quality Strategy in exercising their pollution control functions, for example under the Environmental Permitting (England and Wales) Regulations 2010. However, the air quality objectives in the Air Quality Strategy are a statement of policy targets and, as such, there is no legal requirement to meet these objectives except in as far as these mirror any equivalent legally binding limit values in EU legislation.

There is an EU ambient air quality directive (2008/50/EC). This Directive sets legally binding limits for concentrations in outdoor air of major air pollutants that impact public health, for example particulate matter (PM10 and PM2.5) and nitrogen dioxide (NO_2).

This Directive was made law in England through the Air Quality Standards Regulations 2010, which also incorporates the fourth air quality daughter directive (2004/107/EC) that sets targets for levels in outdoor air of certain toxic heavy metals and polycyclic aromatic hydrocarbons. Equivalent regulations exist in Scotland, Wales and Northern Ireland.

The regulations require the Secretary of State to assess the levels of certain pollutants in air and ensure that they do not exceed the limit values set out in the regulation. The limits values set are for sulfur dioxide, nitrogen dioxide, benzene, lead, PM_{10}, $PM_{2.5}$ and carbon monoxide. The regulations also require the Secretary of State to ensure The regulations require the Secretary of State to ensure that all necessary measures not entailing disproportionate costs are taken to ensure that concentrations of $PM_{2.5}$, ozone, arsenic, cadmium, nickel and benzo(a)pyrene do not exceed the target values set out in the regulation.

References

DEFRA (2010a) Environmental Permitting Guidance. The IPPC Directive. http://archive.defra.gov.uk/environment/policy/permits/documents/ep2010ippc.pdf (last accessed 30 July 2013).

DEFRA (2010b) Environmental Permitting Guidance For the Environmental Permitting (England and Wales) Regulations 2010, Version 3.1. http://archive.defra.gov.uk/environment/policy/permits/documents/ep2010guidance.pdf (last accessed 14 July 2013).

DEFRA (2013) Policy: Protecting and Enhancing Our Urban and Natural Environment to Improve Public Health and Wellbeing, Published 9 April 2013. http://www.defra.gov.uk/environment/quality/permitting/ (last accessed 14 July 2013).

EC (European Commission) (nd) Employment, Social Affairs and Inclusion. http://ec.europa.eu/social/home.jsp (last accessed 14 July 2013).

EC (European Commission) (2000) Commission of the European Communities Brussels, 2.2.2000 COM(2000) 1 Final Communication from the Commission on the Precautionary Principle.

EC (European Commission) (2012a) Application of EU Law, What are EU Regulations? Last Update 25 June 2012. http://ec.europa.eu/eu_law/introduction/what_regulation_en.htm (last accessed 14 July 2013).

EC (European Commission) (2012b) Application of EU Law, Directives – Definitions. Last Update 11 June 2012. http://ec.europa.eu/eu_law/directives/directives_en.htm (last accessed 14 July 2013).

ECHA (European Chemicals Agency) (nd) Guidance. http://guidance.echa.europa.eu/docs/guidance_document/information_requirements_part_e_en.pdf?vers=20_08_08 (last accessed 30 July 2013).

ECHA (European Chemicals Agency) (2013) Registered Substances Page. http://apps .echa.europa.eu/registered/registered-sub.aspx (last accessed 14 July 2013).

EC JRC (European Commission Joint Resesearch Centre) (nd) Reference Documents. http://eippcb.jrc.es/reference/ (last accessed 30 July 2013).

HSE (nd-a) HSE's Role in Europe. http://www.hse.gov.uk/aboutus/europe/roleineurope .htm (last accessed 14 July 2013).

HSE (nd-b) Chemicals. http://www.hse.gov.uk/chemicals/index.htm (last accessed 14 July 2013).

HSE (nd-c) REACH and COSHH Working Together. http://www.hse.gov.uk/coshh/detail/ reach.htm (last accessed 14 July 2013).

HSE (nd-d) Coshh Essentials. http://www.hse.gov.uk/coshh/essentials/index.htm (last accessed 14 July 2013).

HSE (nd-e) Example COSHH Risk Assessments. http://www.hse.gov.uk/coshh/riskassess/ index.htm (last accessed 14 July 2013).

HSE (nd-f) COMAH – Guidance. http://www.hse.gov.uk/comah/guidance.htm (last accessed 14 July 2013).

HSE (2003) Dangerous Substances and Explosive Atmospheres Regulations 2002. Approved Code of Practice and Guidance. http://www.hse.gov.uk/pubns/priced/l138 .pdf (last accessed 14 July 2013).

HSE (2011) EH40/2005 Workplace Exposure Limits. Containing the List of Workplace Exposure Limits for Use with the Control of Substances Hazardous to Health Regulations 2002 (as amended). http://www.hse.gov.uk/coshh/table1.pdf (last accessed 14 July 2013).

HSENI (nd). http://www.hseni.gov.uk/index.htm (last accessed 14 July 2013).

UK Gov (2002) The Control of Substances Hazardous to Health Regulations 2002. http:// www.legislation.gov.uk/uksi/2002/2677/contents/made (last accessed 30 July 2013).

Web References

DEFRA (Department for Environment Food and Rural Affairs) (2007) *The Air Quality Strategy for England, Scotland, Wales and Northern Ireland*, Vol. 1, TSO, London, Presented to Parliament by the Secretary of State for Environment, Food and Rural Affairs By Command of Her Majesty: Laid before the Scottish Parliament by the Scottish Ministers. Laid before the National Assembly for Wales by Welsh Ministers. Laid before the Northern Ireland Assembly by the Minister of the Environment, July 2007. http://archive.defra.gov.uk/environment/quality/air/airquality/strategy/documents/air-qualitystrategy-vol1.pdf (last accessed 14 July 2013).

UK Gov (2010) The Air Quality Standards Regulations 2010. http://www.legislation .gov.uk/uksi/2010/1001/introduction/made (last accessed 14 July 2013).

4

Identification of Substances for REACH – Practicalities

This section sets out an overview of the considerations any producer needs to cover prior to registering a substance. The need for careful definition of the substance in process and compositional terms is explained.

A cornerstone of the Registration, Evaluation, Authorisation and restriction of CHemicals (REACH) process is to correctly and unambiguously identify the substance. All chemical risk management activity is dependent upon this. It is considered by many as the easiest part of the Registration. However, in 2013, the European Chemicals Agency (ECHA) stated that about two-thirds of registration dossiers evaluated had shortcomings related to substance identity (ECHA, 2013).

Guidance on substance identification in 23 EU languages[1] is available online from the ECHA web site (ECHA, nd, 2011, 2012).

Understanding the substance is the first step in REACH and helps to define the whole registration strategy.

The composition of the substance is defined in terms of chemical analysis and/or a description of the manufacturing processes. All registered substances must be defined clearly, using appropriate terms and methods, whether they are highly pure synthesised chemicals, or extremely complex materials extracted from natural sources.

4.1 Substance Identification

REACH distinguishes three broad categories of substance in terms of regulatory status:

1. Phase-in substances derived from the list of existing substances on the European Inventory of Existing Commercial Chemical Substances (EINECS) inventory.

---

[1] The 23 languages are Bulgarian, Croatian, Czech, Danish, Dutch, English, Estonian, Finnish, French, German, Greek, Hungarian, Italian, Latvian, Lithuanian, Maltese, Polish, Portuguese, Romanian, Slovakian, Slovenian, Spanish and Swedish.

Chemical Risk Assessment: A Manual for REACH, First Edition. Peter Fisk Associates Ltd.
© 2014 John Wiley & Sons, Ltd. Published 2014 by John Wiley & Sons, Ltd.

2. Non-phase-in substances derived from the list of New Substances on the European List of Notified Chemical Substances (ELINCS) inventory, which are deemed as 'already registered' under REACH.
3. Substances not already on the market, which must be registered before they are sold or used outside of the site of production; there is some possible exemption for research and development (termed PPORD in REACH, standing for product and process-oriented research and development).

Substance identification is the combination of description of manufacturing processes and chemical analysis in order to define the composition of a substance. All registered substances must be defined clearly, using appropriate terms and methods, whether they are highly pure synthesised chemicals, through to extremely complex materials extracted from natural sources.

The methods of identification and reporting are built on the experiences gained from the New Substances system, but have to accommodate the phase-in substances in a consistent way. Understanding the substance is the first step in REACH and helps define the whole registration strategy. This is discussed further in Appendix B.

4.1.1   Types of Substances

The type of substances can be divided into five main categories.

1. Organic substances.
2. Inorganic substances.
3. Substances of biological origin (often referred to as *Unknown or Variable composition, Complex reaction products or Biological Materials*).
4. Nano-materials.
5. Articles.

4.1.2   Mono-Constituent Substances

Substances classed as 'essentially pure' contain a main constituent at 80% or more; any other constituents should be considered as impurities. Those present at more than 1% should be identified where possible.

4.1.3   Multi-Constituent Substances (MCSs)

These contain more than one well-defined constituent, each present at more than 10% by weight.

Registrants of multi-constituent substances (MCSs) should be aware that ECHA requires the substance to be accurately named. The name used for registration of an MCS substance should accurately describe the constituents in it. This may mean moving away from nomenclature used traditionally, or in other parts of the world.

4.1.4   Substances with Unknown or Variable Composition, or of Biological Origin (UVCBs)

The REACH regulation definition of a UVCB (unknown or variable composition, or of biological origin) is:

substances of Unknown or Variable composition, Complex reaction products or Biological Materials [UVCB substance] cannot be sufficiently identified by their chemical composition, because:

- *The number of constituents is relatively large and/or*
- *The composition is, to a significant part, unknown and/or*
- *The variability of composition is relatively large or poorly predictable.*

These are produced in diverse industrial sectors and a great many existing registrations define the substance as a UVCB. They typically arise from:

- Products based on fractions of crude oil, or petrochemical processes.
- Extracts of biological origin and their derivatives.
- Products derived from technical grade reagents that perhaps needed several chemical reactions to get the desired product, and hence can be complex and difficult to analyse.

Inorganic reaction products where the exact constituents may depend sensitively on the precise molar ratios of reagents.

UVCBs need to be defined rigorously in terms of the process used to make them and in terms of what characterisation can be achieved. The registrant will make the registration easier in creating an effective dossier if every reasonable effort is used to characterise the composition, even if it is understood that variability is inherent. It may be necessary to build up a picture of the composition through analytical methods.

The intrinsic challenge of registration of UVCBs will emerge in subsequent sections; however, it must be realised that a UVCB is not exempt from being assessed for hazard and risk even though the difficulties in doing so are recognised. Guidance about the strategies to adopt and the dossier must include the significance of all the constituents in its scope. Understanding of the constituents present can allow predictive methods to be used to fill in the gaps in property data that inevitably arise due to the difficulties in performing and interpreting experimental studies for these substances.

If it is possible to characterise a substance well enough to allow it to be defined as a multi-constituent substance rather than a UVCB, this is usually desirable. The disadvantages of defining a substance as a UVCB are:

- This is defined by the manufacturing process; therefore, if this process changes, a new registration may be needed.
- There is no such thing as an impurity and all of the (possible very many) constituents need consideration.

4.1.5　Nanomaterials

The use of nanotechnology is rapidly expanding in health care, cosmetics, electronics, energy technologies, food and agriculture. A nanomaterial is defined as being approximately $1-100\,nm$ in at least one dimension. Its nanosize may result in different specific physico-chemical properties from those of particles of a larger size.

There are currently no explicit requirements for nanomaterials under REACH or CLP (Classification, Labelling and Packaging) other than required for a substance. However, this may change in future legislation, as, in 2011, the European Commission released a specific recommendation on a nanomaterials definition that is used in different European regulations, including REACH and CLP (2008). The use of nanomaterials raises questions regarding their potential effects on health and the environment. There is a need to adequately assess and manage the potential risks of these new forms of materials, and even whether manufacturers, importers and downstream users have to ensure the safe use of each substance (whatever its form) under REACH.

4.1.6   Articles

An article is an object composed of one or more substances or mixtures given a specific shape, surface or design. It may be produced from natural materials (e.g. wood or wool) or synthetic materials (e.g. polyvinyl chloride (PVC)). Most of the commonly used objects in private households and industries are articles, for example furniture, clothes, vehicles, books, toys, kitchen equipment, and electronic equipment. Buildings are not considered to be articles, so long as they remain fixed to the land on which they stand.

It is required to identify where possible with the following identifiers.

4.1.7   EC Number

The **EC Number** is the key numerical identifier for substances within European regulatory programmes. It is a unique seven-digit identifier that is assigned to chemical substances for regulatory purposes within the European Union by the regulatory authorities. The EC Number is written as NNN-NNN-R, where N represents integers and R is a check digit calculated using the International Standard Book Number (ISBN) method. The list of substances having an EC number is called the *EC Inventory* and is a combination of three independent and legally approved European lists of substances from the previous European Union chemicals regulatory frameworks:

- European Inventory of Existing Commercial Chemical Substances (EINECS).
- European List of Notified Chemical Substances (ELINCS).
- No longer polymer list (NLP list).

The entries in the EC Inventory consist of a chemical name and a number (EC name and EC number), a CAS (Chemical Abstracts Service) Number, molecular formula (if available), and description (for certain types of substances).

Substance types and analysis – basics of substance definitions, and analytical characterisation

- *Organic substances* – a normal complement normally including at least: UV-visible spectroscopy; IR spectroscopy; ^{1}H-NMR (nuclear magnetic resonance) and/or ^{13}C-NMR spectroscopy; chromatography (HPLC (high performance liquid chromatography) or GC (gas chromatography)), possibly mass spectroscopy.
- *Inorganic* – X-ray diffraction (XRD), X-ray fluorescence (XRF), Atomic absorption spectroscopy (AAS), elemental analysis.
- *Heteroatoms* – Si-NMR; P-NMR may be useful.

If one of the usual methods is deemed not useful, for example if it is decided not to conduct UV–vis spectroscopy, this should be explained in the registration.

As with other technical issues, there are regulatory and commercial considerations behind the methods developed by the authorities, and unsupported statements are not acceptable. The ECHA makes particular scrutiny of information about substance characterisation, for the following reasons:

- Data from different producers may need to be shared, even if the compositions are not identical.
- Impurities or minor constituents may be important or even dominate the regulatory compliance process (i.e. may be substances of very high concern, SVHCs).
- Interpretation of the test data requires insight into the composition.
- Substances which had different EINECS or ELINCS numbers may turn out to be the same and it is desirable to rationalise such inconsistencies.

It should be remembered that, at present, polymers are outside the scope of REACH, although impurities in them may not be, depending on properties.

Several broad groups of substance are identified here. In each case, methods of production are an important part of the information to be provided, but without disclosure of confidential business information. A good description of mass balance is vital: the substance composition information should add up to 100% even if some of this is not fully identified due to technical limitations. The broad types are:

- Essentially-pure substances (single constituent)
- MCS
- UVCB

A fundamental principle affecting all substances is the ability to identify whether any constituents that might meet the SVHC criteria are present at 0.1% or more, whatever the type of substance. For this reason alone, a realistic description of production methods should be provided, because it will support statements concerning the absence of SVHCs which will not have been analysed for.

4.2   Sameness

Experience with the early stages of REACH has shown that the regulation has helped to rationalise many inconsistencies in choice of CAS/EC number in Europe and indeed around the world. Registrants have found that it has been possible (with due respect to confidentiality) to agree a substance identification profile that is acceptable to all. Usually, companies sell different grades of the same substance and sameness is best established as the highest purity commercial grade. With MCS and UVCBs, variability is far more likely, and this should be taken into consideration. Departures from the agreed norm can often be easily accommodated by company-specific additions to the dossier, without requiring the registrant to opt-out of the joint submission. Nevertheless affected registrants should make a careful assessment case-by-case.

4.3   Essentially-Pure Substances

Substances classed as 'essentially pure' contain a main constituent at 80% or more; any other constituents should be considered as impurities. Those present at more than

1% should be identified where possible. The effects of impurities should be considered. Methods of analytical characterisation should be appropriate to the type of substance. For organic substances, ideally all of the following techniques will have been used by analysts:

- Gas chromatography (GC) or liquid chromatography (LC), usually high performance liquid chromatography (HPLC).
- Nuclear magnetic resonance (NMR).
- Mass spectrometry (MS).
- Infra-red (IR) spectrometry.
- Ultra-violet (UV) spectroscopy.

Coupled techniques such as LC-MS may be particularly useful. The analytical data should be accompanied by interpretation and there should be enough description given to allow another laboratory to be able to use the same methods.

Should a substance contain some polymeric impurities, it may be useful to demonstrate this by use of a mass-sensitive technique such as size exclusion or gel permeation chromatography.

Inorganic substances present more difficulties in terms of compliance, and the expertise of producers may be important. Elemental analysis of some form is essential and can be supported by some of the techniques already listed; ion chromatography may be useful. X-ray diffraction results may be useful, to detect individual constituents in a mixed product.

4.4   Approaching the Substance Data Set – Understanding the Substance

A thorough understanding of the constituents present in a substance allows the correct regulatory strategy to be developed. It is important to realise, especially with REACH that compliance is not about an unthinking 'box ticking' mentality. There are fundamental considerations for valid chemical safety assessment that depend on understanding substance composition such as:

- Study design in all areas of data requirements.
- How data are used for risk characterisation.
- Are there sufficient data available for hazard and risk to be characterised fully?
- Are any results, and waiving proposals, consistent with the constituents?

Due to the vital, but complex nature of this topic, further discussion is provided in Appendix B.

References

CLP (2008) CLP-Regulation (EC) No 1272/2008.

ECHA (nd) Guidance on REACH. http://echa.europa.eu/guidance-documents/guidance-on-reach (last accessed 16 July 2013).

ECHA (2011) Guidance in a Nutshell, Identification and Naming of Substances Under REACH and CLP. Reference ECHA-11-B-10-EN, European Chemicals Agency (ECHA).

ECHA (2012) Guidance for Identification and Naming of Substances Under REACH and CLP, ECHA. Reference ECHA-11-G-10.1-EN, European Chemicals Agency (ECHA).

ECHA (2013) Evaluation under REACH, Progress Report 2012. European Chemicals Agency (ECHA), p. 41.

5

Physico-Chemical Properties for REACH – Purpose and Practicalities

This chapter provides a brief introduction before wider discussion of intrinsic properties in Chapters 6, 7 and 9.

5.1 Physico-Chemical Properties

As part of the assessment process, it is necessary to submit certain physico-chemical data on the substance to be registered. The clear understanding of a substance begins with proper and adequate understanding of the substance identity (Section 4.1 and Appendix B of this book) and the physico-chemical properties that are associated with such substance. The information needed depends on the tonnage imported or manufactured in the EU (European Union); however, almost all physico-chemical data are required for substances even at ≥ 1 tonnes/year (Annex VII). Understanding the physico-chemical properties of a substance helps proper assessment of other endpoints, such as toxicology, environmental fate and exposure assessment, and is essential for ecotoxicology. In some cases it may be useful to conduct prediction to aid understanding, even for properties not required at the Annex level.

The basic physico-chemical data endpoints (i.e. types of properties) required at Annex VII of the REACH regulation for all substances are:

- Melting point
- Boiling point
- Density
- Granulometry (particle size distribution)
- Vapour pressure
- Octanol–water partition coefficient

Chemical Risk Assessment: A Manual for REACH, First Edition. Peter Fisk Associates Ltd.
© 2014 John Wiley & Sons, Ltd. Published 2014 by John Wiley & Sons, Ltd.

- Water solubility
- Surface tension
- Flash point
- Flammability, including pyrophoricity and flammability in contact with water
- Auto-flammability or self-ignition temperature
- Explosive properties
- Oxidising properties.

The information requirement for physico-chemical data can be fulfilled either by the use of available data, by conducting a new study and/or by the use of and appropriate data waiver where relevant. The data should be reliable and scientifically valid. In addition, some of the physico-chemical data can be estimated using a validated (Q)SAR ((Quantitative) Structure-Activity Relationship) prediction method (Sections 5.8 and 5.9 of this chapter). Considerable care is needed with hazardous endpoints such as flash point, which is batch specific, where the result can be influenced by the presence of impurities in the sample.

Why is vapour pressure important?

Vapour pressure is one of the physico-chemical properties for which the value determined has significant influence on a number of areas of the chemical safety assessment. Some examples are discussed here.

A very high vapour pressure indicates:

- The substance might tend to volatilise from test systems, so special adaptations to guideline test methods should be considered, for example sealed vessels, no headspace.
- Inhalatory exposure of humans in the workplace, public domain or from consumer use could be significant; it may be important that measured inhalatory toxicity data are available.
- Releases to the air from industrial and wide dispersed uses, including waste-water treatment plants, are likely to be important in the environmental exposure assessment.

A very low vapour pressure indicates:

- Dermal and, in some circumstances, oral exposure pathways for humans may be more relevant than the inhalatory pathway.
- Default releases to air from ERCs (environmental release categories) are likely to be significant overestimates. Some SPERCs (Specific Environmental Release Categories) include variable release rates to air dependent on the vapour pressure of the substance.
- Measured vapour pressure for the REACH data set should preferably use experimental techniques that are less susceptible to influence from volatile trace impurities.

Safety assessment – flash point

The value of the flash point is a measure of flammability of a liquid test substance. A value should be measured for a representative sample and multiple samples from different technical grades may be necessary. This would be, for example, if the composition is variable. Flash point is an example of a property that can be significantly influenced by impurities present, for example residual solvent, which may be more volatile and flammable than the major constituent(s).

Why does octanol–water get mentioned so much?

The octanol–water partition coefficient is an equilibrium constant and is usually expressed as log K_{ow} (or log P). It is significant in various respects in a chemical safety assessment for an organic substance, including multi-constituent and UVCB (unknown or variable composition, or biological origin) substance types. It is not relevant for inorganic substances. In the property data set the value of log K_{ow} represents the substance's affinity for lipid-like media or for aqueous media, and is an important predictor of other properties that are associated with uptake from solution into tissue (bioaccumulation; aquatic ecotoxicity) and adsorption to organic substrates (K_{oc}, and hence distribution between water and sludge in a waste-water treatment plant and fate in the wider environment). In exposure modelling, the value of log K_{ow} has a role in determining exposure of humans and predators via the food chain. Some human exposure models also utilise log K_{ow} as an input.

Results of some of the physico-chemical tests are vital for risk assessment calculations and in the commissioning and designing of the appropriate environmental fate, ecotoxicological and toxicological test packages.

5.2 Strategy in Physico-Chemical Testing Plans

The sequence in which physico-chemical studies are conducted is important because the results of one test can influence how and whether another test should be performed. This is referred to as a tiered testing strategy.

5.2.1 Tier 1 Tests

Pyrophoric properties: This property can be assessed either by the use of theoretical screening assessment (based on experience in handling and use and structural examination) or by conducting a full study. If the result is positive, a flammability study is not required and, in addition, testing for other endpoints becomes extremely difficult to perform.

5.2.2   Tier 2 Tests

Tier 2 tests are divided into groups and tests within the group should be performed in the order shown. In particular, having the melting point result will be useful in selecting the appropriate flammability and auto-flammability studies.

5.2.2.1   *Group 1*

Flammability in contact with water can be assessed either based on theoretical screening assessment (experience in use and handling and also by structural examination) or by conducting a full study using EU Method A.12. If the result is positive, further water-based physico-chemical testing on the substance is not required, but further toxicological or ecotoxicological data may be required on the reaction products.

Water solubility and hydrolysis: Whilst the REACH Annexes and technical dossier place hydrolysis as part of the environmental fate data set, for some purposes it should be considered along with the physico-chemical property data set. To adequately understand and interpret the water solubility result, the hydrolysis potential of the substance should be assessed prior to conducting the water solubility study. Also, information on water solubility is required in order to conduct the hydrolysis study because water solubility is useful when setting the starting concentrations of the hydrolysis study. Therefore, for a hydrolysable substance; the water solubility should be predicted.

Surface tension: This test can be assessed either based on theoretical screening assessment based on structural examination or by conducting a full study using EU Method A.5 if surface activity is suspected. Information on water solubility and time needed to generate a saturated solution are required before conducting a surface tension study. For substances that hydrolyse, the time scale of the test should be such that significant hydrolysis has not occurred.

Octanol–water partition coefficient (also referred to as log K_{ow} or log P): Knowledge of the surface tension is required before conducting the partition coefficient study because surface-active substances interfere with partitioning. However, for substances that are close to the 60 mN/m threshold for surface tension, the shake flask method may be valid, although the validity of the method should be examined by conducting a preliminary partition coefficient. If there is an observation of foaming or formation of an emulsion, then the method is not suitable.

5.2.2.2   *Group 2*

Explosivity: This test can be assessed either based on the theoretical screening assessment (this can be by structural examination, calculating the oxygen balance or evidence from differential scanning calorimetry (DSC) experiments) or by conducting a full study using EU Method A.12. If the result is positive, the need for further test in this group is removed.

Flammability: Flammability tests are not required for liquids; but tests are required for pyrophoric liquids. If the substance is classified for flammability, the test requirement on oxidising potential is removed.

Auto-flammability: The physical form of the substance should be known in order to select the appropriate test method. The test requirement is removed for low melting solids (that is melting point $< 160\,°C$).

Oxidising properties: The test requirement can be waived if based on structural examination, oxidising properties are not expected. The test should not be performed on flammable or explosive substances.

5.2.2.3 Group 3

Melting point: Melting point results can affect both the boiling point and the vapour pressure directly. The melting point does not need to be measured precisely if it is known to be very low (less than $-20\,°C$). Differential scanning calorimetry can be used to evaluate both the melting and boiling point studies in a single test, and gives information about thermal stability.

Boiling point: A boiling point study does not need to be conducted for substances that decompose on melting. This is closely related to vapour pressure and a single test can provide results for both endpoints for some substances.

Vapour pressure: There are several methods available for the vapour pressure measurement (OECD 104 (OECD, Organisation for Economic Co-operation and Development) or EU Method A.4). It is important to select the most suitable method in order to avoid factors which could compromise the measurement and invalidate the test. Information on the melting and boiling point should be available to select the appropriate temperature range.

The vapour pressure is a critical input to exposure assessment models and is also an essential consideration in selecting both appropriate conditions and interpreting existing studies for other endpoints. Examples are its relevance to biodegradation (affecting choice of method), ecotoxicity (affecting the potential for volatile losses from the test system and the need to minimise these), and toxicity (affecting choice of exposure route). Furthermore, it can impact a REACH submission by influencing appropriate emission scenario models. Vapour pressure may be a useful input to prediction of other physico-chemical properties.

5.2.3 Tier 3 Tests

Relative density: There are several methods available for conducting a density measurement (OECD 109). The information on dynamic viscosity of liquid substances should be available prior to conducting a density study because dynamic viscosity affects the choice of method for density determination. For gases, determination of density is not required but should be calculated from its molecular weight using the ideal gas equation.

Particle size distribution: Particle size distribution study is only applicable to solid substances as sold. The result from particle size determination is closely to related human exposure (inhalation toxicity).

Viscosity: This endpoint is relevant only to liquid substances. Certain substances or preparations with low viscosity and low surface tension may present an aspiration hazard in humans. The lower the viscosity, the more easily a liquid disperses into the soil.

Stability in organic solvents: The endpoint is only relevant if stability of the substance in organic solvent is critical to interpretation of study results.

5.3   Difficult-to-Measure Substances

The detailed assessment and understanding of the properties of a substance should be considered when conducting physico-chemical tests on difficult substances. The types of substances highlighted in the following sections can be considered as difficult substances for the purpose of physico-chemical tests. Note that similar issues can affect ecotoxicological testing; this is discussed in Section 7.3.4 of this book.

5.3.1   Multiconstituent or UVCB Substances (Mixtures)

The constituents of a mixture may behave in a different way within the environment and also in a physico-chemical test and, as such, such results should represent each individual constituent. Where the constituents are known and identified, the individual values for the constituents may be presented for physico-chemical endpoints such as vapour pressure, water solubility and octanol–water partition coefficient. Where the hazard profile of a mixture such as a multiconstituent substance can be sufficiently assessed based on the hazard information available for its constituents, there may not be a need to conduct hazard tests on the substance as such.

5.3.2   Poorly Soluble Substances

Poorly water-soluble substances may be difficult to test because it is not always possible to develop analytical methods that are suitable and sensitive enough to obtain an accurate assessment of these substances. A poorly soluble substance will be difficult to test for surface tension and hydrolysis.

5.3.3   Volatile Substances

Minimising the loss of volatile and unstable substances from the test system should be the main point of concern. This can be achieved by the use of closed or covered vessels and conducting the test under dark.

5.3.4   Unstable Substances Either by Hydrolysis, Photolysis or Oxidation

Substances that are unstable in contact with water will be difficult to measure for water solubility. On the other hand, data on water solubility are needed prior to conducting a hydrolysis study and, as such, the water solubility should be predicted using one of the available methods.

5.3.5   Ionisable Substances

Small changes in the pH of ionisable substances can affect some of the physico-chemical tests greatly; an example is the octanol–water partition coefficient. Therefore, the test should be performed on the non-ionised (neutral) form of the substance; this can be achieved by the use of a buffer within the pH range relevant to the environment in the mobile phase or aqueous phase if the HPLC (high performance liquid chromatography) method or shake-flask method is used respectively.

5.3.6   Surface Active Substances

Surface active substances are difficult to test for partition coefficient because surface activity inhibits partitioning. In addition, for substances that degrade or hydrolyse rapidly, the surface tension result is likely to be for a mixture of both the parent and hydrolysis product(s).

5.4   Hazardous Physico-Chemical Data

The hazardous physico-chemical properties (flash point, flammability, explosive properties, oxidising properties and self-ignition temperature) are used mainly for the purposes of safe handling and hazard communication, that is classification and labelling and risk assessment. For liquid substances, the flash point is used to allocate the substance into the appropriate flammability class. The individual registrant is responsible for assessing and identifying the hazards associated with the substance produced. The result from one physico-chemical hazard can influence how the other tests are carried out: for example a flammability test should not be performed for a substance that is found to be explosive. Section 10.6 of this book provides more detail on dealing with the risks that are related to hazardous physico-chemical properties.

5.5   Relationship between Physico-Chemical Tests

The results of the physico-chemical tests are related to one another in a variety of ways; examples are stated below:

- The melting point gives information on the purity of the substance and, as such, it is important in choosing the appropriate hazard test methods for flash point, flammability, auto-flammability, oxidising and explosive properties. Also, substances that decompose during a melting point study will not require a boiling point study. Boiling point is an indication of how volatile a substance is. As a general rule of thumb, the higher the boiling point, the lower the volatility will be. The boiling point is also relevant in placing a substance in the relevant flammability hazard class. The melting and boiling point results should be considered before selecting temperature ranges over which a vapour pressure study is conducted.

- The result for surface tension should be known prior to conducting a partition coefficient study because an increase in the solubilisation and emulsification of the hydrophobic phase will occur as a result of decreased surface tension of water. The surface tension study requires a substance that is stable against hydrolysis and soluble in water at > 1 mg/l. The viscosity (dynamic) of the material should not exceed 200 mPa s.
- Water solubility and hydrolysis are interrelated because the understanding of hydrolysis is needed before conducting a water solubility study on a material. Similarly, information on water solubility is needed in order to select appropriate test concentration for a hydrolysis study. Also, hydrolysis and water solubility results are needed before conducting studies for surface tension, partition coefficient and adsorption. It is also important to have knowledge of dissociation constant before conducting a water solubility and hydrolysis study.

5.6   Application of Physico-Chemical Test Data

Physico-chemical data are applied to many areas of the dossier submission: classification and labelling, toxicology and ecotoxicology and risk assessment.

Substances that are poorly water soluble (i.e. water solubility < 1 mg/l), not readily biodegradable and with a log $K_{ow} \geq 4$ (unless it has an experimentally determined BCF (bioconcentration factor) ≤ 500) are classified (Section 7.1 of this book) for the environment in the absence of further ecotoxicological evidence (Section 7.3.3 of this book). Data on hazardous physico-chemical properties are directly used for classification[1] of substances into a relevant flammable category: flammable; highly flammable; extremely flammable; also explosive or oxidising, along with the associated labelling codes. The partition coefficient is used as an indication of bioaccumulation potential in organisms.

5.7   Can Physico-Chemical Tests Be Omitted?

With only a few exceptions (viscosity, dissociation and hydrolysis), all physico-chemical tests are required for all substances that are imported or manufactured at ≥ 1 tonne/year; but, practically, certain tests for some endpoints may be omitted from the test requirement. Examples are listed below:

- A melting point study should not be conducted below a lower limit of $< -20\,°C$. A boiling point study can be waived if decomposition is observed during a melting point study, or for solids that melt above $300\,°C$.
- A particle size (granulometry) study is not required for liquids or waxy solids, or if the substance is not used or supplied in a solid form.
- A vapour pressure study may be omitted if the melting point is $> 300\,°C$ or if the substance has a boiling point of $< 30\,°C$.

---

[1] Full details on classification and labelling of dangerous substances can be found in Regulation (EC) No. 1272/2008 of the European Parliament and of the Council of 16 December 2008 on classification, labelling and packaging of substances and mixtures, amending and repealing Directives 67/548/EEC and 1999/45/EC, and amending Regulation (EC) No. 1907/2006.

- Water solubility and partition coefficient studies can be waived for substances that hydrolyse rapidly ($t_{1/2} < 12$ hours).
- An hydrolysis study may be waived if the substance is poorly soluble and/or a suitably sensitive analytical method cannot be developed to detect 10% loss of the substance in solution.
- A surface tension study may be waived if water solubility is < 1 mg/l.
- Hazardous properties such as flammability in contact with water, pyrophoricity, explosivity and oxidising potential are routinely waived if, based on structural examination or experience in use and handling, these properties are not expected.

5.8 (Q)SAR and Physico-Chemical Tests

A general introduction to (Q)SAR is given in the next section. It has been established that in many circumstances (Q)SAR can be applied to predict physico-chemical properties where data gaps exist. The (Q)SAR method used should be validated and the test substance should be within the applicability domain of the method. Established (Q)SAR tools for many physico-chemical properties are available off-the-shelf but it is sometimes useful for registrants to develop novel methods.

Where the predicted data significantly vary from the available measured results, it is advisable for the measurement to be conducted.

Table 5.1 indicates physico-chemical endpoints where (Q)SAR can be applied.

5.9 (Quantitative) Structure-Activity Relationships ((Q)SAR)

(Q)SAR[2] is concerned with the relationship between one type of substance property and another. When no measured data are available for a particular property of a substance, data on similar substances can be used to derive a model to predict the property.

The application of methods based on (Q)SAR principles to estimate chemical properties is a topic with a long and distinguished history, but is a 'bag of tools' which still cannot meet every need, despite the growing power of computational chemistry and statistical analysis. Many practitioners prefer the word 'activity' to be replaced by 'property', that is QSPR, but, nevertheless, (Q)SAR persists as the preferred acronym. There have been many types of (Q)SAR throughout its history:

- Early studies of alcohols' toxicity to tadpoles (Hansch and Leo, 1979; Meyer, 1901).
- Prediction of hydrolysis rate ((Ingold, 1930) and followers).
- Prediction of vapour pressure (chemical engineers, 1950s and onwards).
- Prediction of octanol–water partition coefficient (begun in the 1960s) (Hansch and Leo, 1979).
- Modelling pesticide and pharmaceutical properties (numerous examples from the 1960s onwards) using physico-chemical and quantum mechanical descriptors.

---

[2] The Q is in parentheses since, in the ECHA guidance, some models may not be quantitative, that is may answer the question with 'yes' or 'no'.

Table 5.1 *Discussion of the use of (Q)SAR for different endpoints.*

Property area	Discussions
Melting point	It is often difficult to develop good (Q)SAR models for melting point, and the resulting predictions are often associated with large uncertainties. It may be possible to develop models for small subsets of closely related substances. It may also be useful to establish if limit values apply (i.e. < -20 or $> -20\,°C$)
Boiling point	Many 'off-the-shelf' methods are applicable to organic substances
Relative density	A small number of 'off-the-shelf' methods with adequate performance are available for relative density prediction of solid and liquid organic substances
Granulometry	(Q)SAR is not applicable for this endpoint. Tests are recommended except where this endpoint can be waived due to being inapplicable (e.g. for substances which are liquids, gases or amorphous solids)
Vapour pressure	Many 'off-the-shelf' methods are applicable to organic substances. However, since boiling point is an input to vapour pressure determination, it is advisable that one of the two endpoints should be a valid and reliable experimental measurement
Surface tension	Calculation of hydrophilic–lipophilic balance may be useful for some substances; otherwise (Q)SAR is not applicable
Water solubility	Water solubility can be predicted using various freely available methods. It is advisable to input reliable measured melting point data when predicting water solubility of solid substances
Partition coefficient	Partition coefficient prediction can be applied to many organic substances and high performance is typical. Prediction may be particularly useful for substances that decompose, are highly insoluble substance or substances that are surface active. Various predictive methods are applicable
Flash point	Flash point relates to hazard assessment of the substance and it is affected by the presence of volatile impurities. Flash point data relate to the handling and use of the commercial substance and it is not advisable to use prediction for this endpoint
Flammability	(Q)SAR prediction is not advisable for flammability of a substance. Tests are recommended except where the endpoint can be waived
Explosivity	(Q)SAR prediction is not advisable for the explosivity of a substance. Tests are recommended except where the endpoint can be waived based on the oxygen balance calculation, or where the substance does not contain functional groups that are associated with explosivity
Self-ignition temperature	Measurement is recommended for self-ignition temperature
Oxidising properties	(Q)SAR prediction is not advisable for oxidising properties of a substance. Tests are recommended except where the endpoint can be waived
Stability in organic solvent	(Q)SAR prediction is not applicable to this endpoint
Dissociation constant	Some 'off-the-shelf' methods are available to predict the dissociation constant
Viscosity	Several 'off-the-shelf' methods are available to predict the viscosity of some liquid substances

Specific examples follow in later sections but the general principles are set out here. The aim of (Q)SAR is the link of one type of substance property to another, as seen below:

Molecular descriptor Some or all of the following	The link – a mathematical model	Desired property for prediction
A physico-chemical property Spectroscopic measurement Molecular connectivity index Quantum mechanical output Molecular dynamics calculation	→	Physico-chemical endpoint Environmental endpoint Toxicological endpoint

The model is derived using data for a set of substances for which the desired property is known. It can then be used to predict that property for similar substances for which no measured data are available.

Whilst the use of (Q)SAR may be desirable, it is not easy. Every measured property value, of whatever source or reliability, has uncertainty associated with it and so many see (Q)SAR as a kind of paper-based or computer-based 'experiment'.

The REACH legislation follows the stringent principles of validation laid down by the OECD. (Q)SAR predictions may be used instead of test data when the following conditions are met:

- Results are derived from a (Q)SAR model whose scientific validity has been established.
- The substance falls within the applicability domain of the (Q)SAR model.
- Results are adequate for the purpose of classification and labelling and/or risk assessment.
- Adequate and reliable documentation of the applied method is provided.

(criteria taken from REACH Regulation, Annex XI, Section 1.3).

If a prediction fulfils some but not all of the conditions it may still be helpful as part of a weight of evidence approach, to aid interpretation of measured data, or in the design of new experimental studies.

A reliable (Q)SAR model (the mathematical link between substance descriptors and the property to be predicted) must be associated with the following information:

1. A defined endpoint.
2. An unambiguous algorithm.
3. A defined domain of applicability, in terms of both numerical range and chemical structural type.
4. Appropriate measures of goodness-of-fit, robustness and predictivity.
5. A mechanistic interpretation, if possible.

These 'Setubal principles' are well-described in REACH guidance[3] and by the OECD (OECD, 2013) in its method development work.

Note on domain of applicability: The applicability domain of a (Q)SAR model defines the substances for which it can reliably be used to make predictions. This depends on the type of model and the substances used in developing the model. For example, a boiling point model trained using alkanes with boiling points in the range 0–200 °C is not appropriate for predicting the boiling point of an alcohol or of an alkane with an estimated boiling point of 400 °C; and an acute ecotoxicity model for substances with a narcotic mode of action is not appropriate for a substances with a suspected different mode of action. If a prediction is obtained from a valid model and applied to an appropriate substance, its relevance to the endpoint being considered and the completeness of the information provided need to be considered.

Note on robustness of the method: This is of particular note in that it is easy to 'overfit' data, with too many variables chasing too few data points, resulting in a method that describes the data well but which has no predictive value. The answer to this is to have enough data for the model which allows a 'training set' for development of the model and a 'test set' (not used in model development) to trial the model on. This necessarily implies a data set of some size to work with, depending on how narrow the error of the predicted value needs to be.

REACH guidance on (Q)SAR is complex and extensive. In short it describes:

- How to establish the validity of a (Q)SAR model.
- How to establish the adequacy of a (Q)SAR model result for regulatory purposes.
- How to document and justify the regulatory use of a (Q)SAR model.
- Many existing techniques (without granting them formal approval).
- The principles of proportionality and caution (e.g. much more information would be required to justify use of a prediction for long-term toxicity than for a boiling point).

Models may be taken from published sources or developed by the registrant(s). In either case the reporting will need to meet certain requirements:

- (Q)SAR method reporting format (QMRF) in which the method is validated and described; where the method is reported in the public domain it may be possible to waive this requirement.
- (Q)SAR prediction reporting format (QPRF): the way in which the method was applied to the substance under discussion.

Different (Q)SAR predictions methods have been extensively discussed in the literature, but where an 'off-the-shelf' method is directly used for a substance, it is still important to support such a method with a (Q)SAR Prediction Reporting Format (QPRF). It has been seen in the early stages of REACH that registrants simply use an 'off-the-shelf' method and do not fulfil the requirements as laid down in the Guidance.

(Q)SAR methods are commonly used for prediction of the properties shown in Table 5.2.

---

[3] ECHA Guidance on information requirements and chemical safety assessment, Part R.6.

Table 5.2   *Examples of common predictive methods in different endpoints.*

Property area	Some examples of the type of prediction that can be used easily
Physico-chemical properties	Well-established methods exist for a number of important properties, particularly water solubility, vapour pressure, partition coefficient. Refer to Section 5.8 of this chapter
Corrosivity	A property–property relationship based on the pH can be used to predict that a substance is likely to be corrosive (i.e. if pH is ≤ 2 or ≥ 11.5). Refer to Section 9.7.7 of this book
Acute systemic toxicology	The presence of certain features in the chemical structure can be an indicator that a specific mode of toxic action is possible. This is a form of non-quantitative structure–activity relationship
Genotoxicity	Certain structural features are known to be associated with genotoxicity and a variety of no-test methods exist for predicting genotoxicity, from expert systems and metabolic simulators to (Q)SARs
Ecotoxicity	Reliable methods exist for acute and chronic aquatic ecotoxicity. Refer to Section 7.3.5 of this book
Biodegradation	There are good methods for some structural types. Refer to Section 7.2 of this book
Environmental fate properties	Reliable methods for properties such as bioaccumulation and soil adsorption are available. Refer to Section 7.2 of this book

A registrant should take steps to explore what the significance of a data gap might be, as part of validating the use of (Q)SAR to predict data. For example, it could be that a bioconcentration factor test is required by the Annex requirements. If a (Q)SAR that establishes the likely range of values of that property is available, then the risk characterisation ratios (RCRs) affected by the exact value of bioconcentration factor could be explored. Should it be that RCR is less than one, whatever the exact value, it is reasonable to propose to use the (Q)SAR to fill the data gap, even if there is uncertainty in the predicted result.

In conclusion, (Q)SAR can be very powerful and cost effective when working with large numbers of close analogues, or with very well-studied types of substances. They can give Klimisch 2 (Section 6.8 of this book) results which are fully-valid as key studies. More examples follow in later chapters of this book.

References

Hansch, C. and Leo, A. (1979) *Substituent Constants for Correlation Analysis in Chemistry and Biology*. John Wiley & Sons, Inc., New York.

Ingold, C.K. (1930) The Mechanism and constitutional factors controlling the hydrolysis of carboxylic esters. *Journal of the Chemical Society*, 1032.

Leo, A., Hansch, C. and Church, C. (1969) Comparison of parameters currently used in the study of structure-activity relationships. *Journal of Medicinal Chemistry*, **12**, 766–771.

Meyer, H.H. (1901) Zur theorie der alkoholnarkose. Der einfluss wechselnder temperature auf wirkungsstärke und theilungscoefficient der narcotica. *Archiv fur Experimentelle Pathologie und Pharmakologie*, **46** (5–6): 338–346.

OECD (2013) The OECD (Q)SAR Toolbox. http://www.oecd.org/document/54/0,3746,en _2649_34379_42923638_1_1_1_1,00.html (last accessed 20 July 2013).

Overton, C.E. (1901) *Studien über die Narkose zugleich ein Beitrag zur Allgemeinen Pharmakologie*. Gustav Fischer, Jena.

6

Assessing and Documenting the Intrinsic Properties of Substances in REACH

6.1 Introduction to REACH Data Requirements

The data requirements applying for most substances to be registered are set out in four Annexes to the REACH regulation, numbered VII to X; these apply at specified tonnage thresholds unless certain exemptions apply, particularly when the potential for exposure is very limited due to use under 'strictly controlled conditions' (Section 1.2.10 of this book and below). For this reason, data sets may commonly be referred to as (for example) an 'Annex IX dossier' or 'Annex X compliant'.

Literature sources and good research

Published property data can be obtained from a number of readily available sources, many of which may be accessed without charges. Peer-reviewed chemical handbooks including references are reliable sources of physico-chemical property values, although they lack supporting details about the source of a measurement. The online e-chem portal is a single point link to a number of international databases. The ECHA's (European Chemicals Agency's) disseminated data portal is searchable and it may be that data for a similar or related substance exist in a registration already made. Hazardous Substances Data Bank (HSDB) and other databases from the Toxnet network may provide leads in the chemical literature and chemical abstracts can provide data from published journals, although an access charge is normally made for commercial users. For any data that are ultimately owned by companies, a prospective registrant

---

Chemical Risk Assessment: A Manual for REACH, First Edition. Peter Fisk Associates Ltd.
© 2014 John Wiley & Sons, Ltd. Published 2014 by John Wiley & Sons, Ltd.

must establish legitimate data access before using the data in the registration set. This is true even for data from disseminated dossiers and HPV (high production volume) dossiers published by regulators or the OECD (Organisation for Economic Co-operation and Development)/UNEP.

Each Annex presents lists of chemical properties for which data are required, in the areas of physical chemistry, chemical hazards, toxicology, ecotoxicology and environmental fate. In general, the data requirements are more complex and more costly to test, as the annex number increases. This chapter makes reference where appropriate to the requirements at each Annex level (Table 6.1).

Case Study 6.1   Do ALL these tests have to be done?

A registrant is concerned because discussion with a laboratory has revealed that the large costs of undertaking testing to complete all the REACH required endpoints would mean the viability of the registration would be put into doubt.

- Carefully check whether data waiving applies.
- Survey the published literature and check if the substance has been covered in an HPV chemical programme.
- Are chemically similar substances also going to be registered? If so, consider whether read-across between the two is appropriate.
- Consider whether (Q)SAR ((Quantitative) Structure-Activity Relationship) can be used to predict properties.

In addition to identifying the properties, these Annexes also present fundamental reasons why each property or test might be inapplicable in certain cases, which can form the basis of a 'data waiver'. This is a short discussion of scientific reasoning to justify the non-inclusion of data about a particular chemical property that would normally be required at the indicated Annex level.

6.1.1   Strictly Controlled Conditions

A reduced package of registration requirements applies for substances used only under 'strictly controlled conditions'. This was misinterpreted by some registrants early in the process and the ECHA has instigated a new series of checks on dossiers using this claim. In the REACH regulation the concession for strictly controlled conditions applies only when the substances are used *only* as chemical intermediates (used on site or transported isolated) and are managed with an extremely high degree of enclosure and control throughout their life cycle. It is necessary for registrants to provide certain supporting evidence and arguably registrations making such claims will require more continuing maintenance than full submissions.

Table 6.1 Reach data requirements.

	Data requirements at 1–10 tonnes/year	Data requirements at 10–100 tonnes/year	Data requirements at 100–1000 tonnes/year	Data requirements above 1000 tonnes/year
Full registration On-site isolated intermediate (Strictly Controlled Conditions)	Annex VII Existing data only	Annexes VII–VIII Existing data only	Annexes VII–IX Existing data only	Annexes VII–X Existing data only
Transported isolated intermediate (Strictly Controlled Conditions)	Existing data only	Existing data only	Existing data only	Annex VII
Full registration but no CLP applying and no consumer uses or wide dispersed uses	Annex VII (physico-chemical data requirements only)	As full registration	As full registration	As full registration
	Annex VII data requirements	Annex VIII data requirements	Annex IX data requirements	Annex X data requirements
Physico-chemical properties	Physical state		Stability in organic solvents	
	Melting/freezing point Relative density Vapour pressure Surface tension Water solubility Partition coefficient Flash point Flammability		Dissociation constant Viscosity	

(continued overleaf)

Table 6.1 *(continued)*

	Annex VII data requirements	Annex VIII data requirements	Annex IX data requirements	Annex X data requirements
Toxicological data	Explosive properties Self-ignition temperature Oxidising properties granulometry			
	In vitro skin irritation/corrosivity and eye irritation	*In vivo* skin irritation and eye irritation (if required)	*In vivo* mutagenicity tests (if required)	Long-term repeated toxicity (12 months)
	Skin sensitisation	Bacterial and mammalian *in vitro* mutagenicity tests	Repeated dose toxicity	Studies related to particular concerns
	Mutagenicity	Mammalian *in vitro* cytogenicity test	Repeat dose (28-days) toxicity (if not already provided)	Two-generation reproductive toxicity
	Acute oral toxicity	Inhalation and/or dermal toxicity Repeated dose toxicity 28 days (90 days if necessary) Reproductive toxicity	Subchronic (90-days) toxicity (if necessary) Developmental toxicity (two species) Two-generation reproductive toxicity (if indicated by subchronic tests)	Carcinogenicity study
		Toxico-kinetic behaviour		

Environmental data			
Short-term (or long-term) toxicity to *Daphnia* Algal growth inhibition Biotic degradation Ready biodegradability	Fish short-term toxicity Activated sludge respiration inhibition test Further degradation studies (if a need is indicated in chemical safety assessment) Abiotic degradation-hydrolysis as a function of pH Adsorption/desorption	Long-term toxicity to *Daphnia* Long-term toxicity to fish (if required) Degradation testing (surface water, soil, sediment) if required Identification of degradation products Bioaccumulation in fish (if required) Further adsorption/desorption studies (if required) Effects on terrestrial organisms (if appropriate)	Confirmatory testing on biodegradation Fate and behaviour in the environment of substance and/or degradation products Effects on terrestrial organisms (if required) Long-term toxicity to sediment organisms (if required) Long-term or reproductive toxicity to birds (if required)

6.2   Hazards

An important early step in the chemical safety assessment is an assessment of hazards for the substance being registered. Hazards are essentially intrinsic properties of a substance that cause effects which present some sort of danger to humans or the environment, or issues for storage and handling. There is no consideration of the likelihood of such effects being expressed (assessment of risk) (Section 1.1 of this book).

When the REACH regulation was first published, the pre-existing classification system in the European Union, under the Dangerous Substances Directive (DSD, Directive 67/548/EC), was still in place. A subsequent corrigendum has brought the requirements of REACH up to date in law to refer to the system now in force for the classification, labelling and packaging requirements for substances (CLP, Regulation 1272/2008/EC). Many of the rules and criteria within CLP were inherited from the international globally harmonised system (GHS) (UNECE, 2003 as amended).

In CLP (EC Regulation 1272/2008), fixed criteria are set out, by which it can be established whether each of a wide range of specific potential hazards exists for any given substance, based on experimental or other data about its properties.

As well as a general identification of the hazard profile of the substance, the REACH hazard assessment aims to identify whether a substance is an SVHC (substance of very high concern). Substances classified as SVHC are subject to in-depth review by the ECHA.

Some of the most serious carcinogenicity, mutagenicity, and reproductive toxicity effects, associated with specific classifications under CLP, are referred to as CMR (carcinogens, mutagens and reproductive toxins). If a hazard of this type, among others, is present, then the hazardous substance in question has a particularly significant status in REACH, and is considered a 'SVHC'.

For SVHCs, it is not always possible to do fully quantitative risk characterisation but exposure assessment may still be very important.

Criteria for classification and labelling under CLP are presented in the CLP Regulations – EC 1272/2008. There were several important changes, significantly affecting CLP rules for the environment, in the second Adaptation to these regulations – EC 286/2011, which is discussed in detail in Appendix A.

6.3   PBT

As well as assessing the applicable classification and labelling, registrants must evaluate the substance's properties of persistence, bioaccumulation potential and toxicity, again in terms of established threshold criteria of chemical properties. The adoption of these methods in regulatory assessment is a relatively recent development in the European Union compared to other aspects of hazard assessment. The criteria implement a precautionary approach in that conventional risk characterisation is not carried out. In summary, according to the current REACH Annex XIII criteria:

Persistence is defined in terms of half life in water, sediment or soil; some substances may be termed as 'very persistent' or vP.

Bioaccumulation is defined in terms of bioconcentration factor (BCF); some substances may be termed as 'very bioaccumulative' or vB. There are additional factors such as trophic magnification that come into a discussion of weight of evidence.

Toxicity is defined in terms of harm to aquatic organisms or certain mammalian toxicity classifications. There are no criteria for toxicity to terrestrial or sediment organisms, as yet.

A substance meeting screening criteria for P, B and T, or vP and vB only, will be subject to more in-depth assessment.

6.4   Equivalent Concern

Substances which do not meet PBT criteria but are for other reasons considered very hazardous (for example endocrine disrupting substances) are also considered SVHC. In these cases, the premise of the hazard is referred to as of 'equivalent concern'. This term is deliberately open-ended to ensure that rare and unforeseen hazards and risks are covered appropriately in REACH. Thus, there are not well defined criteria, and assessment is a matter for expert judgement on the part of the registrant and the ECHA reviewers. However, new advice concerning endocrine disruption is being issued by various authorities.

An endocrine disrupter is a substance that produces an adverse effect by interfering with the hormone systems of animals or humans.

According to the REACH Regulation, the hazards that are considered of very high concern that may lead to a substance being subject to Authorisation include endocrine disrupters.

However, there is no requirement under REACH to test for endocrine disruption. The reason for this is that endocrine disruption is a subject of much debate and only recently have advances been made towards defining them. The European Commission has recently published a report on the identification and characterisation of endocrine disrupters (JRC, 2013). This report is not specific to any particular legislation, but will in all probability be used in a variety of legislative frameworks to aid in the identification of endocrine disrupters.

The definition of endocrine disrupter accepted and amplified by the report is that agreed by the World Health Organization: 'An endocrine disrupter is an exogenous substance or mixture that alters function(s) of the endocrine system and consequently causes adverse health effects in an intact organism, or its progeny, or (sub)populations' (IPCS/WHO, 2002, cited in JRC, 2013).

6.4.1   Adversity

The effects must be adverse, that is must have a detrimental impact on the individual (for human health) or the population (for ecotoxicological endocrine disruption). For human health, such an effect causes a change in morphology (structure) or physiology (function) that results in impairment of function. For the environment, an adverse effect has the potential to affect the population, including sensitive life stages; these effects can include growth, development, and reproduction.

6.4.2   Mode of Action

For a substance to be identified as an endocrine disrupter, the adverse effects need to be shown to be caused by endocrine changes. Chemicals can affect the endocrine system in a number of ways, leading to changes in levels of hormones by affecting production, metabolism or transport, or by interacting with hormone receptors either simulating the effect of the hormone, or by blocking the receptor. Changes in endocrine systems may be a secondary result of systemic toxicity; such effects are not indicative of endocrine disruption, as they are secondary. An endocrine disrupter produces adverse effects by means of disrupting the endocrine system.

The standard toxicological and ecotoxicological reproductive tests are not adequate to demonstrate endocrine disruption; information on mode of action is needed to identify an endocrine disrupter. It is possible for a test to demonstrate an adverse effect and an endocrine mode of action but more often a weight of evidence approach is need to conclude on the adverse effect and the mode of action separately. There should also be a biologically plausible link indicating that the endocrine disruption causes the adverse effect, and that this is not secondary to a non-endocrine toxicity.

Evidence of mode of action might require:

- Evidence of *in vitro* endocrine activity AND
- *In vivo* evidence of endocrine biomarker and adverse effect AND
- Plausible biological explanation of a relationship between the parameters.

In an OECD fish sexual development assay (OECD TG 234), a change in the ratio of sexes in some species was accompanied by an alteration in the levels of vitellogenin (a protein that is involved in egg yolk formation). Depending on the degree of the change in ratio of sexes, this may be diagnostic of endocrine disruption.

In addition to an adverse effect that is the result of a primary effect on the endocrine system, the effects must be relevant. For human health this means that the effects seen in animal models must be relevant to humans. This is assumed to be the case unless it can be demonstrated not to be. In ecotoxicology, data on all species including mammals are considered to be relevant.

Tests for endocrine disruption are discussed in the OECD conceptual framework (OECD, 2012). There is a need for further development of tools for assessment of endocrine disruption. It will be interesting to see how the REACH guidance (and possibly the Regulation) is adapted in future to reflect the scientific progress on understanding of the identification of endocrine disrupters.

6.5   Test Proposal Rule

Once the requirements of Annexes VII and VIII are met, the ECHA requires that any studies needed to comply with Annexes IX and X are presented as proposals for agreement by the authorities. That does not remove the responsibility for showing that uses are safe (Chapter 1 of this book).

6.6   Availability of Existing Data and Rights of Access

REACH requires that data owners share existing data with other registrants. This leads to sometimes complex legal discussions, but custom and practice about what is a fair cost has built up over the years that REACH has been in force (Chapter 1 of this book).

6.7   Data Reliability

All available studies are assigned a reliability score according to a Klimisch scale (1–4), where reliability 1 and 2 studies are deemed acceptable for fulfilling key criteria (Klimisch *et al.*, 1997). The value of an old study is scaled according to reliability (Chapter 1 of this book).

6.8   Data Gaps – Options for Surrogate Data for Description of Hazard and Risk – Including Read-Across

It is a fundamental principle of REACH that duplication of existing studies should be avoided wherever practicable, once sameness of the substances for registration has been established. Several methods exist to fill data gaps:

- Measurement (discussed throughout Chapters 7 and 9 of this book).
- Claiming a waiver that the test is not needed for a technical reason; the normal reasons are described in the Annexes to the regulation. They are based on the principle of the result being predictable, unethical or unnecessary. These waivers are described in Section 5.7 of this book for physico-chemical properties and in later chapters describing toxicological and environmental properties.
- Claiming a waiver based on lack of exposure making a test unnecessary: that is, the result of a test would not affect hazard classification, risk characterisation or safety advice. This is termed exposure-based adaptation.
- Read-across: this is using a result for more than one substance: it can be one substance to another, in a group of analogues or in a category.
- (Q)SAR, as described in Section 5.8 of this book.

An ***analogue approach*** involves estimating data for a particular endpoint from a single or limited number of compounds. A category has more substances and trends in properties may be identified. The dossier documents need to explain carefully what the methods used are, and justify them with valid science.

A ***category approach*** (where applicable) is considered to be more robust because the basis for evaluation of any individual chemical in the category is greater, and there are usually more measured data points available in such a wider approach. Any gaps may be filled by interpolation, with high reliability. Read-across for one end point does not imply that all endpoints may be read across; however, it is sensible (for clarity) to try to avoid the methods becoming too selective or inconsistent. Again, documentation must be fit for purpose.

Several important principles should be followed when using read-across or (Q)SAR:

1. The guidance allows a substance to belong to more than one grouping, even for the same endpoint.
2. (Q)SARs should be applied where possible. Use of a (Q)SAR implies a 'category' in itself and is subject to quantifiable validation.
3. Breaks in numerical criteria (cut-offs between different interpretations, such as classi-fication endpoints) should follow REACH requirements so as to enhance applicability. For example, it is useful to know whether bioconcentration factor (Section 7.2.2.1 of this book) is less than 100, more than 2000 or more than 5000.

For reference, some key terms are defined here. It can be seen that definition of 'category' and 'analogue' could be uncertain, but that is not particularly important since the ECHA's reporting requirements are similar. Refer also to the fuller definitions in the ECHA guidance (Table 6.2).

Table 6.2 *Key terms.*

Category	A group of substances which can be seen as behaving identically or in accordance with a fixed trend, in respect of the particular REACH endpoint, which would usually cover a range of chemical structures within specified norms, for example a homologous or closely related series sharing key structural features. The Klimisch reliability would reflect the reliability of the associated data, but Klimisch 1 or 2 would usually apply.
Analogue	A substance for which the value of an endpoint can be read across one to another substance due to close structural similarity and reasonable evidence to justify an expectation of equivalency, taking into account physical chemistry and evidence from other available property data, particularly for directly related endpoints. The Klimisch reliability would reflect the reliability of the associated data, but Klimisch 1 or 2 would usually apply.
(Q)SAR	A validated quantitative structure-activity relationship for a defined applicability in terms of the property and the structural class covered. A valid method uses only Klimisch 1 and 2 inputs and the predictions derived can therefore be seen as Klimisch 2. By definition, use of a (Q)SAR implies the existence of a category.
Structural class	A set of substances which contain common structural fragments and do not contain structural differences which would be more significant than the points of similarity. The word class is used so as to avoid confusion with the term 'functional group' which has a specific meaning in chemistry.
Endpoint	A specific property listed as a REACH requirement, for example melting point, ready biodegradation, acute mammalian toxicity and so on.

Table 6.3   *Example of read-across.*

Property area	Some examples of the type of read-across that can be used easily
All areas of data need	Close structural 'relatives' within analogue groups or a category as described in Table 6.2. This is very often applied by stating that the result can be read across because the substances' chemical structures are sufficiently close.
Limiting values of physico-chemical properties	For example, very low or very high melting or boiling point, water solubility, partition coefficient.
Corrosivity	Strong acids or bases may easily be identified and unethical animal studies must be avoided, especially if one substance in the group has proven measured properties.
Acute toxicology	Some functional groups possess well-defined properties so read-across could be used (but may not be appropriate for long term studies).
Acute ecotoxicity	Highly insoluble substances do not usually show acute (short-term) effects.
Biodegradation	Very low degradation – it is appropriate to read across the most cautious result.

6.9   Read-Across

Examples of read-across and science-based methods follow. More examples are given in the technical sections that follow later in the book (Table 6.3).

However, it should be noted that the reporting requirements for read-across and (Q)SARs are onerous, and time costs should be assessed before embarking down any of these routes. Some properties are easier to measure than predict, for example density and flash point; the consequences in error can be severe and should be taken into account. More details are found in later chapters.

The arguments to justify that the result for Substance A can be understood to apply for Substance B must still be fully reported. Registrants should bear in mind that this is a form of estimation and assess the reliability and uncertainty accordingly.

References

IPCS (International Programme on Chemical Safety) (2002) Global Assessment of The State of The Science of Endocrine Disruptors. http://www.who.int/ipcs/publications /en/toc.pdf (last accessed 20 July 2013).

JRC (Joint Research Centre), Institute for Health and Consumer Protection (2013) Key Scientific Issues Relevant to the Identification and Characterisation of Endocrine Disrupting Substances. Report EUR 25919 EN, Report of the Endocrine Disrupters Expert Advisory Group.

Klimisch, H.J., Andreae, M. and Tillmann, U. (1997) A systematic approach for evaluating the quality of experimental toxicological and ecotoxicological data. *Regulatory Toxicology and Pharmacology*, **25**(1), 1–5.

OECD (Organisation for Economic Co-operation and Development) (2012) Guidance Document on Standardised Test Guidelines for Evaluating Chemicals for Endocrine Disruption. OECD Environmental Health and Safety Publications, Series on Testing and Assessment No. 150, Organisation for Economic Cooperation and Development, Paris. http://search.oecd.org/officialdocuments/displaydocumentpdf/?cote=env/jm/mono%282012%2922&doclanguage=en (last accessed 20 July 2013).

UNECE (2003) Globally Harmonized System of Classification and Labelling of Chemicals.

7

Assessing Environmental Properties Data

This chapter deals with the properties or hazard side of doing the chemical safety assessment (CSA) for the environment (human health is dealt with in Chapters 9 and 10). This focuses on the data for the intrinsic properties of a substance, which are used to derive the notional safe level at which no effects are expected, that is the predicted no-effect concentration (PNEC). The data on use pattern are used to derive the concentration of the substance that will end up in different compartments of the environment – the predicted environmental concentration (PEC) (as considered in Chapter 8). The estimation of risk is essentially a comparison of the PEC with the PNEC for each use pattern and each relevant environmental compartment (this is set out in Chapter 11).

7.1 Environmental Properties Data

The purpose of using ecotoxicology in the context of REACH is to conduct a CSA for the environment. A CSA requires data for:

- deriving PNECs for risk characterisation;
- identifying Classification and Labelling requirements;
- assessing the toxicity of the substance against persistent, bioaccumulative and toxic (PBT) criteria.

Only key data that are reliable and representative of the registered substance can be used for these assessments.

7.1.1 PNECs

PNECs are the concentrations at which no effects are expected to occur in the compartment under consideration. They therefore define the concentration at which no hazard is expected to occur with a margin of safety. Under REACH, PNECs are derived

Chemical Risk Assessment: A Manual for REACH, First Edition. Peter Fisk Associates Ltd.
© 2014 John Wiley & Sons, Ltd. Published 2014 by John Wiley & Sons, Ltd.

for the aquatic freshwater and marine, sediment freshwater and marine, soil and avian compartments.

The approach to deriving PNEC from available ecotoxicological data is discussed in Section 7.3 of this chapter.

7.1.2   Classification and Labelling (C&L)

Classification is based on the classification, labelling and packaging (CLP) regulation (EC) No 1272/2008 (Appendix A) Environmental classification is based on both short-term EC/LC_{50} values and long-term NOEC (no observed effect concentration) values. Substances can be classified as Aquatic Acute Category 1 and/or Aquatic Chronic Category 1 (most severe), 2, 3 or 4 (the latter is a safety net classification for substances that do not meet the classification criteria but for which there are still grounds for concern; for example, poorly soluble substances for which only short-term data are available, that do not biodegrade and have the potential to bioaccumulate can be included in Category 4).

Since the second ATP (Adaptation to Technical Progress) amendment effective from April 2011, classification is also divided between substances that degrade rapidly and substances that do not degrade rapidly. This is because the rapid degradation of a substance is a mitigating factor against chronic exposure in the environment. Not rapidly biodegradable substances may be classified in chronic categories on the basis of short-term data when no long-term data are available.

The labels, safety, precautionary, and hazard phrases which must be presented in an extended Safety Data Sheet (eSDS) are defined by each classification category and are presented within the CLP regulation.

7.1.3   PBT

The role of ecotoxicological data in PBT assessment is presented in Section 6.3 of this book.

7.2   Environmental Fate

When a substance is released in the environment it may partition between the environmental 'compartments' of water, sediment, soil and air, and may degrade through biotic or abiotic mechanisms in each compartment. The fate at equilibrium for any specific substance will depend partly on its physico-chemical properties, as discussed in Section 5.1 of this book, and also other properties required in REACH for most substances in the area of environmental fate.

It is vital that the environmental exposure assessment takes into account substance properties in order to model the fate of the substance (or its chemical constituents) in the environment in a realistic way. The most important environmental fate endpoints important for REACH CSAs relate to the substance's distribution (e.g. adsorption and bioaccumulation; physico-chemical properties are also important for understanding this) and removal (biodegradation and abiotic degradation).

Biodegradation and bioaccumulation data are also used (in a simplified way) in the hazard assessment of a chemical (Classification and Labelling, and PBT assessment) as well as in the derivation of PECs.

The potential for toxicity to sewage treatment plant (STP) microorganisms is also important as part of understanding biodegradation, particularly when low levels of removal are seen. This property plays an important role in both the environmental fate and ecotoxicity data sets.

Exposure-based adaptation – environment – principles

For most substances, most of the data requirements listed in Annex IX and X (long-term ecotoxicology; follow up degradation and adsorption studies) are only required if the exposure assessment indicates they are needed. This means that when exposure assessment is conducted using lower tier indicative values, with appropriate consideration of uncertainty, if the conclusion clearly indicates that further exploration of the property is unlikely to significantly affect the conclusion then there would be no benefit in conducting of the test. This can be a basis for waiving the further testing.

7.2.1   Degradation

Degradation of a substance leads to its removal from the system and is, therefore, a critical influence on the environmental exposure; it is thus a key parameter in the risk assessment of a chemical. Degradation rate also has a role in correctly assigning classification and labelling under CLP (Appendix A) and for PBT assessment (Section 6.5 of this book).

Tiers of biodegradation tests

- Baseline information – Ready biodegradability – pass or fail, test for $> 60\%$ (70% in some tests) mineralisation over 28 days by activated sludge.
- Inherent biodegradability – similar but more favourable conditions to demonstrate intrinsic degradability.
- Simulation tests in sewage sludge or natural media (river water, water/sediment, soil) system – longer-term test for mineralisation or removal of parent substance by organisms in natural media.

  The simulation tier of testing is mainly useful to refine in case of uncertainty being significant, mainly for regional scale, or when degradation is exceptionally fast.

Why is degradation in the atmosphere important?

Whilst not a REACH requirement, it may be useful to investigate atmospheric photodegradation (e.g. from reaction with hydroxyl radicals) as this can be an important removal mechanism from substances in the atmosphere. Releases via air and concentrations in air can be important in calculating predicted concentration in soil and indirect intake for humans via the environment.

Degradation of a chemical can occur in a number of different ways:

- Most organic chemicals can undergo a degree of biodegradation, that is degradation of a chemical by bacteria or other biological means. Biodegradation may be complete (resulting in mineralisation) or partial (resulting in a more or less persistent by-product).
- Abiotic degradation may be significant depending on the chemistry of the registration substance. Hydrolysis is the degradation by reaction of a chemical with water (Section 5.1 of this book). Oxidation by free radicals and other species, and photocatalysed degradation, may also occur in aquatic environments, soil, and air. Oxidation and photolysis are not REACH Annex requirements for testing. Information on photolysis is difficult to interpret in the CSA, as its significance depends on local environmental conditions. However, where data exists it can be used on a case-by-case basis.

7.2.1.1   Ready Biodegradation

Ready biodegradation testing is a REACH Annex VII requirement. It can be waived if the substance is inorganic.

Ready biodegradation tests are stringent screening tests. The test is conducted under aerobic conditions. A high concentration solution of the test substance in a mineral medium is incubated with a small amount of inoculum (domestic sewage, activated sludge or secondary effluent). Parallel blanks solutions are run, with inoculum but without the test substance (to determine the endogenous activity of the inoculum); a reference compound is also run in parallel, to check the operation of the procedures and verify the correct activity of the inoculum. Normally, the tests last for 28 days. The degradation is followed throughout the incubation by the determination of parameters such as dissolved organic carbon (DOC), carbon dioxide production and oxygen uptake.

A positive result in a ready biodegradability test is indicated by the passing of a specified threshold: normally 60 or 70% degraded within a fixed time period, depending on the method used. This indicates that the chemical will undergo rapid and ultimate degradation in most environments, including in STPs. Ultimate degradation is the degradation of the substance to carbon dioxide, biomass, water and other inorganic substances such as ammonia.

Physico-chemical properties should be considered when identifying the appropriate studies to conduct; for example, certain biodegradation tests are not applicable for volatile, adsorbing or poorly water-soluble chemicals.

Biodegradability can be predicted if testing is not feasible. For example, the Syracuse Research Corporations Estimation software program BIOWIN, calculates the probability that a chemical will biodegrade rapidly or slowly under aerobic conditions with mixed cultures of microorganisms.

7.2.1.2   Simulation Tests

As well as ready biodegradation tests, more complex simulation-type degradation tests may be used in the hazard assessment and CSA. These tests attempt to simulate degradation in specific environmental compartments, for example water, aquatic sediment or soil.

Simulation tests in surface water, soil and sediment are an Annex IX requirement but are only required if the CSA indicates the need to investigate further the degradation of a substance in a particular compartment, that is if the conclusion of safe or unsafe use may be sensitive to this rate.

Simulation testing can also be waived based on exposure considerations, such as if the substance is readily biodegradable. The water simulation test can be waived if the substance is highly insoluble in water; the soil and sediment simulation tests can be waived if direct and indirect exposure of soil/sediment is unlikely.

If the substance is found to be a PBT/vPvB (very persistent and very bioaccumula-tive) candidate based on screening criteria then it may be useful to consider conducting simulation tests.

7.2.1.3   Special Cases

Special consideration should be given to chemicals that hydrolyse within the time-scale of a ready biodegradability test. This may give misleading results in ready biodegradation tests: for example, if the chemical hydrolyses to one product which biodegrades rapidly, giving a pass result in the test, and also a second product which is not susceptible to biodegradation.

It should be considered whether the substance being assessed may degrade to give stable and/or toxic degradation products. The conventional threshold is a half-life of less than 12 hours. If this is the case, the exposure assessment, C&L (classification and labelling), and PBT assessment should consider the properties of the degradation products.

Unknown or variable composition, or of biological origin (UVCB) substances also need special consideration. Some constituents of the substance may degrade rapidly but others may not degrade at all. For UVCB substances the biodegradation rate of whole substance is required for the purposes of hazard classification. However, for risk characterisation it is appropriate to look at the constituents individually, as consideration of whole substance properties is not relevant once in the environment. Prediction of biodegradation for individual constituents may be more practical.

Biodegradability tests in seawater may be available for a substance. Chemicals gener-ally degrade slower in seawater than in freshwater tests inoculated with activated sludge and sewage effluent (seawater tests use natural seawater both as the aqueous phase and as the source of microorganisms). Therefore, a positive result in a biodegradability seawater test is regarded as evidence of potential for biodegradation in the marine environment. This type of test may be useful for substances with an intended end use in the marine environment, or if the sea is the most likely receiving water.

There are also standard tests for inherent biodegradability (OECD 302 A–C) (OECD, 1981, 1992, 2009), although these are not normally part of the standard set of required new testing under REACH, so the data may exist already. The test procedure in inherent biodegradation tests differs to that in ready biodegradation tests in that the conditions are more conducive to biodegradation, for example a higher concentration of microorganisms and longer test duration. Biodegradation above 20% is regarded as evidence of inherent, primary biodegradability; biodegradation above 70% is regarded as evidence of inherent, ultimate biodegradability.

7.2.2   Bioaccumulation

It is necessary to investigate the potential for a substance to be taken up from contaminated waters and become concentrated in biota. High K_{ow} and K_{oc} values can be indicative of the potential for a substance to bioaccumulate.

Furthermore, depending on whether or how the substance is metabolised, there may be accumulation within the food chain, leading to exposure of predators through their diet. Information of bioaccumulation of a substance in aquatic organisms is used for classification and labelling and PBT assessment. It is also used for modelling of exposure via the food chain in the CSA. It can also be an indicator that long-term ecotoxicity testing is necessary (Section 7.3.6 of this chapter), because if a substance accumulates in an organism it may result in toxic effects over time, even if it is present in the environment at low concentrations.

7.2.2.1   *BCF and Fish Feeding Studies*

A study for bioaccumulation in aquatic species, preferably fish, is required at REACH Annex IX. Normally this is assessed with a bioconcentration fish test. This test determines the bioconcentration factor (BCF) of a substance, which is a useful representation of the potential for a substance to bioaccumulate. Bioconcentration is the accumulation of a substance dissolved in water by an aquatic organism.

The BCF fish test involves two phases: an exposure (uptake) phase (normally 28 and up to 60 days), and a post-exposure (depuration) phase, where the fish are transferred to a medium free of test substance. The BCF at apparent steady state is calculated from the ratio of the concentration of test substance in the fish to the concentration in the water. A kinetic BCF can also be calculated from the uptake and depuration rate constants.

Bioaccumulation is the uptake of a substance via all environmental routes. Bioaccumulation of a substance in an organism can occur via food and sediment as well as via water alone. Uptake via food or sediment is more important for substances that are less water soluble and highly adsorptive. To study bioaccumulation in fish, a test can be carried out using spiked food. For substances with low water solubility, fish feeding tests are more practical to conduct than BCF tests, as it is easier to achieve a constant and high enough dose. For substances with high log K_{ow} (> 6), a fish feeding study is the recommended method.

Bioaccumulation testing can be waived at Annex IX if the substance has a log $K_{ow} < 3$, as this indicates low potential for bioaccumulation. Exposure-based considerations also allow testing to be waived, if direct and indirect exposure of the aquatic compartment is unlikely. However, this waiving consideration implies a low probability of exposure, rather than low level of exposure; therefore, use of exposure-based waiving is limited.

Hydrolysis of a substance can have a marked effect on the bioaccumulation potential. If the half-life for hydrolysis is < 12 hours it can be assumed that the rate of hydrolysis is greater than the rate of uptake by exposed organisms. In some cases it may be necessary to assess the bioaccumulation of the hydrolysis product.

7.2.2.2   *Biomagnification*

Biomagnification is the increase in the concentration of a substance in organisms in successive trophic levels in a food chain. The biomagnification factor (BMF – the increase in concentration from prey to predator) and the trophic magnification factor (TMF – the average increase in concentration per trophic level) are calculated via field studies or modelling.

Substances which bioaccumulate may not biomagnify and trophic magnify, as species at higher trophic levels may be better able to metabolise or excrete the substance.

Fish BCF and BMF values are used to calculate concentrations in fish for secondary poisoning assessment in the CSA.

7.2.3   Adsorption

Adsorption refers to the tendency of a substance to move from free solution to a solid surface (this can be physical or chemical binding). It is important for understanding how a substance will partition when in the environment, and is thus relevant to environmental exposure assessment.

When studies of adsorption are really important

In some cases the tendency of a substance to adsorb does not correlate with its octanol–water partition coefficient (log K_{ow}), or log K_{ow} is not available. The capacity for adsorption to sludge, sediment and soil can be very significant in the environmental exposure assessment, so it is important to identify when this is the case and evaluate adsorption in another way, normally by measurement, to ensure the exposure assessment is realistic. Some important examples are:

- Substances which adsorb to inorganic matter in sludge, soils and sediments (for example some strong acids).
- Surface active substances, for which experimental log K_{ow} values may not be reliable.
- Inorganic substances, for which log K_{ow} cannot usually be determined.
- Substances which react with components of the substrates (for example anilines; some metals can form complexes with humic acids in soil).

Some substances are not amenable to studying in an high performance liquid chromatography (HPLC) system, in which case the full OECD 106 test should be conducted (OECD, 2000a).

Adsorption to sludge in a STP will remove some of the substance from the water and if the sludge from the plant is spread to land it is likely that some amount of the substance will end up in the soil compartment. Once in the soil, desorption of the substance will affect its movement and potential to reach surface or groundwaters. Adsorption to suspended matter in water bodies will mean some of substance will be removed from the water and be present in aquatic sediment.

It can be seen from these considerations that the adsorption potential of a substance is also important for decisions regarding whether it is necessary to carry out toxicity testing to sediment- or soil-dwelling organisms. Decisions regarding whether soil and sediment biodegradation simulation testing is necessary also depend on adsorption behaviour; if a substance is highly adsorptive the rate of degradation will be slower than expected, as it will be less available to microorganisms.

High adsorption also has practical implications for the approach to performing other tests. For example, the substance may adsorb to glassware, which limits achievable aqueous concentrations and, thus, affects testing or supporting analytical work. Also, the substance may adsorb to ecotoxicity test organisms, potentially causing physical effects (Section 7.3.4 of this chapter).

7.2.3.1   Adsorption Coefficient (K_{oc})

There are a variety of media within the environment to which a substance could adsorb; climatic factors such as rainfall, temperature, sunlight and wind can affect the extent of adsorption as well. Therefore, to allow comparison between different chemicals it is often assumed that all adsorption is related to the organic matter content of a medium. This assumption is not valid for ionic substances or complexing agents.

Therefore, adsorption studies first measure the distribution coefficient (K_d): the ratio of the concentration of the substance in the sorbent (sludge or soil) to the concentration in water, at equilibrium. The distribution coefficient (K_d) is then adjusted to account for the organic carbon content of the sorbent. The K_d value is divided by the weight fraction of organic carbon in the sorbent to give the adsorption coefficient (K_{oc}).

7.2.3.2   Adsorption Testing

An adsorption/desorption screening study is required at Annex VIII of REACH and, depending on the results of the screening study, further information (for example a batch equilibrium test) may be required at Annex IX.

The screening study (and further information) are not normally required if the octanol–water partition coefficient of the substance is low (log $K_{ow} < 3$) or if the substance decomposes rapidly, as these two properties would mean that adsorption is unlikely to dominate the CSA.

It is important to note that the log K_{ow} cut-off is not appropriate for ionising substances or surface active substances, as the adsorption for these substances is not expected to be related to organic matter content and a measured adsorption coefficient is necessary.

If no reliable measured data are available, the possibility of using valid read-across or (Quantitative) structure-activity relationship ((Q)SAR) prediction should be considered first. Adsorption can be estimated from K_{ow}, but note that this is only relevant for non-ionic organic molecules and where adsorption is controlled by the affinity for the organic content of the substrate.

However, an experimental value should be measured if prediction is not possible, or if K_{oc} is an important factor in the CSA (i.e. there are risks indicated for sediment/soil compartments based on a predicted K_{oc} value, and log K_{ow} is > 3). The simplest form of screening study for adsorption is the adsorption control within an inherent biodegradation test, should this already be available. A more advanced screening study for adsorption is the HPLC method, which can be used to estimate the adsorption coefficient (K_{oc}) on soil and on sewage sludge. This uses a stationary phase based on a silica matrix. This method is not applicable to all substances and the guideline should be consulted. Other screening methods are available, including some methods using soil and sludge samples.

The full test (OECD 106) is a batch equilibrium method and represents a more realistic scenario (OECD, 2000a). The study uses a number of different soil types, with varying characteristics (organic carbon content, clay content, soil texture, pH), collected from field sites. The ratio, at equilibrium, of the amount of substance in the soil to that in the aqueous solution is calculated for each soil. The kinetics of adsorption and desorption are also assessed.

The K_{oc} value can be used to estimate solid–water partition coefficients for any media (soil, sediment, sewage sludge) by adjusting the value to account for the organic carbon content in that media.

7.3   Ecotoxicology

7.3.1   Introduction

Ecotoxicology is the science of examining the effects of chemicals on organisms, populations, communities and ecosystems. Within REACH CSAs, ecotoxicology underpins the determination of the toxic hazard of substances and the risk they pose to the environment as a whole. In this context the data obtained from ecotoxicology studies are used to:

- Determine classification and labelling requirements for a substance.
- Assess whether a substance meets the 'T' criterion for it to be considered a PBT.
- To set PNECs for risk characterisation.

The overall process is illustrated by the flow chart in Figure 7.1.

It is important to integrate ecotoxicology data for a substance with an understanding of its mode of action and its environmental fate and behaviour properties.

7.3.2   Hazard Assessment and Risk Characterisation

The types of ecotoxicology data required for a REACH CSA are dependent on the Annex, and thus the tonnage (Table 7.1).

Ecotoxicology tests are conducted with organisms from different trophic levels and environmental compartments. The species used for testing are normally selected from a number identified in published test guidelines. These have been chosen to represent different trophic levels within an ecosystem and have been selected on the basis of their susceptibility to toxic substances and their amenability to laboratory culture and handling. The test organisms are typically exposed to a geometric series of substance concentrations for a fixed period of time. Effects on mortality or other endpoints are recorded over fixed time intervals. The data are then plotted graphically and/or statistically analysed to derive the concentration that causes an absolute (e.g. 0 or 100%) or intermediate level of effect (e.g. 50%) on the specified endpoint (e.g. mortality, mobility, growth, reproduction etc.).

The results of short-term (acute) tests conducted over hours or a few days are commonly (but not exclusively) prefixed by terms denoted by the acronyms EC_x (effective concentration) and LC_x (lethal concentration) – where x is the specified level of effect. Similarly, the terms NOEC and LOEC (lowest observed effect concentration) are used to prefix the results of long-term tests.

Short-term tests are normally conducted in advance of long-term tests; if the CSA concludes on the basis of their results that no risk is likely, the need for long-term tests

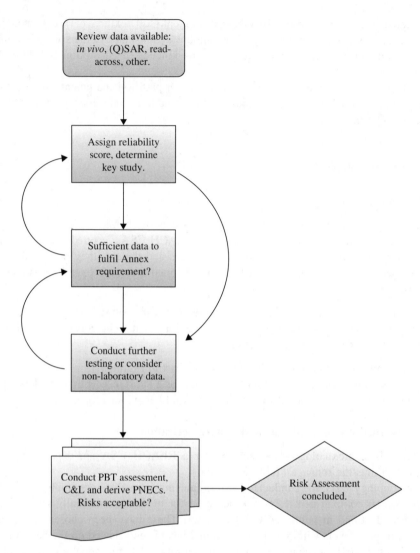

Figure 7.1   *Flow Chart Illustrating the Processes in a Chemical Safety Assessment*

may be waived, depending on the substance. However, there are cases in which it may be more appropriate to directly conduct a long-term study and avoid a short-term one. For example, long-term studies are more appropriate with hydrophobic (log $K_{ow} > 3$) and low water solubility substances, as it may take longer for the equilibration state to be reached (refer to discussion on bioaccumulation, Section 7.2 of this chapter) and for toxicity effects to manifest. Poorly soluble substances are considered to be ones with water solubility < 1 mg/l according to guidance R7.b (ECHA, 2008a).

In some circumstances it is permissible to conduct what are termed 'Limit tests'. In this case the response in a single exposure concentration is compared with that of a control.

Table 7.1   *Studies required for each REACH Annex level.*

Compartment	Standard information requirement	REACH Annex	Study duration	Guidance
Pelagic	9.1.1 Short-term toxicity to invertebrates (preferred species *Daphnia*)	VII	48–96 h	R.7.8.1–7.8.6
	9.1.2 Growth inhibition study aquatic plants (algae preferred)	VII	72–96 h	
	9.1.3 Short-term toxicity testing on fish	VIII	96 h	
Sediment	9.5.1 Long-term toxicity to sediment organisms	X	28 d	R.7.8.7–7.8.13
Terrestrial	9.4.1. Short-term toxicity to invertebrates	IX	2–4 wk	R.7.11
	9.4.2. Effects on soil microorganisms	IX	6 h to 28 d	
	9.4.3. Short-term toxicity to plants	IX	14–28 d	
	9.4.4. Long-term toxicity testing on invertebrates	X	21 d to 8 wk	
	9.4.6. Long-term toxicity testing on plants	X	*Not defined*	
Avian	9.6.1. Long-term or reproductive toxicity or birds	X	Egg incubation time + 10 wk	R.7.10.14

This might apply where no effects are expected at the limit of solubility of a substance or at a maximum concentration specified under regulatory criteria (e.g. 100 mg/l in short-term tests). Such limit tests minimise the number of organisms that need to be used in the tests and are therefore desirable on animal welfare grounds.

7.3.2.1   Pelagic

Tests with aquatic organisms are required at and above Annex VII, because at this tonnage level there is a high probability that some of the substances will be released to the environment via waste-water treatment plants (WWTPs) outflows, for example. Within the pelagic compartment three trophic levels are covered: primary producers (aquatic plants), primary consumers (invertebrates) and predators (fish). Exposure of organisms to chemicals in this compartment is considered to be via the water only: direct ingestion, by for example feeding, is not considered under REACH. The aquatic plant tests are normally conducted with unicellular algae and are multigenerational. Consequently, both a NOEC and an EC_{50} can be generated and used in the assessment of short-term and long-term effects. An EC_{50} based on algal growth rate (E_rC_{50}) is the preferred statistic for hazard assessment because biomass is dependent on study length, design and growth rate. Toxicity to aquatic microorganisms is covered in Section 7.3.7 of this chapter (Guidance R.7b, ECHA, 2008a)

7.3.2.2   Sediment

The sediment compartment can be an important sink for chemicals through deposition and sorption. It can also be a source of chemical exposure through re-suspension.

Sediment studies are required only at Annex X. However, they should also be considered for substances with a log K_{oc} or log $K_{ow} \geq 3$ or for those that can significantly bind to sediment via other means, such as ionic binding: such substances have a significant potential to adsorb onto suspended material and, thus, be deposited in sediment. Testing with benthic organisms may also be preferable for substances with low water solubility.

The living and feeding habits of sediment organisms need to be taken into account when setting up a study (Annex I of R.7.b) (ECHA, 2008a); exposures can be conducted via spiking of the sediment or of the overlying water. When spiking the sediment, it is important to allow sufficient time for equilibration of the substance with the various phases of the sediment matrix. Long-term sediment studies are most relevant for risk characterisation because once the substance is incorporated into the sediment matrix it is likely to stay there for a relatively long time.

7.3.2.3   Terrestrial

The terrestrial compartment is complex; exposure to contaminants can occur through air, soil pore water, food and bulk soil. Terrestrial ecotoxicology data are required at Annex IX and above, and testing should be considered for all substances that have the potential to adsorb onto soil or that are persistent. Test data are seldom available at the commencement of a CSA and alternative approaches to obtaining alternative data or the development of testing proposals may be needed to address data gaps.

Organisms that live in the soil can be divided by ecological process and route of exposure: primary producers (plants, exposed through soil pore water), decomposers of organic matter (invertebrates, exposed through all routes) and re-cyclers of nutrients (microorganisms, exposed through soil pore water). Under REACH only non-vertebrates living the majority of their life in soil are considered part of the terrestrial compartment. The soil type and age can influence the bioavailable fraction of a substance. Results of studies with terrestrial organisms should be expressed with reference to organic matter or clay content to give an indication of the bioavailable fraction.

7.3.2.4   Avian

Avian toxicity data are only required at the tonnage level corresponding with Annex X. The purpose of avian toxicity data is to assess the risks of secondary poisoning of top predators through ingestion of fish or earthworms derived from aquatic and terrestrial food chains. The correlation between short- and long-term study data is poor for this compartment. As a consequence, only long-term data are considered suitable for a CSA (guidance Part B) (ECHA, 2008a). Mammalian data are extensively considered for human health assessment purposes and can be used to assess risk from secondary poisoning (PNEC$_{oral}$). Particular consideration should be given to substances that have structural similarities with known avian toxicants, such as organophosphates, heavy metals and pesticides (e.g. DDT). Testing is not considered necessary for substances with a log $K_{ow} < 3$.

7.3.3   Data Review

Reliable data must be used to characterise the hazards and the risks of substances. The Klimisch scoring system provides a framework to assign reliability scores for ecotoxicological studies (Section 6.8 of this book). Studies must fulfil the key validity criteria set out in the testing guidelines in order to be considered reliable. Common validity criteria relate to water quality (e.g. dissolved oxygen concentration and pH) and control response. A study with no analytical monitoring of the test substance in the test medium and/or no good laboratory practice (GLP) cannot be assigned a reliability score better than two, even if the study has been well documented. Monitoring the test substance in solution is particularly important when testing difficult substances, for example when testing a volatile substance (defined as a substance having a Henry's Law Constant of > 100 Pa) (OECD, 2000b).

The choice of key study in REACH reflects a worst-case scenario, that is the lowest reliable result.

7.3.4   Testing of Difficult Substances

Testing in aquatic media can be problematic for some substances. Properties of a substance that can make it difficult to test include:

- *Low water solubility*: The highest exposure concentration should be equal to or less than the water solubility limit. Above this, it is possible that physical effects, such as gill clogging or surface film entrapment, are responsible for the observed effects.
- *UVCB*: See Case Study 7.1.
- *Hydrolysis*: For substances that hydrolyse, testing should be optimised for assessing the hazard potential of the parent substance or the degradation product. The degradation half-life under ambient temperature and neutral pH conditions is critical: the cut-off is considered to be 12 hours (R.16.4.4.1, ECHA, 2012), above which effects of the parent substance are considered instead of the product(s). However, if the end product is more toxic than the starting material, testing with the former may be more appropriate, and vice-versa.
- *Vapour pressure*: As already mentioned, effects observed with substances that can escape the solution may lead to an underestimation of toxicity. Reduction of headspace, using a closed test vessel choice of exposure regime, can help to retain the substance in solution.
- *Surfactants*: Surfactants or detergents can form emulsions or dispersants that can lead to overestimation of the bioavailable fraction and entrapment. Study concentrations should not exceed the critical micelle concentration (CMC), rather than solubility limit, of a substance.
- *Metals*: The speciation, thus bioavailability and solubility of metals, is affected by water quality parameters such as pH, alkalinity and water hardness.

Ionisable, complexing, coloured, and adsorbing substances can also present difficulties, as can those that degrade (biotic and abiotic), oxidise and photolyse. The R7 guidance describes a decision process for identifying difficult to test substances and designing appropriate tests. OECD 23 (OECD, 2000b) and R7.b (ECHA, 2008a) provide guidance on testing with difficult substances and on presentation and interpretation of results.

Endocrine disrupting chemicals are not substances difficult to test *per se*. However, this endpoint is not part of the Annex VII to X assessment and the standard short- and long-term toxicity tests will not detect them. The definition of an endocrine disrupting substance adopted by the OECD expert groups is: *'an exogenous agent that causes adverse health effects in an intact organism, or its progeny, consequent to changes in endocrine function.'* (ECHA, 2008a). Mammalian toxicological data must be used where available, and screening assays must be conducted where there is the potential for concern. This area is being developed further in the current testing programs and further information on endocrine disruption in the context of REACH can be found in Section 6.5 of this book.

7.3.5   (Q)SARs, Data Waiving and EPM

The results of reliable laboratory toxicity tests are preferred as the basis for CSA because of their direct relevance. However, in the absence of such data it is permitted under certain circumstances to use data obtained by other means. (Q)SARs (as discussed in Section 5.9 of this book) are a tool for estimating toxicity using established correlations between chemical structures and measured toxicity. The use of well-validated (Q)SARs in ecotoxicology can speed up the process of obtaining useful data and reduce the need for animal testing.

Property prediction may also be useful in the case of substances for which the conduct of a test is made technically difficult by its chemical properties. For example, a hydrophobic chemical with low solubility for which it is difficult to maintain stable test concentrations.

The need for sediment and soil *in vivo* testing can be assessed on the basis of the more widely available aquatic data. The equilibrium partitioning model (EPM) uses aquatic toxicity data in conjunction with environmental and physico-chemical properties, such as K_{oc}, K_{ow}, water solubility and vapour pressure, to derive PNECs and RCRs (risk characterisation ratios) to establish whether a substance is likely to pose a risk in these compartments. However, for substances that are likely to accumulate in sediment and soil, further testing should be considered even if the equilibrium partitioning model indicates no risk.

Work is continuing to develop *in vitro* alternatives to *in vivo* ecotoxicological tests. There is still generally a poor correlation between the two, but *in vitro* screening of endocrine disrupting chemicals is generally recognised.

REACH recognises that it is not always necessary to conduct a test to fulfil an Annex requirement and permits data waivers under the following circumstances:

- Where data already exist for higher tier tests.
- Adequate information already exists for classification and labelling.
- Mitigating factors indicate that toxicity is unlikely to occur.
- The RCR based on existing data already indicates no risk (i.e. RCR < 1).
- No direct or indirect exposure is likely.
- It is not technically feasible or scientifically justified to conduct a study, for example the substance exists only in gaseous form.

7.3.6   Further Testing

Whilst unnecessary further testing should be avoided, it may be justifiable to make proposals to carry it out under certain circumstances. For example:

- To fulfil an Annex requirement.
- To refine the risk characterisation (i.e. RCR > 1).
- To establish hazard in the compartment considered to be the most likely fate of a substance.

The choice of a short-term or long-term test is determined by Annex requirements and the other criteria outlined in the previous sections. REACH guidance provides information on the accepted harmonised and national testing guidelines available for each ecotoxicological compartment. When proposing further testing the trophic level or species that is most likely to be susceptible to a substance should be selected. The design of a new test must take into consideration physico-chemical properties of the substance that will determine its behaviour in the environmental compartment under consideration.

7.3.7   Toxicity to Sewage Treatment Plant Microorganisms

Assessing the toxicity of a substance to STP microorganisms is important for ensuring that municipal and industrial STPs exposed to the substance are able to function normally.

Some substances may cause adverse effects on microbial activity in STPs. Therefore, it is important to calculate a $PNEC_{stp}$ and assess the risk by comparison with PEC_{stp}, that is calculate a RCR for the STP.

7.3.7.1   STP Microorganism Toxicity Testing

Assessment of toxicity to microorganisms is a REACH Annex VIII requirement. The test type specified is activated sludge respiration inhibition (ASRI) testing (e.g. OECD 209 (OECD, 2010)). Though this is the preferred test type for new testing under REACH, other types of microbial inhibition testing (e.g. growth inhibition) can also be used, particularly if such studies have already been carried out.

In an ASRI test, test mixtures containing water, activated sludge, synthetic sewage feed, and test substance are incubated under aeration for 30 minutes or 3 hours. Subsequently, the oxygen consumption of the test mixtures is measured and compared with that of a control.

A nitrification inhibition test, which assesses the effects on a subpopulation of nitrifying organisms, can be used instead if there are indications that the substance may be toxic to nitrifying bacteria.

The activated sludge used in the tests should be from a municipal STP, as microorganisms in sludge from an industrial STP are assumed to be adapted to the substance being tested, and $PNEC_{stp}$ values obtained would not be protective of municipal STPs.

It is also possible to use the results from ready biodegradation testing to calculate a $NOEC_{stp}$. If the substance degraded well in the test, or did not inhibit the degradation of a positive control, the relevant concentration can be used as the $NOEC_{stp}$ value.

If the operators have plant performance data and chemical emission/exposure information, field data from industrial STPs can also be used to derive a $PNEC_{stp}$.

7.3.7.2 Other Considerations

Use of (Q)SARs for toxicity to STP microorganisms is not recommended, as validated models are limited and testing is simple.

It is preferable to test mixed inoculums for toxicity, to represent the entire microbial community in STPs, but inhibition data on individual bacterial species may be available, for example on *Pseudomonas putida*. Results from such tests should only be used if no other test results are available. If PNECs are derived from results with single species, a lower assessment factor is used. This is partly because the single species is more sensitive to toxic effects than whole populations, partly because tests with *P. putida* measure growth inhibition, which is a long-term endpoint (respiration inhibition is a short-term endpoint).

REACH Annex VIII states that testing is not necessary if:

- There is no emission to a STP.
- There are mitigating factors indicating that microbial toxicity is unlikely to occur, for instance the substance is highly insoluble in water.
- The substance is found to be readily biodegradable and the applied test concentrations are in the range of concentrations that can be expected in the influent of a STP.

As for toxicity testing with higher organisms, microbial toxicity testing at concentrations above the water solubility of a substance should be avoided. This is also an unrealistic scenario as insoluble chemicals will likely be removed in the STP treatment stages prior to exposure to activated sludge. However, where there is an existing test at concentrations above the water solubility, a valid PNEC can be derived. The presence of an undissolved test substance does not have as much effect on microorganisms as it can on higher organisms.

7.4 Turning Intrinsic Properties into 'No-Effect' Concentrations

As discussed in Section 7.3.2 of this chapter, laboratory studies are used to investigate the ecotoxicology of chemicals and observed effect concentrations (EC_{50} or LC_{50} from short-term tests and NOEC or LOEC from long-term tests) are determined on the basis of these studies. In order to carry out risk characterisation, these concentrations are used to estimate the level of exposure that is safe for the environment. This notional safe level – a predicted no-effect concentration – is defined as the concentration below which no adverse effects on a particular compartment are expected. The PNEC is lower than the observed effect concentration in order to give a margin of safety that accounts for the differences between laboratory studies and the real world, and uncertainties in the assessment.

The derivation of PNECs is discussed in detail in REACH Guidance document R.10 (ECHA, 2008b); key points are summarised here.

PNECs must be calculated where possible for substances manufactured or imported into the EU at > 10 tonnes/annum. If it is not possible to develop a particular PNEC, this must be fully justified. PNECs are calculated for several different compartments as conditions differ widely between the different compartments:

- Water (freshwater and marine).
- Sediment (freshwater and marine).
- Agricultural soil.
- Predators exposed via the food-chain (secondary poisoning).
- Microorganisms in the STP.
- Atmosphere (for this compartment, the guidance not yet well developed and it will usually not be possible to develop a PNEC at this stage).

7.4.1   Selecting a Suitable Starting Point for a PNEC Calculation

The first step in calculating a PNEC is to pool the available NOECs and EC/LC$_{50}$s from the toxicological data and select a suitable starting point. All available data should be considered and the PNEC will need to be reviewed if further data become available. The selection of a key result for a particular compartment and trophic level is discussed in Section 7.3.3 of this chapter. Expert consideration of the reliability, relevance and completeness of the data is essential. For PNEC derivation, long-term data are generally preferred over short-term data. Where data for more than one trophic level are available (for example for the aquatic compartment data for fish, invertebrates and algae will often be available), the lowest result is selected.

As discussed in Sections 7.3.5 of this chapter and Section 5.9 of this book, values from (Q)SAR predictions may be used in the absence of laboratory measurements if certain validity conditions are fulfilled. These values may then be used in PNEC derivation in the same way as in measured results. Prediction may be particularly useful for substances for which testing is difficult.

A special case is where all available studies show no effects at the concentrations tested; commonly, for short-term aquatic studies, this would mean three results of L(E)C$_{50}$ > 100 mg/l. It is possible to derive a limit value for the PNEC in this case; however, the true L(E)C$_{50}$ may be much greater than 100 mg/l. Therefore, the true PNEC is much higher than the limit value obtained. If exposures are low, and the RCR obtained using the limit PNEC is < 1, the limit value may be considered sufficient. However, if risk characterisation indicates that the RCR may be above one, it is advantageous to obtain a better estimate of PNEC. One approach may be to use a (Q)SAR prediction as the starting point for PNEC derivation.

7.4.2   Calculating a PNEC Using Assessment Factors

The assessment factor method is suitable for use when a standard data set is available and is the most commonly used method for calculating PNECs for REACH. The selected starting point from the toxicity data is divided by the appropriate assessment factor. Default assessment factors are available for water, sediment, soil, STP organisms and predators. These include factors to account for some or all of the following uncertainties:

- Intra- and inter-laboratory variation of toxicity data.
- Intra- and inter-species variations (biological variance).
- Short-term to long-term toxicity extrapolation.
- Laboratory data to field impact extrapolation.

The choice of assessment factor depends on the available data set. The size of the assessment factor is designed to reflect the extent and type of data available (e.g. short or long term) and to provide a margin of safety against uncertainties. For example, for freshwater aquatic organisms a default assessment factor of 1000 is applied to the lowest $L(E)C_{50}$ from the relevant studies if only short-term data are available. A lower assessment factor is applied to the lowest NOEC from a long-term study. The assessment factors are dependent on the length of the studies used for the assessment, the trophic levels for which data are available and the number of species tested within each trophic level.

In circumstances where releases are 'intermittent', the receiving environment has more time to recover as exposure is not sustained. PNECs are derived from short-term aquatic data using a reduced assessment factor (Section 7.4.5 of this chapter).

The PNEC for marine aquatic organisms will often be based on data for freshwater organisms, as studies with marine organisms are not part of the standard data set. A higher assessment factor (10 000 if only short-term data are available) is applicable due to the greater species diversity present in the marine environment. This means that there is a greater sensitivity distribution and so a greater uncertainty in deriving a safe level.

In some circumstances it may be appropriate to use a non-default assessment factor. This may be higher or lower than the default and should always be fully justified on a case-by-case basis. Examples of situations where it may be appropriate to use a non-default assessment factor include:

- Evidence from structurally similar compounds indicates that a higher or lower factor is appropriate.
- The mode of action for toxicity is known. A lower assessment factor might be appropriate for a substance with a non-specific (narcotic) mode of action; a higher factor might be appropriate for certain known specific modes of action, such as endocrine disrupting effects.
- Data are available from a variety of species covering the taxonomic groups of the base-set species across at least three trophic levels, providing that multiple data points are available for the most sensitive taxonomic group.
- Data are available from a wide range of species from additional taxonomic groups other than those represented by the base-set species.
- Substances that bioaccumulate may in some cases require a higher factor.

Example 7.1

Substance A has the following results from short-term toxicity studies:

Fish 96 hours $LC_{50} = 32$ mg/l
Invertebrate 48 hours $EC_{50} = 20$ mg/l
Algae 96 hours $EC_{50} = 17$ mg/l; NOEC $= 1.8$ mg/l
STP microganisms (ASRI) NOEC $= 80$ mg/l.

The starting point for the calculation of PNEC$_{\text{freshwater aquatic}}$ is selected as the lowest $L(E)C_{50}$ from these studies: 17 mg/l. As the studies are all short term, a default assessment of 1000 is appropriate. Therefore, PNEC$_{\text{freshwater aquatic}}$ is $17/1000 = 0.017$ mg/l.

A NOEC from an ASRI study is available; therefore, a default assessment factor of 10 is applied to this NOEC to obtain $PNEC_{STP} = 80/10 = 8\,g/l$.

No marine data are available, so $PNEC_{marine\ aquatic}$ is calculated from the freshwater data and a default assessment factor of $10\,000$ is appropriate as only short-term data are available. Therefore, $PNEC_{marine\ aquatic}$ is $17/10\,000 = 0.0017\,mg/l$.

At a later date, long-term aquatic testing is carried out for this substance and the following results become available:

Fish 21 days NOEC 2.4 mg/l
Invertebrates 21 days NOEC 1.1 mg/l.

Long-term results are now available for three species (the standard algal study is multigenerational and so may be used to assess both short- and long-term effects). A default assessment factor of 10 is now applied to the lowest NOEC, giving $PNEC_{freshwater\ aquatic} = 1.1/10 = 0.11\,mg/l$.

7.4.3 Calculating a PNEC Using Sensitivity Distribution

Sensitivity distribution methods for calculating a PNEC aim to determine a concentration at which a certain percentage (e.g. 95%) of the population of an ecosystem is protected against toxic effects. This usually requires a larger data set, including experimentally determined NOEC values for a number of species from different taxonomic groups, so is not often applied for REACH purposes.

7.4.4 Calculating a PNEC Using Equilibrium Partitioning

It will often be the case initially that no data are available for sediment and soil organisms. PNECs for these compartments can provisionally be calculated using the equilibrium partitioning model. This method may result in an over- or underestimation of the toxicity to sediment and soil organisms; it is thus is only used as a rough screening in deciding whether further testing is needed. If the risk characterisation for sediment and soil based on the equilibrium partitioning method indicates that the substance may pose a risk to these compartments, testing will usually be required.

The method uses $PNEC_{freshwater\ aquatic}$, the suspended matter/water (for $PNEC_{sediment}$) or soil/water (for $PNEC_{soil}$) partition coefficients as input. These partition coefficients may be calculated from the octanol–water partition coefficient (log K_{ow}) if no further information is available.

Water and sediment/soil dwelling organisms are assumed to be equally sensitive to the substance. In sediment/soil, the substance will be both adsorbed to the sediment/soil particles and dissolved in the interstitial water. It is assumed that only uptake from the water phase is important and that the concentration dissolved in water can be calculated from the partition coefficients. Uptake of the substance could also occur by other pathways, such as ingestion of soil or sediment, and this is particularly important for adsorbing chemicals (log $K_{ow} > 3$). For substances with a log $K_{ow} > 5$ or that are likely to adsorb strongly due to other mechanisms (such as ionisable or surface active substances), the

---

Case Study 7.1   Assessment approach for a UVCB substance

A plant-derived substance comprising of more than 50 constituents has been the subject of a registration. The complex and variable composition of the substance meant that it was considered to be a UVCB.

The environmental hazard data for the substance were obtained for fish, invertebrates, algae and waste-water treatment organisms in tests conducted with the water-soluble fraction of the substance. This was necessary to avoid complications arising from the presence of undissolved test material. The test results were expressed in terms of the nominal treatment levels at which the water accommodated fractions were prepared so that the test results could be directly interpreted with respect to hazard classes and labelling criteria.

No-effect concentrations for risk characterisation were set for individual constituents or groups of constituents (referred to as 'blocks', Section 11.3.1 of this book) that make up the substance. Toxicity data for the constituents were obtained from published studies or from validated predictive models and used in conjunction with the default assessment factors to derive the no-effect concentrations for the constituents and blocks. These concentrations were used in conjunction with exposure concentrations determined in exposure assessments to obtain risk characterisation ratios for each constituent and block. Summing of the block-specific RCRs gave an overall RCR from which the acceptability of risk is determined.

---

risk characterisation ratios for soil and sediment are increased by a factor of 10 to account for the uptake via ingestion of soil/sediment.

7.4.5   Intermittent versus Continuous Releases

The standard PNEC values derived for fresh and marine waters are applicable for continuous release of a substance into the environment. It is reasonable to assume that populations can tolerate higher concentrations when exposed to a substance for a short-time rather than continuously over a long period. Therefore, when discharges are for short periods (for example from some batch processes) a separate $PNEC_{water, intermittent}$ may be derived. The starting point for this PNEC is always the short-term $L(E)C_{50}$ values. An assessment factor of 100 is applied where data are available for three trophic levels (compared to a factor of 1000 for PNEC for continuous releases).

References

ECHA (2008a) R7.b: Endpoint Specific Guidance. Guidance on Information Requirements and Chemical Safety Assessment, May 2008. European Chemicals Agency.

ECHA (2008b) Chapter R.10: Characterisation of dose [concentration]-response for environment. Guidance on Information Requirements and Chemical Safety Assessment, May 2008. European Chemicals Agency.

ECHA (2012) Chapter R.16: Environmental Exposure Estimation. Guidance on Information Requirements and Chemical Safety Assessment, October 2012. European Chemicals Agency.

OECD (1981) Test No. 302A: Inherent Biodegradability: Modified SCAS Test. OECD Guidelines for the Testing of Chemicals, Section 3. OECD Publishing.

OECD (1992) Test No. 302B: Inherent Biodegradability: Zahn-Wellens/EVPA Test. OECD Guidelines for the Testing of Chemicals, Section 3. OECD Publishing.

OECD (2000a) Test No. 106: Adsorption – Desorption Using a Batch Equilibrium Method. OECD Guidelines for the Testing of Chemicals, Section 1. OECD Publishing.

OECD (2000b) OECD Series on Testing and Assessment, Number 23. Guidance Document on Aquatic Toxicity Testing of Difficult Substances and, OECD Publishing, Mixtures.

OECD (2009) Test No. 302C: Inherent Biodegradability: Modified MITI Test (II). OECD Guidelines for the Testing of Chemicals, Section 3. OECD Publishing.

8

Environmental Exposure

The methods used for modelling environmental exposure are largely unchanged since the pre-REACH requirements for new and existing substances. The ECHA (European Chemicals Agency) has issued guidance on conducting the environmental exposure assessment (Chapter R.16) (ECHA, 2012).

Registrants must cover production (if this occurs within the European Union) and use of the substance they supply. Full risk assessment (in which the exposure associated with multiple producers and their supply chains) may follow, as part of the ECHA evaluation (Section 1.2.3 of this book).

Environmental exposure scenario development should follow a 'tiered' stepwise approach, in which registrants should refine as far as is useful to demonstrate safe use. Concentrate on the most important issues.

It is not useful for registrants to treat their exposure assessment as an advertisement for their company's standards of control. If the risk characterisation ratio (RCR) is < 1, then this is enough.

Estimating exposure is a balance between definitive knowledge and guesswork. A registrant combines several concepts to arrive at a quantitative estimate of final concentration:

- How much of the substance is released to the aquatic, terrestrial, atmospheric and biotic compartments?
- What are the sizes and properties of the compartments themselves? (defaults exist for these)
- How does the substance behave in a compartment, and move between compartments?

8.1 Substance Identity and Approach to Exposure Assessment

The approach to assessment of high-purity single-constituent organic substances is straightforward. However, registration substances are often not so simple. The nature

Chemical Risk Assessment: A Manual for REACH, First Edition. Peter Fisk Associates Ltd.
© 2014 John Wiley & Sons, Ltd. Published 2014 by John Wiley & Sons, Ltd.

of the substance has a detailed bearing on approach taken to exposure assessment. The registrant must make a careful assessment case-by-case. Some examples are:

- *Lower purity single-constituent substances*: consider the potential impact of the impurities present to affect test data. For example, a vapour pressure value measurement that represents a volatile impurity, a biodegradability test result that represents a more degradable impurity. It may be useful to conduct a literature search or property prediction for the major constituent and impurities. Consider whether impurities could be contributing to observed toxicity, in which case multiple assessments may be appropriate (next bullet point).
- *UVCB and complex multi-constituent substances*: in the environment the constituents will partition, degrade and cause effects based on their own intrinsic properties. The assessor should are separate and group constituents for assessment based on similarity in their properties and effect levels. Each is then assessed separately and the resulting RCRs summed to give an overall conclusion for the UVCB substance. This is termed the 'hydrocarbon block' approach.
- *Ionisable substances and salts*: consider the likely speciation in the wider environment, which may be dominated by prevailing pH conditions and common dissolved ions. The log K_{ow} value, among others, may require adjustment to derive a value applicable at pH 7.

8.2   Characterising Releases

8.2.1   Evaluating Use Pattern

For the environmental aspect of exposure assessment, it is important to be able to account for the breakdown of various uses of the substance.

- Track the volume manufactured from production (or import) through subsequent uses
- Establish how many users and locations as far as reasonably practicable
- Develop realistic relationships with downstream users.

REACH requires an assessment of releases associated with waste, for every step of the life cycle. This must normally be a quantitative assessment, unless it is demonstrably insignificant compared to the environmental release from the local site for that life cycle stage itself.

Example 8.1

A registrant is aware that the registration substance has a large number of potential uses globally, but is uncertain which uses are the most significant within EU, and a proportion of the manufactured tonnage is sold through a distributor company.
Points to consider:

- Business and marketing departments are likely to have access to sales data
- Patent literature, public info (e.g. chemical encyclopedias and databases) and sales literature can be useful but may not reflect the typical real uses
- Some EU countries have product registers which record uses

- Whilst customers can be reluctant to discuss detailed use information, it is in their interests to have their uses covered by the manufacturer's chemical safety assessment.
- An impartial third party can be useful to collate data in case of confidentiality concerns. A consultant; consortium secretary; or even the distributor may be able to act as an interface. The public use mappings from industry sector groups may also help in this regard.

8.3   Evaluating Releases

The loss rate may also depend upon such factors as the vapour pressure, water solubility, degree of control imposed and tonnage handled. It is helpful to understand batch sizes and handling procedures. Larger sites may operate with better technology and a smaller fractional release rate than smaller sites.

If departing from environmental release category (ERC) release defaults, which tend to be very conservative, then estimates of fractional releases must be justified in a scientific way. The ECHA looks for valid recording and reporting of measured data to a high standard, using representative information and with the sources of information properly described.

Support from published sources will help considerably. Table 8.1 presents some of the most useful types of sources to refer to.

Table 8.1   *Useful sources of published information for environmental exposure.*

Document type	Summary of key features	Remarks
ECHA part R12	Default release fractions and site sizes within the official regulatory guidance	These can be very conservative
OECD emission scenario documents	OECD web site. Industry sector-specific documents designed to support environmental exposure assessment and covering chemical functions, releases and site tonnages	These documents are ideally suited for the purpose of exposure assessment. It is nevertheless necessary to properly document the applicability of the emission scenario document (ESD) to the substance and exposure scenario in question, case by case
Industry codes of practice	Relevant industry codes of practice may have information on recommended handling procedures and control methods	May represent the best case rather than the worst case

(*continued overleaf*)

Table 8.1   *(continued)*

Document type	Summary of key features	Remarks
IPPC BREFs	Regulatory industry sector-specific reference documents in support of the EU IPPC legislation (Chapter 3)	Generally include a lot of detail about the modern industry. Useful source documents may also be available for download. By definition these would tend to represent the better pollution control technologies in general use
Past published EU risk assessments under the preceding legislation	An EU-wide approach was used. These documents were peer reviewed in depth	The approach to environmental exposure assessment was essentially the same
Detailed guidance from industry groups, including Specific Environmental Release Categories (SPERCs)	Release rates, site sizes, WWTP details, produced by industry sector-focussed working groups and published through CEFIC	The parameters are in a very useful and directly-applicable format. However the figures are sometimes not well referenced or supported by documented evidence. SPERC fact sheets, published to support the figures, are variable in quality
CEFIC RMMs (risk management measures) library	Further guidance on information from regulatory and other sources	Further sector-specific RMM information is expected to follow in future
General literature	Little specific evidence, but can be useful leads on applications, methods and technology, and to answer specific questions	Sources such as chemical encyclopaedias can be useful
Patent literature	As above	It must be recognised that patent literature is not evidence of current common practice
Scientific know-how from industrial chemists	Invaluable real-life evidence on relevant matters	Will need to be suitably documented. It is important to demonstrate that a sufficient proportion of the tonnage is covered
HERA programme	Established methods for assessing environmental exposure for detergents and cleaning products	Related to EU methods but adapted factors for widely dispersed end use. Does not cover entire life cycle but only end use

8.3.1 Reality Checking – Top Down and Bottom Up

In the majority of circumstances, the individuals conducting risk assessment are office based – whether consultants or staff from the company's own health, safety and environment department. Unless they are highly trained or have first-hand industrial experience, there is a significant risk that the resulting model will lack realism. A good description of the use of a substance should be clearly recognised by users at the facilities described, as well as to the regulator reviewing the assessment.

The numbers should be realistic relative to the market and the application. An assessment must be made both from the 'top down' and also from the 'bottom up'.

Top-down method

Total volume sold, fraction in each application and life cycle stage and number of users (distribution of substance used across a realistic number of sites) leads to a quantity used per day and a release rate per day for each life cycle stage throughout the life cycle.

Bottom-up checks

For each life cycle stage, consider this estimated quantity used per day in terms of what it implies about the volume of formulated product in which the substance is contained, for the scale of operations in which the substance is used. Is it realistic?

Consider for the case in question what the sources of release might be. If a realistic estimate is made for each and summed up, how does this compare to the total released amount being used in the exposure model?

Example 8.2

A registrant of an industrial process additive considers that the ERC default release from the processing plant is very high and wishes to collect information from the site to attempt to check whether it is unrealistic.

What processes need to be considered?

The plant manager is normally best placed to set out all processes any one substance goes through on the site. The following are likely to be relevant:

- Unloading delivery vessels such as tankers and drums, and cleaning them.
- Transfer and storage on site.
- Charging of mixing vessels.
- Filling of drums with product.
- Routine cleaning practices for equipment and surfaces.
- Washings from stack scrubbers.
- Volatile emissions from local exhaust ventilation (LEV).

What kind of data and evidence are required?

- Each of these processes has the potential to be a source of low level releases or to lead to wastes containing the substance.
- Consider the degree of enclosure, available technological controls and other substances used in the same process.
- General information on everyday practice will often be supported by company records anyway.
- Check if the work has been done already as part of site compliance with integrated pollution potential and control (IPPC) legislation (refer to Chapter 1 of this book) or for a local permit.
- Site inspectors from the regional regulatory authority may wish to audit site-specific data.

8.4   Documentation for the Registration

The chemical safety report (CSR) documents the outcomes of the chemical safety assessment (CSA). The CSR does not need to contain details of separate iterations, only the finalised scenario.

Each individual scenario will need to be set out in detail in Chapter 9 of the CSR. If any confidential information has been used in assessing the exposure and risk, it may be appropriate to document this separately in a confidential annex.

8.4.1   Uncertainty

It is important to bear in mind areas of uncertainty in the exposure assessment and its possible effect on the resulting risk characterisation. Any possible effect on persistent, bioaccumulative and toxic (PBT) assessment should also be discussed clearly. The significance of the uncertainty on individual inputs to the exposure assessment (chemical properties; environmental fate; releases factors and predicted no-effect concentration (PNEC)) may require expert assessment.

8.5   Local Scale Releases

Environmental exposure from industrial locations (e.g. manufacture, formulation, many types of end use) is modelled based on a whole-site emission (Box 8.1). Releases to air, water and, in some cases, industrial soil are estimated in the first instance based on standard defaults.

8.5.1   Site Size

For the purpose of environmental exposure assessment, the site size is defined in terms of the processing volume of the substance of interest. Depending on the case, it is quite possible for this to be very small, even if the installation is very large in other terms (or vice versa).

Box 8.1   Spatial scales summary

- *Local scale*: industrial or wide-dispersed alternative definitions. An industrial instal-lation or a town or borough, served by a municipal waste-water treatment plant (WWTP) and a river receiving its effluent.
- *The region*: a notional area of $200 \times 200 \, \text{km}^2$ with population 20 000 000 in which industrialised urban areas and farmland are closely intermixed.
- *The continent*: the remaining EU.

8.5.2   Site Inspections

Inspectors from the national regulatory authorities are already beginning to undertake validation checks during site visits. It is important to maintain adequate monitoring records in support of releases figures, if the exposure assessment is reliant on justification of release levels well below the ERC default values.

The requirements of REACH and the way it affects an individual industrial installa-tion overlap, or at least approach closely, the requirements of other legislation. This is discussed in Section 3.8 of this book.

The distribution of releases over time is also significant. It is important to consider a realistically relevant volume processed and handled per day (Section 8.3 of this chapter). Intermittent releases (releases occurring not more frequently than once per month and for less than 24 hours) can be assessed slightly differently and use what may be a more favourable PNEC value in risk characterisation (Section 7.4.5 of this book).

8.6   Exposure Assessment – Models or Measurements?

As explained in Chapter 1 of this book, risk characterisation is a key part of REACH chemical safety assessment, in which exposure of a substance is compared to effects. In approaching risk characterisation, it is usual to use a combination of estimates, models and measurements of exposure, which have different merits in flexibility, general appli-cability and specificity. This section uses terms set out in Table 8.2 and expanded upon in later sections.

8.6.1   Using Measurements

The REACH guidance on this topic is extensive. In summary, it addresses such ques-tions as:

- Are the numbers and timings of samples representative of the pattern of use?
- Are temporal variations included? This should be considered in respect of daily practice (for example, releases on working weekdays compared to days when the plant is closed) and seasonal variation (for example, dilutions may be very different during dry compared to wet weather).
- Are all sites covered?
- Are findings analysed statistically, and reported?

- Is the spatial scale of the environmental measurement accurately defined, that is does the measurement represent a local scale measurement or a background concentration? This concept also applies to human exposure measurements (often termed occupational hygiene studies). It is necessary to know whether measurements are the background concentration in a facility or represent the working environment of a worker performing a particular task.
- Are all sources of the substances truly known?

The guidance sets high standards on all these topics, whether the item being analysed for is release level or concentration. The outcome is that measurements meeting all the validation criteria are hard to come by.

Table 8.2 *Definitions of some terms relevant to modelling or measuring exposure.*

Term	Discussion
Environmental exposure	REACH considers the environment in terms of notional 'compartments' such as surface water, soil, air and so on. The concentration of the substance (all constituents individually) in each of the compartments is termed here 'exposure'
Human exposure	In qualitative terms, exposure of individual workers, consumers or members of the general population would first be made by considering the potential for a substance to come into contact with the skin, be ingested or breathed in. Then the concentration within the body can be considered
Measurement	This is the application of analytical chemistry methods to determine the amount of substance per unit volume or mass, in an environmental compartment or in a person
Release	This is usually expressed as the mass of substance escaping from equipment, containers or articles during the normal life cycle of the substance, over a known period of time
Model	In this context, an estimate of the amount of substance released is combined with a mathematical description of the behaviour of the substance to determine the concentration expected to be experienced by an environmental compartment or a person
PEC	Predicted environmental concentration: applied to one compartment at a time this can be based on measurements or models, or a combination. It is expressed as mass per unit volume or weight of the substrate
Validation	Analytical methods should be validated as fit for purpose: are they reliable, and able to achieve the desired concentration measurement? Measurements of concentration need to be further validated as being fit for purpose Models need to be validated in terms of the quality of the science behind them and their ability to do what they are intended to Measurements of concentration need to be further validated as being fit for purpose Models need to be validated in terms of the quality of the science behind them and their ability to do what they are intended to

8.6.2   Using Models

In estimating exposure, any model may be used, but acceptability by the authorities will be most likely when well-known and widely validated models are used. Details are given in Sections 8.6 and 10.2.5 of this book. The REACH guidance never gives unreserved recommendations but there is clear 'custom and practice', which are described in this book.

The benefit of using reasonable worst-case models is largely that of cost and time. This principle means not that every site is at the modelled level but that some might be. Models of releases can be very cautious, unless there are well-researched exposure scenarios available, such as those published by the OECD.

Models of how the release turns into an exposure concentration are largely about dilution and partition into other compartments. The commonly used ones are simple screening-level tools, termed Tier 1 in the Guidance. A higher tier model will include progressively more complex description of the work place or of the environment.

8.6.3   Models or Measurements – Recommended Approach

Models are more widely applicable than measurements but need good information about use pattern. Using measurements to validate an exposure model is likely to be either very costly or not accepted by the authorities. Therefore, the following general strategy is recommended:

- Measure releases on a representative basis where possible, because the ECHA defaults are very high. Make sure to give a good description of the industry, one that 'insiders' would recognise, but remember that the CSA/CSR are neither a handbook nor a publicity brochure.
- Justify all statements about release; avoid saying that release or concentration 'is zero' – this cannot be proven and antagonises the reviewer. Instead, make a realistic effort to calculate limit values even if they are very low. It is better to end up with a defendable 'less than' estimate.
- Use published emission scenario documents where they are available, since they have been developed by regulators.
- As a registrant, do not expect much information in advance from downstream users; why should a customer tell you commercial insights? However, there is much information in public sources.
- Use standard models of exposure but try to do better than a 'Tier 1' approach, even if it gives acceptable RCRs, so as to minimise the possible need for further testing. It is not easy to retract once a CSR is finalised.
- Learn from downstream users' response to the extended Safety Data Sheet (eSDS).
- Use experienced personnel to do the work, remembering that exposure calculations are in the end as important as hazardous property values when an RCR is calculated.

8.6.4   Tools

It is possible to conduct environmental exposure assessments manually but this is very challenging unless the life cycle is extremely simple. Most registrants choose to use one of the available tools to simplify the task.

8.6.4.1   *ECETOC TRA for Environment*

The organisation ECETOC (European Centre for Ecotoxicology and Toxicology of Chemicals) has published several documents associated with exposure assessment. The ECETOC TRA (targeted risk assessment) tool, however, is a simplified tool using Microsoft Excel to apply standard algorithms for the EU assessment method.

8.6.4.2   *CHESAR*

The CHESAR (CHEmical Safety Assessment and Reporting) plug-in tool undertakes a very basic assessment of exposure and risk in the International Uniform Chemical Information Database (IUCLID) programme (Appendix C). It is equivalent to a Tier 0 assessment, in which ERC defaults are applied and no modifications can be applied for complex behaviour. (For example, in the case of a rapidly-hydrolysing substance, the exposure assessment of the final product substance should be conducted. CHESAR cannot do this.)

8.6.4.3   *EUSES*

EUSES (European Union System for the Evaluation of Substances) is the software that was most commonly used for environmental exposure and risk modelling prior to REACH. Like ECETOC TRA for the environment, it applies the standard methods required. EUSES has the advantage of being highly adaptable when used by experts. It is, therefore, a better choice in the case of substances that have non-standard behaviour of which the chemical safety assessment needs to take account.

EUSES is published through the Institute for Health and Consumer Protection, a division of the European Commission's Joint Research Centre (JRC, 2013).

8.6.4.4   *CHARM*

In the case of some types of offshore applications, the end use life cycle stage takes place in an open sea setting rather than coastal waters. For these steps in the life cycle, it may be justifiable to deviate from the normal marine modelling methods. A model such as Chemical Hazard Assessment and Risk Management (CHARM) produced by the European Oilfield Speciality Chemicals Association (EOSCA), which is more closely designed for this kind of context, may be more useful.

8.7   Water

8.7.1   Release via Waste-Water

The aquatic environment is principally exposed through releases via the waste-water system. The baseline assumption is that the significant majority of waste water is treated in the municipal system along with domestic sewage.

8.7.1.1   *WWTP*

The standard models assume that the substance in waste water passes through a moderately-sized secondary biological WWTP.

Within the WWTP, various partitioning and rate mechanisms are accounted for:

- Ready or, if available, inherent biodegradation results.
- Sewage simulation test results.
- Henry's Law constant.
- Adsorption constant for sludge (normally derived from organic carbon partition coefficient).

Note that removal by hydrolysis is not automatically taken into account by the models. If the extent of hydrolytic removal in the WWTP needs to be investigated, this must be done by manually setting the rate in the degradation field. This could be important to explore, for example if high levels of adsorption could restrict the potential for degradation.

Note that when the mass balance is calculated, the programme calculates the concentration of substance in treated effluent. The programme will indicate if this figure exceeds the limit of water solubility but makes no automatic corrections. In almost all cases, the registrant will need to check for this and make adjustments so as to ensure the solubility limit is not exceeded. The correct approach is to adjust the mass balance in favour of the sludge.

Typical time line of waste-water treatment

Drain to WWTP: < 1 hour to several hours
Primary settler: 2 hours
Activated sludge tank: about 18 hours (sludge retention about 9 days)
Solid/liquid separation: about 6 hours
The total elapsed time of normally approximately 24 hours is to be expected.

Many installations have on-site treatment of different types. Biological treatment normally requires municipal sewage to be collected from nearby residential areas. Physico-chemical treatment conducted on-site can include settlement tanks, oil water separators or chemical treatments such as acids or oxidising agents.

The standard default is a treatment plant with a flow of $2000 \, m^3$/day. The treated effluent is modelled to be released to a freshwater river with a flow of $18\,000 \, m^3$/day (a dilution factor of 10). Many municipal waste treatment plants in Europe are larger than the default size and, also, dilution into the receiving water may be more than the default, particularly if located on a major river.

In the WWTP secondary treatment tanks, the biodegradation is performed by a mixed population of many different strains of bacteria and protozoa known as 'activated sludge'. The functioning of the treatment is biologically complex, which is why for all substances at 10 tonnes/year and above, microbial inhibition is tested. The purpose is to examine the effect of the substance on the function of whole biological community. In the absence of better data, toxicity controls from ready biodegradability tests (in which the test substance is tested in a mixture with degradable reference substance and the impact on its removal checked) can provide useful evidence but are not normally considered highly reliable.

Following sewage treatment, the effluent is released to a receiving environment. Depending on the case, either the freshwater or marine scenario is likely to be applicable. EUSES automatically runs both for all local scenarios but it is important to realise that the marine scenario defined as a result is not downstream of the receiving river. It is an alternative receiving environment to the river.

It is important to be aware that it is normal in the first instance to assume if the receiving environment is marine, then the waste water is released and diluted into the marine water without passing through the WWTP.

Tip

If risks are found for marine compartments for industrial use scenarios, check if the risks are acceptable when wastes are treated in a WWTP. This could be a useful risk management measure.

8.7.2   River Environment

Simulation biodegradation results may be used in adapting the river environment but it is rarely justifiable to do this at the local scale. This is more a matter for the regional scale. The resulting action is to use measured rate constants for degradation in environment, rather than extrapolated values.

8.7.2.1   *Consider*

Is there susceptibility to reaction that will remove the parent substance, but not degrade it completely (i.e. mineralisation to carbon dioxide, water, ammonia etc.). Factors such as hydrolysis, oxidation and so on would suggest this. If so, then the exposure and risk strategy probably needs to be considered carefully by an expert.

8.7.2.2   *Background Concentrations*

Environmental concentrations in the local scale, associated with the emission period, are added to another predicted environmental concentration (PEC) which represents the background. Background concentrations at the regional level and in the wider continent are modelled. A steady-state model is used for these background concentrations. This type of modelling takes due account of possible build-up and slow degradation of the substance in the environment over time. One consequence is that any degradation of the substance has the potential to show significance in the assessment: even slow rates of (bio)degradation and photo-degradation in air could be important in allowing the models to reach a steady state over realistic time scales.

Another consequence is that the standard models are well suited to modelling the long-term scenario without significant changes to the pattern or volume of use and releases. For some purposes other types of modelling, such as dynamic fugacity models, can be very useful. It is most unusual to apply such methods in a REACH CSA.

8.7.3   Marine Environment

The detailed quantification of exposure and risks in the marine environment is a relatively recent development in environmental risk assessment and has only been required since 2003. The marine environment under consideration using the standard models relates to coastal waters. ECHA guidance and general expert opinion is that other models such as CHARM are more useful for cases where the open sea specifically must be assessed.

8.7.4   Sediments

In fresh water, partition coefficients are defined for the partitioning processes for suspended sediment and settled sediment. On the whole, the major route to the sediment compartment from water is through partitioning to suspended sediments.

8.8   Soil

8.8.1   WWTP Sludge and Agricultural Soil

For the majority of substances, the primary route of exposure to soil in standard modelling is through the assumption that sludge from a WWTP is spread on agricultural land as fertiliser. This practise is indeed widespread and has largely been considered a beneficial form of recycling. The SimpleTreat model, developed by the National Institute for Public Health and the Environment (RIVM) in the Netherlands, derives the concentration in WWTP sludge and this leads to the concentration of the substance in dried sludge at the time of sludge spreading.

The PEC in topsoil is estimated based on a 10-year cycle of sludge application plus continuous aerial deposition, with degradative removal mechanisms taken into account. PEC local soil can represent the 30-day or 180-day average following the tenth sludge spreading event.

Experience suggests that some WWTPs that accept industrial waste streams specifically avoid sludge spreading to avoid chemical contamination, and landfill or incineration may be used for disposal of excess sludge. WWTPs have high standards of water quality to meet and it is vital for the normal functioning of the WWTP to avoid poisoning of the active sludge. For local scale site-specific assessment, see Box 8.2.

Box 8.2   Tip for refinement of soil PECs

Consider checking the actual fate of WWTP sludge by contacting the treatment plant. WWTP sewage sludge is spread to agricultural land only in about 50% of cases.

8.8.2   Deposition

Deposition of the substance from the air is taken into account. In general, this is a very small contribution to soil PECs. If it is necessary to refine a risk associated with the soil compartment, it is usually a better strategy to consider first the sludge spreading exposure pathway.

8.8.3   Biodegradation in Soil

Agricultural and grassland soils are teeming with microorganisms and the standard models take due account of biodegradation processes. The key assumption that the models make is that, in soil, degradation takes place in water (channels or films) in the soil matrix, that is on surface of solid particles. A consequence of this is that substances with very high organic carbon binding tend to have much higher PECs.

However, data tend to show that the assumption is incorrect and that degradation can take place on the soil surface. The consequence is that measuring soil degradation may be useful.

8.8.4   Crops and Grassland

Exposure of agricultural soil can lead to uptake of the substance into growing crops. Similarly, cattle grazing on exposed grasslands can accumulate the substance due to eating contaminated grass. This can lead to meat and milk containing a concentration of the substance which the models can quantify. The typical quantity of crops, meat and milk in a consumer's diet are taken into account in the models. Along with the concentration in air, these are the key contributors to the indirect exposure of consumers.

8.8.5   Industrial Soil

In the case of some uses, a release factor exists for industrial soil. This means a direct release of the substance to the land immediately surrounding the sites which are sources of the substance. This contributes to the wider modelling at the regional scale only.

8.9   Air

8.9.1   Air in the Standard PEC Models

The releases to air from both the local site itself, and volatilised from the WWTP processing aqueous wastes from the local site, are evaluated by a simple Gaussian plume model. PEC local air is based on the larger of the two. The calculation of PEC takes account of degradation in the air (photo-degradation by hydroxyl radicals) and deposition by both dry (via aerosol adsorption) and wet (via wash-out by rain) mechanisms.

It is important to note, when entering parameters for a photo-degradation removal mechanism, that this should always be entered in the form of a reaction rate. The standard assumptions for the concentration of the OH radical in the atmosphere differ in the EU compared to other parts of the world, particularly the USA. Therefore, the half-life derived by EUSES from a given value of k, will differ from the half-life output from (for example) the AOPWIN estimation software.

8.9.2   Ozone Depletion and Other Specific Effects

For several particular effects associated with the atmospheric compartment (ozone depletion, photochemical ozone creation potential, strong odour and tainting), the normal methods of risk characterisation do not apply. REACH makes provision that these must be dealt with on a case-by-case basis. The exposure calculation of atmospheric concentration can be useful and should be reported, even if none of these effects apply. The concentration in air contributes to indirect exposure daily exposure doses.

8.9.3   Long Range Pollutants

Substances which are persistent in the wider environment may have the potential for long-range transport via air, as well as any localised pollution effects. Standard models exist but it is not part of the normal requirements for a REACH registrant to attempt to model this. However, EU regulatory authorities have responsibilities under the UN convention on long-range trans-boundary air pollution and the registration data set will facilitate their activity to identify any new candidates.

8.10   The Food Chain

Predators and birds can be exposed to the substance via the food chain. Organisms at the lower level of the food chain take up the substance directly from the environment; higher organisms eat them and so acquire an additional dose in their diet.

8.10.1   Biomagnification

Some types of substances are metabolised to a degree but in the initial screening level assessment a biomagnification factor of at least one is assumed. A value in the range $1-10$ is set, selected on the basis of value of log K_{ow} and fish bioconcentration factor (BCF). If the metabolic pathway is well defined, potential for accumulation of metabolites should also be considered.

Methods for understanding metabolism are improving (see also Section 7.2.2 of this book on BCF property discussion). A number of factors can influence bioconcentration. Some structures could be identified and facilitated by active biochemical transport mechanisms, for example specific interaction with cell membranes and active transport across them. Some substances are maintained in homeostasis by specific mechanisms. Physical chemical factors can have an influence, for example large bulky molecules and charged species may be less likely to be taken up. Surface-active substances in solution may form micelles, with very different behaviour than might be expected from modelling based on a single molecule.

If rapid, rate of hydrolysis could exceed rate of uptake. Hydrolysis products could theoretically accumulate, but often tend to be more water soluble than the parent structure, making this unlikely.

The rate of uptake could exceed rate of biodegradation, so it is very possible for a substance to have high BCF even if concentration in the environment low.

8.10.2   Secondary Poisoning

It is necessary to consider secondary poisoning in the chemical safety assessment if certain criteria are fulfilled. If the substance has a very low bioaccumulation potential then secondary poisoning need not be considered further. If the potential to bioaccumulate exists then it must be checked which hazards are present, based on self-classification as well as harmonised classification. If there are other indications (e.g. endocrine disruption) this would also be a factor to consider.

The main protection targets in the standard assessment are:

- *Fish-eating birds and mammals*: this relates to avian and mammalian predators on freshwater fish. The constituents of their diet are modelled on defaults to allow a daily 'dose' via lower trophic level organisms to be estimated.
- *Worm-eating birds and mammals*: note that the bioconcentration factor for earthworms is almost never tested experimentally. The BCF is derived based on other data.
- *Marine food chain*: two trophic levels are considered, marine predator and marine top predator.

References

ECHA (2012) Guidance on Information Requirements and Chemical Safety Assessment, Chapter R.16: Environmental Exposure Estimation, Version: 2.1, October 2012.
JRC (European Commission Joint Research Centre, Institute for Health and Consumer Protection) (2013) EUSES. Last update 5 July 2012, http://ihcp.jrc.ec.europa.eu/our _activities/health-env/risk_assessment_of_Biocides/euses (last accessed 21 July 2013).

9

Assessing the Hazards to Human Health from Chemicals

9.1 Mammalian Toxicology

The science of toxicology is the study of the inherent capacity of a chemical to cause adverse health effects to living organisms. Toxicity testing uses animals as surrogates to establish what effects chemicals are likely to have on humans (should exposure occur).

Toxicokinetics – overview of importance

Toxicokinetics is the assessment of adsorption, distribution, metabolism and excretion of a given substance entering the human body. This is significant in understanding uptake levels, clearance times, and identifying target organs or tissues and is important in defining strategy for the human health exposure assessment and planning further testing programmes. In REACH, the main exposure pathways (oral, inhalatory, dermal) should all be considered. A toxicokinetics assessment based on knowledge of the chemical properties is normally adequate; the assessor must properly consider behaviour under the relevant conditions of pH and temperature. Experimental studies in mammals can be very important and useful but can be very expensive to conduct, particularly if using radiochemical analysis.

9.2 Exposure Routes and Local/Systemic Effect Types

In order for toxicity of any description to occur there has to be exposure, or chemical contact. There are three main ways by which this can take place: these are inhalation, skin contact and ingestion. In some cases chemical exposure could conceivably occur

---

Chemical Risk Assessment: A Manual for REACH, First Edition. Peter Fisk Associates Ltd.
© 2014 John Wiley & Sons, Ltd. Published 2014 by John Wiley & Sons, Ltd.

by all three routes. In the work place the most likely routes of exposure are inhalation and skin contact, although ingestion can never be totally ruled out.

Chemical exposure could result in local or systemic effects, or even both. Some chemicals will only cause local effects; these are typically reactive chemicals such as skin irritants and corrosive chemicals. Systemic effects relate to adverse effects occurring elsewhere in the body away from the original site of exposure. However, unlike local effects absorption of the chemical in question has to take place for the possibility of systemic effects to occur. For example, exposure to sodium hydroxide will cause local effects, but will not cause systemic effects. In contrast, exposure to benzene will primarily cause systemic effects should inhalation occur but can also cause local, degreasing effects to the skin upon prolonged exposure by this route.

The use of chemicals is a prevalent concern in contemporary society and questions regarding whether or not a given chemical could be less toxic than another are frequently asked. In the sixteenth century the German-Swiss physician Philippus Aureolus Theophrastus Bombastus Von Hohenheim, or Paracelsus, concluded that everything had the potential to be toxic. The only thing that differentiated between something being toxic or non-toxic was the dose. In other words, 'it is the dose that makes the poison'. For many people today, the idea that 'chemicals' are in their vicinity can cause great concern. However, it is the dose at which exposure occurs that will determine the toxicological effect. In fact, it should be noted that arsenic was a very effective medicine in the combat against syphilis before penicillin was discovered. There are, of course, some exceptions to this theory laid out by Paracelsus, amongst which are included genotoxic carcinogens and sensitisers, as it is thought that these types of effect can occur at any dose.

9.3   Acute and Chronic Effects

Toxicity, or the inherent ability of a chemical to cause harm, can be broadly subdivided into two main types: *acute toxicity* and *chronic toxicity*. These two terms are associated with the duration of exposure and resulting adverse effects.

Acute toxicity describes the effects arising from a single or limited number of exposures over a short time period. In contrast, chronic toxicity arises as a result of intermittent or continuous exposure over a lifetime. Both these types of toxicity manifest themselves in different ways, giving rise to different effects. It is, therefore, not possible to predict the acute effects that may occur from knowledge of the chronic toxicity of a substance, and vice versa. There are also other divisions of toxicity, which lie between acute and chronic toxicity. These are known as subacute and subchronic toxicity.

9.4   Influences on Toxicity

There are a number of factors which will influence the extent to which harm may occur as a result of chemical exposure. These are, in part, related to the chemical in question and include the dose and chemical structure. In addition, factors such as exposure route and host factors (age, sex, lifestyle, nutritional and health status) will influence the toxicological outcome.

9.5   How Chemicals Cause Harm

There are numerous ways in which chemicals can cause harm; these are summarised in Table 9.1. These common ways in which chemicals can cause harm are discussed in more detail here.

9.5.1   Asphyxiants

These are substances that reduce the level of oxygen in the body; they can be divided into two subgroups: chemical asphyxiants and simple asphyxiants. Chemical asphyxiants interact at a biological level in the body, thereby preventing the oxygen from being used and causing interference with normal cellular metabolic processes. Examples include carbon monoxide and hydrogen cyanide. Simple asphyxiants, such as methane and nitrogen, cause asphyxiation indirectly by displacing the oxygen that is present the air.

9.5.2   Narcotics

These are substances that depress the normal functioning of the central nervous system. Organic solvents and ethanol are examples of chemicals that can cause this effect.

9.5.3   Irritants and Corrosives

In terms of handling chemicals in the workplace, probably the most common way in which chemical exposure occurs is by accidental splashing or spillage: in other words by skin contact. Chemical irritants will cause a reversible inflammatory response at the site

Table 9.1   *Definitions of types of toxicity relevant in REACH and CLP.*

Nature of harm	Description	Local or systemic effect?
Asphyxiant	A substance which causes a reduction in the availability of oxygen in the body	Systemic
Narcotic	A substance which depresses the normal functioning of the nervous system	Systemic
Irritant	Causes reversible inflammation in living tissue at the site of contact	Local
Corrosive	Causes irreversible destruction in living tissue at the site of contact	Local
Sensitiser	A substance which causes an allergic response	Systemic
Carcinogenic	Ability of a substance to cause cancer	Systemic
Genotoxic	Causes heritable mutations	Systemic
Reproductive effects	Effects on different aspects of the reproductive cycle, including fertility	Systemic
Developmental effects	A substance which causes adverse effects on the developing foetus	Systemic
Target organ effects	Specific effects on organs usually away from the initial site of exposure	Systemic

CLP – classification, labelling and packaging.

of contact. Corrosives, however, cause irreversible destruction to the living tissue; another expression for corrosive effects is 'chemical burns'. In most cases the concentration of a chemical will be a factor in whether or not it is a corrosive or an irritant. For example, a 5 M sodium hydroxide solution will be corrosive to skin, whereas a 0.1 M sodium hydroxide solution will be irritating but not corrosive. The pH is also another consideration: in general, chemicals which have a pH of 2 or less or greater than 11.5 are, from a regulatory and safety perspective, automatically considered as being corrosive.

In terms of exposure by skin contact, chemical irritants cause what is known as 'irritant contact dermatitis'. This is a local effect and the severity of the symptoms will depend in part on the chemical and also factors intrinsic to the individual person who is exposed.

9.5.4   Sensitisation (Allergic Reactions)

The process of development of an allergic reaction involves the cells of the immune system and is, in fact, a systemic effect. Common allergies include hay fever and nickel allergies. It is important to note that the immune system is specific (to a particular allergen) and has memory. This means that once the immune system has been stimulated, any subsequent future exposures will result in a rapid response.

In terms of the development of an allergic response, there are two defined stages:

1. *Sensitisation*: this occurs upon the first encounter with a relatively large exposure to an allergen and results in the immune system being stimulated as a response.
2. *Elicitation (of the allergic reaction)*: this second stage arises as a result of exposure to what can be a much lower level of allergen compared to the sensitisation step. The onset of the symptoms of allergy (such as itching, redness, pain etc.) is subsequently seen.

In the workplace, the most common sensitisation effects are related to skin and respiratory exposures. Known skin allergens, such as nickel, give rise to the condition known as 'allergic contact dermatitis'. This is known as a 'delayed hypersensitivity reaction' as the symptoms (redness, itchiness, rash) usually occur hours after exposure. Respiratory sensitisation is of great concern in the workplace, as it causes a wide range of effects in the respiratory tract, which include rhinitis and difficulties in breathing. This condition is often referred to as 'occupational asthma'. Common examples of respiratory sensitisers include isocyanates, animal proteins, flour dusts and colophony. The immunological responses that follow exposure to a respiratory sensitiser are immediate, meaning that the allergic symptoms occur rapidly. Unlike exposure to chemical irritants and the development of irritant contact dermatitis, not everyone who is exposed to a known skin sensitiser or respiratory sensitiser will develop an allergic response. This is because the development of an allergic response is, in part, based on a number of host factors, which include atopy. However, the consequences of developing an allergy as a result of work can have devastating effects and in many cases the only solution is to remove the individual completely from the workplace.

9.5.5   Carcinogenicity

Carcinogenicity is defined as the ability of a substance to cause cancer. Carcinogens are those agents that can cause cancer. Chemical carcinogens include benzene, formaldehyde, asbestos and so on. The process of carcinogenesis is characterised by abnormal cell division and growth. This may result in a tumour (or neoplasm) that is either malignant or benign. Malignant tumours have the ability to spread (metastasise) to other areas of the body, whereas benign tumours are typically localised and do not spread. Chemical carcinogens can be categorised into two types: genotoxic and non-genotoxic. Genotoxic carcinogens are believed to act by interference with DNA, causing damage and mutation. This may bring about changes to normal cell division, resulting in uncontrolled cellular proliferation and the development of a tumour. As it is believed that the development of cancer is due to mutations, any substance that is genotoxic is believed to have the potential to cause cancer. However, although this is true in many cases, there are also exceptions. In other words, not all carcinogens are genotoxic in nature, for example asbestos or phthalate esters. These act via a non-genotoxic mode of action and either enhance another pre-existing carcinogenic process or induce cancer as a result of other cellular effects.

9.5.6 Genotoxic Effects

Genetic toxicology is used to identify chemicals that could cause heritable effects (or mutations) and also identify potential carcinogens. This topic is covered in more detail in Section 9.8 of this chapter.

9.5.7 Reproductive and Developmental Effects

Reproduction is the way in which genetic material is passed on from one generation to another: that is parent to offspring. Adverse effects on the reproductive process may be caused by chemicals. This could be, for example, by decreasing fertility, thereby making conception more difficult. Known chemical reproductive toxicants include lead and glycol ethers.

Developmental effects are those which arise due to damage in the developing foetus, resulting in physical abnormalities; they do not, however, cause any adverse effects in the mother. Chemicals which cause this are known as 'teratogens'; they include methyl mercury and thalidomide.

9.5.8 Target Organ Effects

These are adverse effects that are caused by chemical exposure to different organs within the body. This includes the kidneys, liver, cardiovascular system, respiratory system, skin and so on.

9.6 Toxicokinetics

Not all chemical exposures will result in adverse health effects, because the effect will depend on the concentration present within the body. This is dependent on the toxicokinetics: the study of the rate of absorption, distribution, metabolism, and excretion of a chemical.

In order for a chemical to cause a systemic effect it has to be absorbed. The extent of this occurring will depend on a number of factors, including physico-chemical properties and location of exposure (such as inhalation, skin contact etc.). If there is no absorption into the body, then there will be no risk of systemic effects, although local effects such as irritation could occur. Once a chemical has been absorbed it will be distributed around the body via the blood and lymphatic systems. The extent of distribution and storage within the body will again depend on the physico-chemical properties. Substances which are highly lipophilic ('fat loving') are more likely to remain within the body for much longer periods of time than those which are highly water soluble. The process of metabolism (or xenobiotic metabolism) is related to the removal of unwanted chemicals from the body and primarily occurs in the liver. Lipid solubility is an important factor in both absorption and storage of chemicals. Metabolism is concerned with:

- Making such chemicals more water soluble, so that they may be eliminated from the body via the kidneys in the urine.
- Making chemicals less toxic.

Metabolism breaks down the parent compound into metabolites. The problem, however, is that in some cases these metabolites may be more toxic, such as the case with *n*-hexane. This is metabolised to 2, 5-hexanedione, which causes peripheral neuropathy.

Excretion is the process by which the body eliminates the substance from the body. Rapid removal from the body will reduce the potential for toxicity, whereas the converse is that the longer the substance remains in the body the higher the potential for toxicity. The main routes of excretion from the body are via the kidneys, bile, and lungs.

9.7   Toxicological Testing

For toxicology testing a wide range of animals can be used, ranging from rodents such as rats and mice, to dogs and even primates, depending on the purpose of the study. Ideally, all studies would be conducted using a wide range of species, the logic being that if no adverse effects were noted in any of the species tested then it would be unlikely that any adverse effect would be seen in humans. However, this is not done due to ethical and economic reasons, and hazard assessments are usually made from studies where only one or two species have been used. The choice of exposure route used in the study will be partially determined by what is known to be the most common route of exposure. For example, in the workplace this is likely to be skin contact and inhalation. However, the physico-chemical properties of the substance will also be a determining factor in terms of choice of exposure route. Throughout this section reference will be made to the Organisation for Economic Co-operation and Development (OECD) test guidelines, which can be found online at the OECD web site (OECD, nd-a).

Toxicity testing can be categorised into three groups:

1. Acute toxicity studies
2. Short-term (repeated dose) studies
3. Long-term (chronic) studies.

These studies are, in effect, evaluating the potential for a chemical to cause systemic effects, although information may also be gleaned on local effects such as irritation. In addition to these, there are testing methodologies available for assessing other toxicological endpoints, such as carcinogenicity, genotoxicity, reproductive, and developmental effects, irritation/corrosive potential and sensitisation effects.

Under REACH the level of toxicological testing, as described in the REACH Regulation (1907/2006, Annex VII–X inclusive) will be based on the tonnage level of the substance being either manufactured or imported in to Europe, and whether or not the substance is used as a strictly controlled intermediate. That is, the higher the tonnage of substance being manufactured/imported the greater the level of toxicological testing that will be required.

9.7.1   Data Gaps

In the absence of data, there are a number of options that can be considered in lieu of animal testing:

1. Relevant data from a structurally analogous substance and (Q)SAR ((Quantitative) Structure-Activity Relationship) may be available and used as read-across or prediction.
2. Valid human data which could be used to support a weight of evidence argument.
3. Data from validated *in vitro* studies.

9.7.2   Data Waiving

Data waiving may be used if the toxicological endpoint result would be predictable, unethical or unnecessary. For example, if the test substance was known to be corrosive to skin, an acute toxicity study should be waived on ethical grounds. Similarly, there would be no reason to conduct a 28-day repeated dose toxicity study if a reliable and relevant 90-day repeated dose study already existed.

9.7.3   Acute Toxicity Studies

The main objective of these studies is to study the effects of exposure to a given chemical over a relatively short period of time, with lethality being the end point. In principle, all three exposure routes can be studied, although historically the oral route is by far the most common.

9.7.3.1 LD_{50} Tests

This is a very common test and also a requirement of most chemical control regulations worldwide, including REACH. The term LD_{50} means 'lethal dose 50' and describes the dermal or oral median dose lethal dose: in other words, the dose which when administered will kill 50% of the test population. The result is usually described as mg/kg body weight, which means milligrams (of test substance) per kilogram of animal. For the inhalation route, the test result is described as LC_{50} or 'lethal concentration 50' and can be expressed as mg/m^3 or ppm. The results are plotted as a dose-response curve from which the LD_{50}, or the dose administered that will cause lethality in 50% of the test population, can be obtained. Due to ethical issues and animal welfare concerns, the LD_{50} test (OECD 401) was deleted from OECD test guidelines in 2002. As alternatives, the Acute Toxic Class Method, the Fixed Dose Procedure and the Up and Down Procedure were introduced. The Fixed Dose Procedure (OECD 420) does not use lethality as an endpoint. Instead, it relies on the observation of clear signs of toxicity at one of the series of fixed dose levels used. Both the Acute Toxic Class Method (OECD 423) and Up and Down Procedure (OECD 425) still have mortality as the endpoint, but fewer animals are used.

9.7.3.2   *What Information Can Be Derived from Acute Studies?*

The results can be used to provide a rough measure of relative toxicities and are used for classification and labelling purposes. However, it should be noted that comparison of the results can only be loosely made between different substances as the results will depend in part on the methodology used, species studied and the choice of exposure route.

9.7.4   Short-Term, Repeated Dose Studies

These studies are designed to investigate the effects that arise from repeated exposure to smaller doses of test chemical compared to the acute studies over a larger part of the test organism's lifespan; typically in the region of 10% of the animal's lifespan.

9.7.4.1   *What Information Can Be Derived from Short-Term, Repeated Dose Studies?*

Essentially the objective of these studies is to find the lowest dose that produces a detectable adverse effect. This could, for example, be changes in kidney function or body weight. These studies enable a dose-response curve to be plotted, from which it is then possible to derive the 'no observed adverse effect level' (NOAEL) and also the 'lowest observed adverse effect level' (LOAEL). The results of such studies can help identify any potential target organ effects and also the potential for any accumulative effects and resulting 'delayed' toxicity.

9.7.5   Long-Term (Chronic) Studies

Long-term or chronic toxicity studies investigate the adverse effects that may arise from prolonged exposure to a relatively low level of test chemical over the greater part of a lifetime. These studies are similar to the short-term repeated dose studies, with the main difference being the number of animals used, duration, and dosage. Due to the cost of these studies, it is not uncommon for these to be combined with carcinogenicity studies, with the methodology being such that the objectives of both studies are fulfilled.

9.7.5.1   *What Information Can Be Derived from Chronic Studies?*

Chronic toxicity studies provide information similar to short-term repeated dose studies: that is, a dose-response curve from which a NOAEL and LOAEL can be derived. In addition, they can give information on those effects arising from prolonged exposure, which may not have been seen in the shorter-term studies, as well as for those types of effect which have a long latent period before being seen.

Dose-response curves can provide information on whether exposure to a chemical causes a given effect, regardless of whether or not there is a threshold for the effect and how rapidly the effect is seen upon increasing doses (the slope of the dose-response curve).

9.7.6   Other Systemic Effects

9.7.6.1   Carcinogenicity

These studies are similar to chronic studies in that they involve exposure to the test substance over the greater part of the organism's life. However, the main difference is that these studies are designed to detect the probability of one person in a million developing cancer as a result of exposure to a given test substance. In order to achieve this, these studies use high doses of the test substance in order to compensate for the relatively small number of animals that are used. These studies aim to investigate whether or not there is an increased incidence of cancer as a result of exposure to a given test substance.

9.7.6.2   Genotoxicity

Genotoxicity studies are commonly used to indirectly ascertain whether or not a test substance could have carcinogenic potential. This is because the majority of chemicals that are carcinogenic also demonstrate genotoxicity. The background to these studies has been covered in Section 9.8 of this chapter.

9.7.6.3   Reproductive and Developmental Effects

Groups of rats or mice are dosed using up to three doses prior to mating, and then during the mating period and pregnancy. Effects on fertility, conception, and successful pregnancy are assessed and, in some cases, the study can continue through the offspring being allowed to reproduce (whilst being dosed). These studies provide information on the dose-response effects and the NOAEL. Specific studies are undertaken to investigate the effects of a chemical on the developing foetus, called developmental toxicity. Rabbits or rats are typically used and the study involves dosing post-implantation during the critical first trimester of the pregnancy up until parturition. Again, these studies provide information on the dose-response effects and NOAEL.

9.7.6.4   Sensitisation (Allergic Reactions)

Although the symptoms of allergy can be seen at the site of contact, such as with poison ivy causing a skin rash, it is in fact a systemic effect. This is because the immune system is affected and the symptoms can be seen in other areas of the body. In terms of toxicology testing, the two routes of exposure that are considered are skin and inhalation. Because there is currently no widely accepted test for respiratory sensitisation, most assessments are either made based on structural activity relationships or by human experience. For skin sensitisation studies both the guinea pig maximisation test and the Buehler test are the ones that will be most commonly seen (OECD 406). Both studies use the guinea pig as the test species. The main difference between the two types of test is that the guinea pig Maximisation Test involves subcutaneous injection of the test compound and an adjuvant in order to maximise the likelihood of an effect being seen even for weak sensitisers, whereas the Buehler Test involves topical application of the test chemical without the adjuvant. Both studies are designed to mimic both stages of the development of an allergy: sensitisation and elicitation of the symptoms. It is the elicitation of the symptoms of allergy that are visually scored.

With consideration to animal welfare, a great deal of research has taken place over the past decade or so to reduce the number of animals that are used in these studies. The murine local lymph node assay (LLNA) is such a study that has considerably reduced the number of animals that are used, and under REACH is the 'first choice test method'. Instead of the elicitation of allergic symptoms, this study quantitatively measures sensitisation potential by the measurement of lymphocyte proliferation at the draining lymph nodes (OECD 429).

9.7.7 Local Effects

9.7.7.1 *Irritation and Corrosivity*

Localised reversible inflammation upon contact with the respiratory tract, skin or eye can be caused by some chemicals. However, some chemicals demonstrate an even more severe effect, causing non-reversible burns (or corrosion) at the site of contact. Prior to any assessment for potential irritation it is, therefore, necessary to ascertain whether or not the chemical could in fact be corrosive. This can be achieved with knowledge of the acid/alkaline reserve: typically, chemicals with a pH of ≤ 2 or ≥ 11.5 will automatically be classified as corrosive, and thus not tested.

Traditional studies for irritation use rabbits as the surrogate; classification of irritation is based on visual development of symptoms of irritation, such as redness (erythema) and swelling (oedema). A grading system is used to assess the severity of the effects in a semi-quantitative manner. The results are then used for classification purposes. There are also a number of validated *in vitro* alternatives that have been developed over the past decade for assessment of irritancy to skin. These include the TER test (rat skin transcutaneous electrical resistance test) (EC B.40, OECD 430), EpiSkin™ and EpiDerm™ (EC B.40bis, OECD 431), and SkinEthic™ (EC B.40bis, OECD 431) tests. There are currently no validated eye alternatives available, all that are available are tests which can be used as a screen to eliminate corrosives. These are the BCOP (bovine cornea opacity permeability) (OECD 437) and ICE (isolated chicken eye) (OECD 438) tests.

9.8   Genetic Toxicology

9.8.1   Introduction

Genetic toxicology (also called genotoxicology) assesses the effects of chemicals on the genetic material of cells or organisms. 'Mutagenicity' is the general term used to refer to the induction of transmissible changes in the genetic material of cells or organisms. Most of the information on the mutagenicity of a substance is derived from studies that investigate the genotoxicology of a substance. Assessment of the potential for mutagenicity is part of the hazard assessment of chemicals and data used in the hazard assessment are also used to determine whether classification as mutagenic is appropriate.

9.8.2   Hazard Assessment

The data that are needed to assess the mutagenic potential of substances for REACH are based on a set of three *in vitro* tests (tests using cells grown in culture rather than live animals). Testing in animals (*in vivo*) may be indicated if the results from any of the *in vitro* studies are positive. The *in vitro* studies required have been chosen to give information on gene mutations (changes in the base sequence of DNA), changes in the structure of chromosomes (breaks or re-arrangements) and changes in the number of chromosomes. Table 9.2 sets out the data requirements for genetic toxicity at each Annex level.

The following terms are used in describing tests for genotoxicity:

• Mutagenicity studies investigate damage to genes in DNA (note that this is a more specific use of the word than the general use described in the introduction to this section).
• Cytogenicity studies look at induction of damage to chromosomes (chromosome aberration).
• Clastogenic means causing chromosome aberrations. A substance which gives a positive result in a cytogenicity study is said to be clastogenic.
• Aneuploidy indicates changes in the number of chromosomes.

The Guidance on information requirements and chemical safety assessment Chapter R.7a: Endpoint specific guidance (ECHA, 2008) includes a useful flow chart indicating the testing strategy for mutagenicity.

There are several points to note about the information requirements.

Table 9.2   *Studies required for each REACH Annex level.*

Standard information requirement	Brief description of study[a]	REACH Annex	More information
8.4.1. *In vitro* gene mutation study in bacteria	Bacterial mutagenicity	VII	Further studies should be considered in case of a positive result
8.4.2. *In vitro* cytogenicity study in mammalian cells or *in vitro* micronucleus study	*In vitro* cytogenicity	VIII	Not required if there is information from an *in vivo* cytogenicity test, or the substance is the substance is known to be carcinogenic category 1A or 1B, or mutagenic category 1A, 1B or 2
In vitro gene mutation study in mammalian cells	Mammalian mutagenicity	VIII	Not required if there is information from an *in vivo* mutagenicity test, or the substance is the substance is known to be carcinogenic category 1A or 1B, or mutagenic category 1A, 1B or 2

[a]The brief descriptions are used to refer the information requirements in this section.

9.8.2.1   Results

Results that may be obtained from genetic toxicity testing are:

- Negative (no evidence for induction of mutations or of chromosome aberrations, depending on test type).
- Positive (evidence that the substance induces mutations or causes chromosome aberrations).
- Equivocal (it is not possible to judge whether the result is positive or negative).

The responses to these results will depend on the Annex requirement (Table 9.2). It should be noted that where a report concludes that a substance is clastogenic (positive in cytogenicity assay, structural aberrations) only at overtly toxic concentrations, this is not considered to be an indication that the substance induces chromosome aberrations, because damage to chromosomes occurs during apoptosis (cell death).

Negative Results.   Negative results will need to be followed up with the next test needed. If all test data required in REACH are available (just bacterial mutagenicity at Annex VII, *in vitro* cytogenicity and mutagenicity testing in mammalian cells at Annex VIII and above), and all results are negative, then the conclusion is that the substance is not genotoxic, and no further testing is needed.

Equivocal Results.   Equivocal results should be followed up by repeating the test using the same method, but varying the conditions to obtain conclusive results. It is not always possible to obtain conclusive results, in which case the next step is to continue testing in accordance with REACH requirements.

Positive Results.   If a positive result is obtained in a bacterial mutagenicity study, even if the tonnage band is Annex VII, further *in vitro* (Annex VIII) studies should be considered.

If an *in vitro* cytogenicity study gives a positive result, a mammalian mutagenicity study will not be needed.

If either the *in vitro* cytogenicity study or the mammalian mutagenicity assay gives a positive result, then appropriate *in vivo* studies should be considered.

Conflicting Results.   The results obtained in bacteria may not be reproduced in mammalian cells. This may be due to differences in uptake mechanisms, genetic material or metabolism. Sometimes different results are obtained from the same study conducted at a different laboratory or on another occasion. It should be considered whether there are problems of test substance purity or pH. Positive results at high or low pH are not considered to be indicative of mutagenicity.

Bacterial Mutagenicity.   The current EU and OECD guidelines for bacterial mutagenicity (EU Test Method B.13/14 (2000)/OECD TG 471 (1997)) specify that the strains of bacteria used should include one that is capable of detecting cross-linking mutagens. The strains of bacteria usually used are: *S. typhimurium* TA 98, TA 100, TA 1535, TA 1537, with *S. typhimurium* TA 102, *E. coli* WP2 *uvr*A or *E. coli* WP2 uvrA (pKM101) as the fifth strain. The 1983 guideline had different requirements, and many older studies do not include such a strain. The European Chemicals Agency (ECHA) has made it clear that if the only evidence for lack of mutagenicity to bacteria is derived from studies that do not include an appropriate fifth strain, then additional information is needed (ECHA, 2010).

Table 9.3 In vivo *testing in REACH.*

REACH number	Test type	Example tests (guidelines)
8.4	Somatic cell *in vivo* test to investigate structural or numerical chromosome aberrations	*In vivo* micronucleus test (erythrocytes) EU: B.12 OECD: 474s *In vivo* chromosome aberration test (bone marrow) EU: B.11 OECD: 475 Comet (no guideline)
8.4	Somatic cell *in vivo* test to investigate gene mutations	*In vivo* unscheduled DNA synthesis[a] (EU: B.39 OECD: 486) Gene mutation in transgenic mice[b] (no guideline) Comet assay (no guideline)
8.4	Germ cell *in vivo* test to investigate gene mutations	Comet assay[c] Gene mutation in transgenic mice[b] Rodent dominant lethal test (EU: B.22 OECD: 478) Mammalian spermatogonial chromosome aberration test (EU: B.23 OECD: 483)

[a]Unscheduled DNA synthesis: identifies chemicals that induce DNA repair in liver cells.
[b]Gene mutation in transgenic mice: this test can measure gene mutations. Any tissue may be used, including site of contact tissues and germ cells.
[c]Comet assay: this test can give evidence of DNA strand breaks, and any tissue may be used, including site of contact tissues and germ cells.

The range of strains tested is particularly important for Annex VII registrations, where the only genetic toxicity information available may be from a bacterial mutagenicity study. If this is the case, and the study tested only four strains, or a different range (e.g. *S. typhimurium* TA 98, TA 100, TA 1535, TA 1537 and TA 1538), then testing of a fifth strain is needed to meet the REACH requirements.

Waiving In Vitro *Tests.*   Testing in mammalian cells is not needed if there are results available from appropriate *in vivo* studies. REACH distinguishes between mutagenicity and cytogenicity endpoints in *in vivo* studies. Table 9.3 summarises the ECHA guidance on *in vivo* studies and the endpoint to which they relate. The Annex VII requirement for *in vitro* mutagenicity testing cannot be waived on the basis of *in vivo* studies investigating chromosome aberrations (e.g. *in vivo* micronucleus assay), which can, however, be used to waive cytogenicity testing. Similarly, the Annex VII requirement for *in vitro* cytogenicity testing cannot be waived on the basis of *in vivo* studies investigating gene mutations (e.g. *in vivo* unscheduled DNA synthesis), which can, however, be used to waive mutagenicity testing.

In Vivo *Testing.*   In order to prevent unnecessary animal testing, *in vivo* studies are proposed for ECHA, and only carried out after a period of public consultation. Proposals for *in vivo* testing should be made where there is indication potential for genetic toxicity from one or more of the *in vitro* mammalian assays (Section 9.8 of this chapter).

The type of *in vivo* study should be chosen based on whether there is evidence for mutagenicity or cytogenicity or both. Suitable tests can be found in Table 9.3.

Positive results from tests in somatic cells may need to be followed up by further testing in somatic or germ cells. Germ cell testing is very rarely required.

Other Mutagenicity Endpoints.   There may be data available for other genetic toxicity endpoints that are not part of the REACH requirements, for example DNA binding and sister chromatid exchange. These data should be included in the REACH dossier but if they conflict with the results of the required studies the latter are, generally, of higher significance.

Prediction of Genotoxicity.   Genotoxicity involves interactions between toxicants and molecules involved in replication of cells, so is closely related to the presence of structural features of toxicants. The concept of 'structural alerts' for genotoxicity may be used in assessing genotoxic potential (Benigni and Bossa, 2006; Benigni *et al.*, 2008). Much effort has gone into the development of systems for the prediction of genotoxicity, with the result that there are a range of tools available that are able in some instances to predict potential for genotoxicity.

Many of the predictive tools are available on-line, and links can be found to them from the OECD (Q)SAR Toolbox web site (OECD, nd-b). A review of (Q)SAR methods and software tools has been published by the JRC (Serafimo *et al.*, 2010), which concluded that 'the assessment of model predictions requires a reasonable amount of (Q)SAR knowledge' and in addition, more work is needed on their use for regulatory purposes.

The REACH Technical guidance (Part R.7a, p. 380–382) (ECHA, 2008) discusses the use of prediction in assessing potential mutagenicity and suggests the use of prediction in a weight of evidence approach. As with all (Q)SAR predictions, the use of modelling in genotoxicity must be valid, adequate and properly documented to meet regulatory requirements.

9.8.3   Risk Assessment

9.8.3.1   *Mutagens and DNELs*

If the genetic toxicity assessment concludes that a substance should be classified for mutagenicity, this should, of course, be taken into account in the risk assessment. A substance classified as a mutagen is then regarded as a genotoxic carcinogen, unless there is scientific evidence that this is not the case.

It may not possible to assign derived no-effect levels (DNELs) for mutagenicity, as it is not considered that there is a threshold below which a mutagenic substance will not exert an effect. Risk characterisation is, therefore, based on a derived minimal effect level (DMEL) (Section 9.9.1 of this chapter).

9.9   Turning Intrinsic Properties into 'No-Effect' Levels

When a substance causes adverse effects in mammalian toxicity studies it is necessary, for human health risk characterisation purposes under REACH, to compare exposure levels with a dose that is not hazardous to humans. This value is known as the derived no-effect level. In this context adverse effects can mean effects that lead to classification and labelling (Appendix A) but can also apply to less serious effects that do not trigger classification according to EU criteria but are, nevertheless, considered to be of toxicological significance.

Risk characterisation ratios (RCRs) for humans are calculated by dividing the exposure level by the DNEL for the relevant route of exposure; as with the environment the target is to achieve an RCR value below one.

Other regulatory schemes use a different approach where the no-effect level is divided by the exposure level to determine a 'margin of safety' (MoS). The underlying principles of the MoS approach are identical to the DNEL approach but the target is to achieve the greatest MoS possible.

The basis for setting DNELs for the purposes of the REACH Regulation are laid out in detail in REACH Technical Guidance Chapter R.8 (ECHA, 2010). This document should be consulted and ideally expert evaluation of the available data undertaken before attempting to calculate DNELs in all but the most straightforward of cases.

When calculating DNELs it is necessary to consider:

- The potential exposed population(s), for example workers, general population, sensitive populations such as pregnant women, the elderly or children.
- The potential routes of exposure, that is oral, dermal or inhalation.
- The pattern of exposure, for example repeated low levels of exposure over a period of months or years, or a short exposure to a high concentration on one day per year.

It is necessary to calculate separate DNEL values for each relevant situation. Under most circumstances for REACH registration purposes worker exposure is considered only for the dermal and inhalation routes since oral exposure should be prevented through use of good hygiene practice. For the general population oral exposure may need to be assessed either if there is intentional or likely swallowing of the substance due to use pattern, or if the substance requires assessment for exposure to humans via the environment in drinking water and foodstuffs.

The most normal starting point for calculation of a DNEL value is the no observed adverse effect level or concentration (NOAEL(C)) from a long-term mammalian toxicity test such as a repeated-dose toxicity study, a reproductive toxicity study or a developmental toxicity study (refer to Sections 9.8.3 and 9.9 of this chapter). When adverse effects are seen in more than one type of study DNEL values should be considered for each case and risk characterisation based on the worst-case outcome or leading health effect.

Usually, a DNEL can only be quantified for substances which produce a 'threshold effect', in other words there is a clear cut-off dose below which the adverse effects are not observed. For non-threshold effects, DNELs cannot usually be quantified and alternative approaches must be considered for risk characterisation (Section 9.9.1 of this chapter). DNELs may be derived for both local effects (tissue damage at the site of contact, such as necrosis) and systemic effects (for example organ damage, central nervous system effects, developmental effects follow systemic absorption and distribution throughout the body).

One or more of the following corrections may need to be applied to the experimentally-determined NOAEL(C) value to obtain a 'corrected NOAEL(C)' that is appropriate for humans:

• Extrapolation of experimental study exposure period (e.g. 6 hours per day) to the duration of exposure that is relevant for humans (typically 8 hours for workers, 24 hours for the general population).
• Extrapolation of the study dosing regimen (e.g. 5 days per week) to the relevant human situation (which may be 7 days per week).
• Correction for differences in respiratory rate and volume between experimental animals and humans, and between different exposure conditions (e.g. test conditions relate to 'at rest' conditions whereas a worker may be undertaking light or heavy activity and thus have increased respiration rate compared to the resting value).

A number of assessment factors must then be applied to the corrected NOAEL to take account of the following uncertainties:

• Metabolic differences between experimental animals and humans (known as allometric scaling).
• Extrapolation of duration (to take account of the duration of exposure compared to the lifetime of the subject).
• Remaining interspecies differences.
• Intraspecies differences (to take account of potential differences within the human population).

For the assessment factors described above, guidance document R.8 (ECHA, 2010) provides default values; non-default values can, of course, be applied when there is sufficient evidence to do so. Similar assessment factors are also the basis of other safety assessments such as World Health Organization principles for risk assessment of chemicals in food (WHO/FAO/UNEP, 2009).

In some cases, additional assessment factors might also be applied:

• Route-to-route extrapolation (when systemic uptake of a substance may differ between the experimental exposure route and the route that is relevant for humans).
• Dose–response (applicable, for example when a NOAEL has not been established due to effects at all dose levels but it is likely to be a threshold effect).
• Quality of the data set.
• Use of screening data.

While guidance document R.8 (ECHA, 2010) provides some suggested values or ranges for such assessment factors, these most often require detailed expert judgement.

Where suitable data exist, and the need arises, it is also possible to quantify DNEL values for other types of effect such as acute (short-term) toxicity or sensitisation. Similar corrections and assessment factors apply for these too although there is no specific guidance available.

Alternative approaches such as the benchmark dose and toxicological threshold of concern are also described in the guidance.

9.9.1 Special Cases

For some substances, for example genotoxic carcinogens or mutagens, it is usually assumed that there is no threshold for a given effect. This means that in theory exposure to an infinitesimally small amount of that substance has the potential to cause cancer or genetic damage. Specific legislation exists (the Carcinogens and Mutagens Directive (2004/37/EC)) to control releases of and exposure to such substances. Nevertheless, it is still necessary to perform risk characterisation yet a conventional DNEL cannot be established.

An alternative approach for non-threshold effects is to determine the DMEL, which is considered to be a 'tolerable' level of risk, for example cancer risk levels of 10^{-5} or 10^{-6} for workers and the general population, respectively. DMEL calculation and risk characterisation for non-threshold effects should only be attempted by experts.

References

Benigni, R. and Bossa, C. (2006) Structural alerts of mutagens and carcinogens. *Current Computer-Aided Drug Design*, **2**(2), 169–176.

Benigni, R., Bossa, C., Jeliazkova, N. *et al.* (2008) The Benigni/Bossa Rulebase for Mutagenicity and Carcinogenicity. JRC Scientific and Technical Report EUR 23241 EN, Publications Office of the European Union.

ECHA (2008) Guidance on Information Requirements and Chemical Safety Assessment, Chapter R.7a: Endpoint Specific Guidance, May 2008.

ECHA (2010) Guidance on Information Requirements and Chemical Saftey Assessment, Chapter R.8: Characterisation of Dose [concentration] – Response for Human Health. Version 2, December 2010.

OECD (nd-a) Home page. http://www.oecd.org/ (last accessed 31 July 2013).

OECD (nd-b) Donors to the (Q)SAR Toolbox. http://www.oecd.org/chemicalsafety/ assessmentofchemicals/donorstotheqsartoolbox.htm (last accessed 31 July 2013).

Serafimo, R., Gatnik, M.F. and Worth, A. (2010) Review of (Q)SAR Models and Software Tools for Predicting Genotoxicity and Carcinogenicity. JRC Scientific and Technical Reports EUR 24427 EN, Publications Office of the European Union.

WHO/FAO/UNEP (2009) International Programme on Chemical Safety. http://whqlibdoc.who.int/ehc/WHO_EHC_240_8_eng_Chapter5.pdf (last accessed 23 July 2013).

10

Human Exposure to Chemicals

This chapter discusses the assessment and quantification of human exposure; it covers workers in both industrial and professional use situations, consumers, exposure of people to chemicals in the environment and risks due to physico-chemical hazards such as explosivity.

10.1 Exposure

The assessment of exposure to chemicals in the workplace covers workers in industrial settings, which could be the chemical industry or other manufacturing-type situations. In these situations, the employer has a legal responsibility to provide safe working conditions. Exposure in the workplace is discussed in the European Chemicals Agency ECHA Guidance Part R.14 (ECHA, 2010b).

The population to be considered covers only healthy people of working age: that is adults. Inhalation and dermal exposure are considered; oral exposure is not regarded to be significant in workplace situations, as it can be prevented by basic workplace hygiene measures, such as not eating in working areas. Both acute (peak exposure over a short period) and long-term exposure are considered. The estimates are intended to cover a reasonable worst-case for typical use; accidents, malfunction or deliberate misuse are not covered.

Potential ways to be exposed to chemicals in an industrial setting include:

- Fugitive emissions from vessels
- Direct emissions during use
- Sampling, and so on
- Minor splashes/spills and
- Handling contaminated equipment.

Chemical Risk Assessment: A Manual for REACH, First Edition. Peter Fisk Associates Ltd.
© 2014 John Wiley & Sons, Ltd. Published 2014 by John Wiley & Sons, Ltd.

Industrial workplaces can include both highly automated environments where exposure to substances is very minimal and situations where workers are involved in manual handling of chemicals.

The level of exposure may be estimated using models or measurements; these are each discussed further below. In general, factors to take into consideration include:

- The nature of the substance. For example, whether it is a solid, liquid or gas, the particle size if it is a solid, or the vapour pressure if it is a liquid.
- Whether the substance is used on its own or as part of a mixture or article, and its concentration in the mixture or article.
- The processes and techniques used in handling the substance and the level of containment.
- The duration and frequency of exposure.
- The location: for example, whether the work takes place indoors or outdoors and the room volume if indoors.
- The presence of ventilation systems and their efficiency.
- The personal protective equipment (PPE) recommended. This should usually only be used as a last resort after other control options are used as far as possible, and the exposure estimation is usually performed first without considering this before the effect of PPE is added in.

For initial screening (Tier 1) assessments the level of information needed may be limited, for more sophisticated (higher tier) assessments many additional details may be required.

The process category (PROC) deserves special mention. This forms part of the system of Use Descriptors defined in the REACH guidance, to set a framework for communication of key information on nature of exposure through the supply chain and to guide exposure modelling (refer to REACH guidance part R.12) (ECHA, 2010a). The PROC descriptor codes describe the application techniques or process types; for example PROC 1 represents a closed process with no likelihood of exposure and PROC 7 represents industrial spraying. The categorisation depends on:

- The amount and form of energy applied in a process (e.g. heat, mechanical energy, radiation).
- The surface of the substance available for exposure (the dustiness of the material or the thickness of layers of the material).
- The principal level of containment and engineering controls to be expected.

The PROCs therefore reflect the general level of exposure expected to result from a process or technique. A particular exposure scenario may require consideration of several PROCs. For example, manufacturing of a chemical may involve a batch process (PROC 3 or 4, depending on the level of containment) or a continuous process (PROC 1 or 2, depending on the level of containment) as well as transfer of a substance for packaging/transport (PROC 8a/8b or 9, depending on the context and volumes).

The selection of the correct PROC(s) to describe a particular activity is important and requires some experience in occupational hygiene. When describing a use pattern it is best to start with a brief description in words, using language appropriate for the sector, before selecting PROCs. This description should be included in communications up and down the supply chain; however it is rarely possible to adequately describe a use pattern using **only** the use descriptor codes.

10.2   Exposure to Chemicals in the Workplace

10.2.1   First Tier Models

First tier models are intended to provide a quick screening tool to identify substances that may be of concern. The estimates are intended to be conservative and may significantly overestimate exposures. If the risk characterisation using these models indicates a risk, the exposures estimates can be refined using higher tier models or measurements.

The ECETOC (European Centre for Ecotoxicology and Toxicology of Chemicals) targeted risk assessment (TRA) allows exposures of workers, consumers and the environment to be calculated. For workers, a PROC, a broad sector of use (industrial or professional), the state of the substance (solid or not), the vapour pressure (for a liquid or gas) or dustiness (for a solid) are entered. This leads to estimates of inhalation and dermal exposure. These estimates are then modified according to whether the activity takes place indoors or outdoors, the presence of local exhaust ventilation (LEV, for indoor activities only), the duration of the activity, the type of respiratory protection used, whether the substance is used in a preparation and the concentration range of the substance in the preparation. An example of the use of the ECETOC TRA is shown in Case Study 10.1.

For inhalation exposure, the EMKG-Expo-Tool may be used as an alternative to the ECETOC TRA. This is a generic tool developed by the German BAuA (Federal Institute for Occupational Safety and Health), which applies a banding approach and allows non-risky workplace situations to be filtered from those requiring more detailed attention.

Case Study 10.1   Example of the use of ECETOC TRA

Substance A is formulated in a multistage batch process, which provides the opportunity for significant contact; this is best described by PROC5. The substance is a liquid with a vapour pressure of 1.1 Pa at 20 °C; therefore, it falls into the low fugacity category. The default inhalation exposure (8 hour time weighted average) for indoor use is predicted to be 1 ppm. LEV with default efficiency of 90% is present, so the exposure prediction is modified to 0.1 ppm. Each worker is exposed once per day for between 15 and 60 minutes; therefore, a modifying factor can be applied for the duration of exposure, giving an estimated exposure of $0.2 \times 0.1 = 0.02$ ppm. No respiratory protection is used and the substance may be used neat, so no further modifying factors can be applied.

For PROC5 with LEV, a default dermal exposure of $10\,\mu g/cm^2/day$ is predicted based on a contact surface area of $480\,cm^2$ (palm of both hands). Based on the default worker body weight of 70 kg, the overall daily dose via dermal exposure is 0.07 mg/kg/day.

It requires three input parameters: volatility or dustiness, amount of the substance used and control strategy.

The Tier 1 tools are not yet fully validated and further experience may lead to updates being made.

10.2.2   Higher Tier Models

A Tier 2 assessment is generally much more detailed than the Tier 1 assessment and is specific to the situation under consideration rather than being based on generic processes. The assessment needs to be carried out by an experienced person who will normally need access to much more detailed information on the workplace situation. Gathering this information can present a significant challenge, particularly when downstream sites are being considered.

The assessment can be carried out using any suitable method that can be shown to be valid and sufficiently accurate. The assessor needs detailed information on the specifications of the model and the level of confidence associated with the estimates. Several tools are being developed by industry and European institutions, including the Stoffenmanager exposure model, the RISKOFDERM dermal model and the Advanced REACH Tool (ART) for occupational exposure assessment.

10.2.2.1   Measurements

The development of exposure scenarios for REACH will not usually require new exposure monitoring to be initiated. However, if the Tier 1 exposure estimation indicates a risk, carrying out exposure measurements may in some cases be a good alternative or supplement to carrying out higher tier estimations (Section 10.4.3 in this chapter). Careful design of such measurements is crucial to ensure that the data obtained is robust and suitable for use in the assessment.

Any existing information should be taken into account in the assessment and may be preferred to estimates but will need careful evaluation. Measured data should be:

- Representative of the exposure scenario they are applied to: for example data from a single site may not be representative of a scenario that covers several sites.
- Reliable: sufficient details about the data collection should be available to allow the assessor to determine its reliability.
- Robust in terms of sample size.

Measurements may also be helpful in evaluating the effectiveness of risk management measures and allowing downstream users to assess the validity of the exposure control advice received from their suppliers.

10.2.3   Risk Management Measures

If the risk characterisation indicates a risk after all reasonable refinement has been carried out, risk management measures will need to be put in place. The exposure estimation is then modified to take into account the effect of these measures. For workers, these may include the following:

- Modification of the process, for example by using safer equipment or greater automation.
- Changing the form of the substance or limiting the concentration of the substance in a preparation.
- Installing LEV.
- Limiting the duration of exposure.
- PPE (respiratory protection, gloves, goggles) where exposure cannot be prevented by other means.

Chapter 11 of this book gives a more complete discussion of risk management options.

10.2.4   Exposure in the Professional Use Setting

Exposure in a professional use setting again relates to occupational exposure, but in a non-industrial setting. This may include public domain settings, for example hospitals, schools, institutional buildings or office blocks, but it does not usually consider exposure of the general public, only employees. The exception to this is substances that fall within the scope of the Chemical Agents Directive.

The following examples illustrate professional use settings:

- A farm worker may use a variety of different chemicals, for example cleaning/sterilising products, fertilisers, and pesticide spray adjuvants. Pesticides are not assessed under REACH as they are covered by other legislation. The work may take place indoors or outdoors for varying lengths of time. The types of processes involved may include mixing/formulation, transfer of substances between different vessels, spraying, and roller application/brushing. Both inhalation and dermal exposure may be significant depending on the type of process.
- Hospital cleaners are involved in tasks including spraying of cleaning fluids, wiping of surfaces, and transfer of cleaning solutions between different containers. Therefore, they may be exposed via both inhalation and skin contact to the variety of substances in the cleaning products. The work is indoors and exposure could be for several short periods throughout the working day.
- An automotive mechanic may come into contact with lubricants, greases, engine fluids, coolants, cleaning/degreasing products, fuels and fuel additives in the course of his or her work. There may be significant skin contact with some substances and processes may be carried out at high temperature. The work may take place indoors or outdoors and exposure could be for long periods.
- A large number of tasks involving exposure to chemicals, both in preparations and in articles, can take place on building sites. For example, mixing and laying of cement, weather sealing glazing units, painting, welding or cutting metal, cleaning/degreasing, and spraying masonry treatment solutions. The work may take place outdoors or indoors.
- A small furniture maker may use adhesives as well as polishes, varnishes, and paints in their work. Significant skin contact with some substances, for example when polishing, can be expected. This may only be for short periods but inhalation exposure could occur throughout the day as coatings dry.

- Fuel delivery drivers can be exposed to substances (the fuel itself and any additives) on transfer to and from the tankers. This will commonly take place outdoors, for short periods, and exposure may be by inhalation or by splashes onto the skin.
- A carpet fitter may use adhesives/sealants in affixing or joining carpet. He or she may also be exposed to substances such as flame retardants, glues and dyes evaporating from carpet or underlay or dust/particles released on cutting (exposure from articles). The work is likely to be indoors, the area may be poorly ventilated and exposure may be throughout the working day.

Employers have the same legal obligations as at industrial sites and appropriate working conditions and PPE must be provided. However, this is harder to control than in an industrial setting, for example a tradesperson working in clients' homes away from his or her employer may choose not to wear the gloves that are provided.

As for industrial exposure, the assessment covers healthy adults being exposed over the short and long term via inhalation or skin contact. A reasonable worst-case for typical use is considered, whereas accidents, malfunction or deliberate misuse are not covered.

10.2.5   Models

The process of estimating exposure of professionals to substances is very similar to that for workers at industrial sites. The initial Tier 1 estimation provides a screening level indication of the degree of risk associated with a substance. Both the ECETOC TRA tool and the EMKG-Expo-Tool can be used for professionals; ECETOC TRA has different estimates for industrial and professional use situations.

If risks are identified, refinement of the exposure estimates using higher tier models may be appropriate. Any model can be used to assess professional exposure, provided that the assessor has the necessary information about exposure parameters for input into the model and knowledge of how the model works in order to assess its suitability and uncertainty in the results.

Some of the main exposure tools are now discussed; there are others.

10.2.5.1   ECETOC Targeted Risk Assessment (TRA)

The ECETOC TRA comprises three separate models for exposure estimation for workers, consumers and the environment (ECETOC, 2012). The software is available online for download for free and has been developed and provided by European Centre for Ecotoxicology and Toxicology of Chemicals, to aid with REACH registration-related exposure assessment. The software provides the option of manual or batch mode, where the user can choose between assessing single or multiple exposure scenarios at a time. To facilitate batch operation the tool allows, for both consumers and workers, up to 15 exposure scenarios/uses to be entered for exposure estimation using the relevant descriptor codes, operating conditions and modifying factors.

The ECETOC TRA tool uses the REACH use descriptor codes as a starting point for exposure assessment. Of the different use descriptors, the majority of available PROCs are input parameters for Tier 1 worker exposure estimation. Several article categories (ACs) can also be used as input parameters for consumer exposure modelling in the standalone ECETOC consumer exposure estimation tool. The ECETOC TRA tool has two tables of options, professional and industrial, wherefrom values can be selected that correspond to the use descriptors applicable for the selected exposure scenario. For example, the use descriptor for industrial use (SU3) and professional use (SU22) are suitable as the starting point for Tier 1 exposure estimation in the TRA tool. Similarly the consumer exposure estimation tool is based on selected chemical product categories (PCs). In practice, the user interface of the ECETOC TRA tool allows for the selection of the PROCs and modifications from drop down menus in a spreadsheet.

The results of the ECETOC TRA are expressed as estimated exposure values and risk characterisation ratios (RCRs). Required inputs include physico-chemical properties such as vapour pressure and molecular weight. These values are processed in the spreadsheet to calculate both dermal and inhalation exposure estimations. In order to derive a RCR an indicative reference value, such as the derived no-effect level (DNEL), are required.

10.2.5.2   Advanced REACH Tool (ART)

The ART is a free, web-based higher tier exposure assessment tool, designed for the estimation of inhalation exposure at the workplace (ART, nd). ART is currently able to assess inhalable dusts, mists and vapours, while it is not suitable for fumes, fibres, gases or dust emissions from hot metallurgical processes. This mechanistic model of inhalation exposure makes use of precise estimates which the user is required to input into the tool and which thereby allow for more accurate exposure estimation. The current ART tool does not allow for use of respiratory protective equipment in the assessment of inhalation exposure. The tool is based on a Bayesian statistical framework which integrates the mechanistic model with the available exposure database. In practice, the user goes through a stepwise process where data are entered into the tool on the product type, exposure scenario, user activities (including duration), emission potential (e.g. dustiness, volatility), activity class, localised controls, surface contamination, substance dispersion and measures which are in place to separate the worker from the substance during user pattern. The user has the option of selecting between the 50th and 99th percentile, to reflect the assumed variability of the exposure scenario under assessment. The final exposure estimate is expressed as mg/m^3 with a confidence interval, which reflects the level of uncertainty of the estimated percentile.

10.2.5.3   ConsExpo

ConsExpo is a consumer exposure estimation tool (NIPHE, nd). It uses a set of exposure models with an associated pre-existing database of known exposure scenarios, which is suitable for higher tier estimation of human exposure to indoor consumer products via the oral, dermal, and inhalation routes. The pre-existing database comprises default values based on known product categories.

ConsExpo offers a wide range of outputs ranging from acute to semi-chronic for the different exposure routes. The software operates on a stepwise process and the user is required to input data on substance type and physico-chemical properties, as well as answer a set of questions on parameters relevant to each exposure route and scenario, including the population group for which the exposure estimation is being modelled.

For inhalation, models are available for spray and vapour exposure, where relevant parameters on exposure duration, room dimensions, ventilation rate, volatility, particle distribution, and so on are required for the exposure assessment. Similarly, for dermal exposure estimation, five models are available for instant application, constant rate, rubbing off, migration or diffusion, which can be assessed by entering information on parameters such as the exposed area, contact rate and time, compound concentration and layer thickness. For oral exposure modelling direct oral intake, constant rate, migration and migration from packaging material can be assessed by providing data on the ingested amount, rate and time, migration rate or parameters to do with the type of packaging.

The user is required to understand the supporting factsheets which define the default values and form the basis of the database. This is essential in order to objectively evaluate the relevance of the predetermined parameters assigned to each individual exposure estimation. Where more accurate information is available, this should always be used to substitute the default values.

The results of the ConsExpo model can be displayed as point values, graphically or in text form with the option of a sensitivity analysis of a given exposure estimation. The sensitivity analysis allows for the manipulation of a single parameter, to see what changes occur as a result in the calculated exposure.

10.2.5.4   Stoffenmanager

Stoffenmanager is a validated web-based higher tier exposure assessment tool designed for the evaluation of worker health risks (Stoffenmanager, nd). The software comprises inhalation and dermal exposure models, where hazard information and exposure assessment are combined to calculate a risk score. The risk assessment and relevant control measures are thereafter presented as an 'action plan'.

For inhalation, an assessment can be made for dust and vapour; however, fibres, gases or emissions from metallurgical processes are not suitable for assessment at present. The Stoffenmanager assessment tool is able to estimate a conservative, worst-case scenario based on the 90th percentile of the exposure distribution, or alternatively a more refined assessment can be carried out by defining a lower percentile.

The exposure tool offers two different models for inhalation exposure assessment; a REACH specific Tier 1 inhalation exposure tool and a higher tier option are available. The toolkit is linked to an existing database of 700 known measurements. There are no restrictions for the use of the inhalation models; however, the dermal tool is not suitable for the assessment corrosive of very toxic substances.

Comparative evaluation of the various models is under way through research sponsored largely by national authorities.

10.2.6   Measurements

As discussed in Section 10.2.2.1 in this chapter, measured data may be preferred to models if they are representative, reliable and robust. However, professional use of a particular product often takes place in a wide variety of settings over which the manufacturer of the product has little control. Therefore, it is difficult to ensure that measurements of exposure of professionals are representative for all users.

10.3   Risk Management Measures

If the best possible estimates of exposure indicate a risk when compared to no-effect levels (Section 9.9 of this book), risk management measures need to be put in place. The most effective measures for professional users are often those involving changes to the product, such as reducing the concentration of a hazardous substance or changing the form of the substance. Workplace measures, such as recommending installation of ventilation systems or limiting the duration of exposure, and PPE, such as gloves, can be appropriate in some situations. However, it is not appropriate to assume that such measures are followed in all professional use situations. Chapter 11 of this book gives a more complete discussion of risk management options. These are termed 'operational conditions' for safe use.

10.4   Consumer Exposure

Consumer exposure is discussed in the ECHA Guidance Part R.15. A consumer is any member of the public and the assessment therefore includes 'vulnerable' populations such as the sick, elderly or children (ECHA, 2012). A consumer product or article is considered to be something that can be purchased from a retail outlet by members of the general public.

Consumer exposure may occur in several situations:

- Direct use of a substance or product.
- Post-application exposure: the consumer does not leave the work area and, therefore, exposure may be over a 24-hour period rather than a standard working day.
- Release of a substance from an article or a reacted/dried preparation during use/service life. This can be driven by water or saliva contact, skin contact, elevated temperatures, mechanical abrasion or slow emission from a matrix.
- Exposure to substances in public spaces/buildings or exposure to substances used by professionals (for example decorators) in the home.
- Removal/cleaning of a product.

Exposure via the environment (for example a substance released from a chemical manufacturing plant and then inhaled by a member of the public) is not included; this type of exposure is assessed separately and is discussed in Section 10.5 in this chapter.

Examples of typical product types to be assessed would include:

- Cleaning products
- DIY products: paints, adhesives, sealants
- Hobby materials: paints, adhesives
- Household articles: carpets, furniture, curtains, toys.

If the same person may be exposed to a particular substance from more than one different product/article, the exposures should be added together. The assessment does not include products such as cosmetics or personal care products that are outside the scope of REACH because their safety for consumers is assessed under other legislation.

10.4.1   General Considerations for Exposure Estimation for Consumers

A reasonable worst-case for typical use is considered; for consumers this includes both the intended uses and other reasonably foreseeable uses. For example, consumers may use a larger volume of washing-up liquid than specified, or chew the end of a pen. The assessment does not include deliberate misuse of a substance or product. However, differentiating between deliberate misuse and reasonably foreseeable uses can be difficult in some cases and the decision about what to cover in the assessment should be fully justified.

In addition to the inhalation and dermal exposures considered for workers, oral exposure is relevant for consumers. This may arise, for example, from the use of cleaning products on food preparation surfaces or from the hand-to-mouth and mouthing behaviour of young children. In special cases, in may be appropriate to consider other routes of exposure, such as splashes to the eyes or intradermal exposure arising from piercings.

Exposures of different subpopulations should be considered and appropriate values for parameters such as body weight and skin surface areas selected. For example, a crawling child may be exposed to residues of cleaning products on the floor and their high skin surface area to body weight ratio compared to an adult can have a large impact on the outcome of the exposure estimates.

Key inputs for exposure estimation include the duration and frequency of exposure, the different routes of exposure, the characteristics of the product or article (in particular, the amount of product used and the concentration of the substance in the product or article) and the different ways in which the product or article is handled. For consumer uses, there is often little control or monitoring of how products and articles are used beyond their point of sale and limited information is available for the assessor to use.

10.4.2   Tier 1 Models

The REACH guidance on consumer exposure estimation (ECHA guidance part R.15, section 15.3) (ECHA, 2012) sets out algorithms for initial Tier 1 estimates of exposure. The assessor can either implement the algorithms manually or make use of one of two tools which implement them, ECETOC TRA and ConsExpo Tier 1. Tier 1 tools are designed to be easy to use and require little information; they apply conservative default values and produce a worst-case estimate of exposure. If the risk characterisation based on these initial estimates does not indicate a risk (all RCRs are less than 1) this initial estimate may be sufficient. If a risk is identified, the exposure estimates can be refined by over-riding some of the defaults with more realistic values, using higher tier models, or making measurements.

The key inputs for the Tier 1 estimates are the volume of product used and the concentration of the substance in the product. As a first estimate, it is assumed that all of the substance in a product is available for intake by the route being considered. For example, for inhalation exposure all of the substance is considered to be released into a standard sized room as a gas, vapour or airborne particulates. Defaults for the respirable fraction of the inhaled substance, the ventilation rate of a person, body weight, the duration of exposure and the number of exposures per day are then used to calculate the intake of substance by inhalation per day and body weight.

The ECETOC TRA consumer tool is largely based on the Tier 1 algorithms described in the REACH guidance, with some minor modifications. It incorporates default values for parameters such as exposure time and amount of product per use for more than forty product and article types. These product/article types are linked to the product and ACs in the REACH use descriptor system (ECHA guidance part R.12) (ECHA, 2010a). The user has to enter a process/AC and the volatility of the substance. They may also choose to over-ride the defaults for the fraction of substance in the product or article, the amount of product used per application, the dermal contact surface area and the 'mouthed' surface area. The default values were taken as far as possible from the RIVM ConsExpo fact sheets (NIPHE, nd-a), which give background information and quantify exposure parameters that are important for use of the ConsExpo tool. Expert judgement was used to derive default values for ECETOC TRA where a ConsExpo fact sheet was not available.

ConsExpo is a well-known tool for consumer exposure estimation available from the Dutch National Institute for Public Health and the Environment (RIVM) (NIPHE, nd-b). It includes several different models of varying complexity for each exposure route. It is possible to select Tier 1 models that are comparable to the algorithms given in the REACH guidance and those used by ECETOC TRA.

The two Tier 1 models therefore give comparable results. ConsExpo gives greater flexibility to alter defaults and select alternative models. However, this means that its use requires a greater level of expertise and there is also no direct link to the PC/AC in the REACH use descriptor system.

10.4.3   Refinement of Initial Exposure Estimates, Higher Tier Models and Measurements

A first step to refining the Tier 1 exposure estimates will usually involve a 'reality check' on the input parameters and the outputs they generate. The default inputs to the Tier 1 tools are intended to represent a worst-case scenario and consideration of the values may show that either individual values or the combination of inputs are inappropriate for the specific use being considered. For example, for inhalation exposure it is assumed that 100% of the product is released to the air and for dermal exposure it is assumed that 100% of the product is in contact with skin. Any changes to the default values should be fully documented and justified.

Several higher tier tools for consumer exposure estimation are available. These models are more sophisticated and detailed than Tier 1 models and give more realistic outputs. They should be conducted by an expert assessor, who will require detailed information about the use of the product.

It is possible that measured values are available for exposure model inputs such as migration rates of a substance from an article, percentage of a substance permeating through the skin and air exchange rates. It would not usually be necessary to make new measurements for this purpose, although this could be considered if a parameter was particularly important and difficult to estimate accurately. In all cases, measurements should be reliable, robust and representative for all consumers covered by the assessment.

10.4.4   Risk Management Measures – Consumers

If a risk is still identified after all possible refinement of the exposure estimates, risk management measures will need to be applied. For consumers, the only risk management measures that are usually effective are those that can be controlled by the manufacturer of the product. These include measures such as reducing the concentration of a substance in a product, limiting the total amount of product used by using individual dose packaging, limiting contact with the product by design of the packaging, supplying the product as granules or tabs rather than powder to reduce exposure to dust, or limiting migration from an article by design of the matrix. It is good practise to make recommendations such as using gloves or working in a well-ventilated area on consumer packaging. However, it cannot be assumed that consumers will follow these instructions and they should not be taken into account in quantifying exposures. Chapter 11 of this book gives a more complete discussion of risk management options.

10.5   Indirect Exposure (Humans via the Environment)

As described in Chapter 8 of this book, models exist to quantify how a substance is distributed throughout the environmental compartments. Through these media, humans may be exposed to the substance indirectly, through food, drinking water, and air. The diet of humans naturally varies considerably between individuals and there may be trends over time or between countries. The models use a standard modelled diet.

Indirect exposure of humans is required in REACH to be assessed for substances at >1000 tonnes/year, or at lower tonnage when certain types of human health hazards apply (toxic or carcinogenic, mutagenic, reprotoxic). Tools such as EUSES (European Union System for the Evaluation of Substances) can model this type of exposure.

In some circumstances a registrant may need to consider including the total dose received by consumers or workers in the exposure assessment, accounting for exposure indirectly via the environment combined with workplace or consumer exposure.

Exposure of agricultural soil can lead to uptake of the substance into growing crops. Similarly, cattle grazing on exposed grasslands can accumulate the substance due to eating contaminated grass. This can lead to meat and milk containing a concentration of the substance which the models can quantify. The typical quantity of crops, meat, and milk in a consumer's diet are taken into account in the models. Along with the concentration in air, these are the key contributors to the indirect exposure of consumers.

The concentration in fish can be estimated from the modelled concentration in water and uptake parameters.

Concentration in drinking water is calculated based on leaching of substance from soil into groundwater.

Exposure of air is described in Section 8.9 of this book. Indirect exposure is based on annual average.

10.6   Risk due to Physico-Chemical Hazard

The physico-chemical properties of some substances mean they can present an immediate hazard for handling, transport, storage or use if special measures are not taken. The relevant properties are tested for as part of the Annex VII data set. REACH requires assessment for at least flammability, oxidising properties and explosivity. Where one of these hazards is present, it is necessary to conduct an assessment of the risk posed. The guidance (R9) recommends a quantitative method to assess exposure and evaluate risk (ECHA, 2008).

Substances classified as explosives are controlled under separate legislation (Dangerous Substances and Explosive Atmospheres Regulations 2002) and for these cases, where a risk assessment has already been made, it is not necessary to conduct a full assessment under REACH.

This method is approached using a tailored questionnaire to all end users. In most circumstances when the supply chain has a large number of end users this would be extremely challenging and is rather unrealistic for most registrants. Unless the substance is used as such, then there is no clear benefit in exhaustive attempts to resolve this issue.

A qualitative assessment, combined with appropriate labelling and handling guidance to communicate the findings through the supply chain, may be adequate.

Whichever form of risk assessment is used, the considerations should include such matters as:

- Type and severity of the hazard(s)
- Handling conditions
- Storage conditions
- Packaging materials
- Reactivity hazards
- Temperature
- Enclosure
- Pressure relief
- Use of monitoring systems
- Ventilation to avoid flammable or explosive atmospheres forming
- Avoiding potential ignition sources, potential for leakage, and so on.

If the substance is commercially supplied only in the form of formulations, in which the substance is present at a low proportion, and the other components are non-hazardous, this can mitigate the hazard. This may be the most suitable method for controlling hazards for consumer uses. Classification of the formulations supplied is a responsibility of the supplier but is not part of the REACH registration of each substance as such.

References

ART (nd) Advanced Reach Tool, Version 1.5. www.advancedreachtool.com (last accessed 24 July 2013).
ECETOC (2012) TRA, Version 3. www.ecetoc.org/tra (last accessed 24 July 2013).
ECHA (2008) Guidance on Information Requirements and Chemical Safety Assessment, Chapter R.9: Physico-Chemical Hazards.
ECHA (2010a) Guidance on Information Requirements and Chemical Safety Assessment, Chapter R.12: Use Descriptor System, Version 2, March 2010.
ECHA (2010b) Guidance on Information Requirements and Chemical Safety Assessment, Chapter R.14: Occupational Exposure Estimation, Version 2, May 2010.
ECHA (2012) Guidance on Information Requirements and Chemical Safety Assessment, Chapter R.15: Consumer Exposure Estimation, Version 2.1, October 2012.

NIPHE (National Institute for Public Health and the Environment) (nd-a) ConsExpo, Fact Sheets. http://www.rivm.nl/en/Topics/C/ConsExpo/Fact_sheets (last accessed 24 July 2013).

NIPHE (National Institute for Public Health and the Environment) (nd-b) ConsExpo. www.rivm.nl/en/Topics/Topics/C/ConsExpo (last accessed 24 July 2013).

Stoffenmanager (nd) Stoffenmanager Version 5.0. www.stoffenmanager.nl (last accessed 24 July 2013).

11

Managing Hazard and Risk

11.1 Characterisation, Assessment and Management of Risk

Chapter 6 of this book examined in some detail how chemical hazard is assessed, with the REACH regulation as the main focus. REACH sets out a formalised framework for the characterisation and communication of risks; indeed, this can be seen as one of its primary goals. Risk in the REACH framework was defined in Chapter 7 of this book. This chapter looks at how *characterisation* of risk is refined. It starts from a point of understanding what risk assessment means, and its management is then considered.

Risk assessment is a phrase that occurs minimally in the REACH Regulation, whereas it was used throughout the predecessor regulations. Risk assessment is taken to mean a complete assessment of all sources and exposures to a substance. This was the intention of the previous regulations; however, these focussed on a very limited set of high tonnage existing substances and on new substances. Under REACH many more substances are involved (perhaps 30 000 existing substances compared to about 150 under the ESR, Existing Substances Regulation). The old 'new substances' regulation called upon much effort on a few thousand substances, most of which were on the market at less than 10 tonnes/year. These could be assessed fully in that system.

Risk assessment	Risk characterisation
A process of considering combined risk from all sources of a particular molecule, including all substances that contain it.	A process of considering one substance from one registrant, although combined risk from different sources is included, for that registrant.

Chemical Risk Assessment: A Manual for REACH, First Edition. Peter Fisk Associates Ltd.
© 2014 John Wiley & Sons, Ltd. Published 2014 by John Wiley & Sons, Ltd.

In REACH, risk characterisation is the same basic process as risk assessment but implies an assessment of risk by one registrant only. Of course, if a consortium working on a substance pools all its use information, and it represents a high fraction of the substance producers and importers, then a consortium chemical safety report (CSR) amounts to a true risk assessment.

Therefore, the phrase 'risk characterisation' is used to imply that a registrant has characterised risk by the required methods but a full risk assessment needs more to be done, should that be required. The Authorities can initiate performance of a full assessment.

11.2   What Is 'Risk' under REACH?

In very few instances it is possible to say *for certain* that a risk exists. One set of circumstances could be when there is direct measurement of concentration in the workplace of a substance with known effects on human health from human studies or records of effects. Even in such cases the susceptibility of individuals varies and exposure concentrations are not fixed over time. Despite the simplicity of the scenario, immediately on analysis of the circumstances surrounding a measured exposure, it is necessary to have to bring in some degree of knowledge and interpretation in order to understand the event or events. REACH amplifies many of these considerations, based on systems of regulation and interpretation that have evolved over the last 50 years. The process of risk characterisation accounts for uncertainties in exposure and effects.

Risk is characterised quantitatively (i.e. numerically) on the basis of comparison of a no-effect concentration or exposure compared to the actual or predicted exposure. The methods should not be seen as some kind of absolute truth but as a system that sets priorities for action.

For human health risk characterisation, there are many uncertainties, some discussed already in Chapter 9 of this book. These include:

- The basis on which a derived no-effect level (DNEL) is set.
- Duration of exposure.
- Exposure route (air, skin, oral).
- Absorption, distribution, metabolism and excretion of the substance may not have been studies.
- Variation in susceptibility between individuals.

Even so, systems have been developed to account for uncertainty for the individual.

Environmental risk characterisation differs from the health characterisation in that under REACH the goal is the protection not of individuals but of ecosystems. The kinds of uncertainty now may include:

- Susceptibility of species when only a few representative ones have been tested.
- The basis on which a predicted no-effect concentration (PNEC) has been set.
- Duration and continuity of exposure.
- Organisms in the environment differ from laboratory organisms in susceptibility.
- Exposure to many different substances.

11.3   What Are Risk Reduction and Risk Management?

11.3.1   Risk

For any REACH protection goal, the legal systems set out what is considered to be unacceptable. When that applies it is necessary to reduce risk and then put systems in place to ensure that the risk is controlled on a long-term basis. At the simplest level, users of chemicals are advised by producers on the basis of how an acceptable level of exposure can be achieved. Remember, the properties of the substance cannot be modified. At the more serious extreme, it may be considered that no safe level can be established; then the only way forward is to stop production and use totally. In most cases simple controls are sufficient.

Some people in the industry can be tempted to say 'why do we need REACH? We would know if we have a problem!' That is naïve, because REACH ensures that all important endpoints, environmental and health, are characterised, and many of these endpoints, or protection goals, are not visible. Therefore, it is quite normal to find that a default model of exposure and effects can give rise to an apparent risk. This chapter sets out responses to such a finding, and describes the kinds of iterations that may be necessary to ensure that a use is safe, as defined in the REACH paradigm. This may well include the need for changes in working practices.

REACH aims to protect human health and the environment by assessment of the properties of the substance that may make it dangerous, and the potential of contact between people or the environment to the chemical in ways that may cause harm (exposure). Chapter 1 of this book provides a general discussion of the aims of REACH. REACH requires that risks are 'adequately controlled'. What does this mean and how is it achieved?

The chemical safety assessment (CSA) that must be carried out for REACH will identify any hazardous properties of the chemical being registered. The most severe hazards for health and for the environment are identified and used to set a hazard threshold. For example, a chemical may be more toxic to fish than to invertebrates or algae, in which case a PNEC will be set based on the level of toxicity to fish (Section 7.3 of this book). For human health, DNELs are set for exposure of workers and for consumers for single and repeated contact with the chemical by different routes (e.g. skin contact, oral exposure and inhalation toxicity). For some hazards that cannot be quantified, (e.g. irritation, mutagenicity), a DNEL cannot be calculated. Instead, a derived minimum effect level (DMEL) will be set.

To find out the level of risk posed by hazardous properties, exposure is assessed. Exposure assessment quantifies the amount of chemical reaching different environmental compartments and the extent to which workers and consumers are exposed to the chemical, both directly and by consideration of secondary poisoning in the food chain.

To assess whether risks are adequately controlled, risk characterisation is carried out. This involves comparing the PNECs and DNELs or DMELs with environmental and human exposure levels. If the exposure is greater than the hazard threshold, the risk characterisation ratio (RCR) will be greater than one, which demonstrates that the chemical poses an unacceptable risk to human health or the environment. Adequate control is demonstrated if the RCR is below one, that is the exposure is below the hazard threshold.

This does not mean that there are no risks; it is not possible to eliminate risk completely. The assessment of adequate control of risk assumes that the exposure is steady state; it does not take into account fluctuations in exposure. It also assumes that the potential for harm is captured adequately in the PNECs and DN(M)ELs.

It is a requirement of REACH that risks are adequately controlled. As the hazardous properties of a chemical cannot be altered, reduction of risk is achieved by control of exposure. The ways in which exposure can be controlled are referred to in the context of REACH as risk management measures (RMMs). The CSR documents not just properties and exposure, but also the RMMs needed to ensure safety of workers, consumers, and the environment.

Risk Characterisation for Complex Substances – Environment

The 'Hydrocarbon block method' allows a substance with multiple constituents to be assessed. In brief:

1. The first step is to consider the constituents' physico-chemical properties, degradability and ecotoxicity, using experimental data for the pure constituent where available (e.g. from published literature or handbooks) and predicted data.
2. Constituents should be grouped into 'Blocks' according to these properties.
3. Perform exposure assessment and characterise risk separately for each Block, using representative property inputs.
4. Risk characterisation for the substance as a whole is performed by adding together the RCR values for all Blocks.

Risk Characterisation for Complex Substances – Human Health

Complex substances (multi-constituent substances and UVCBs, unknown or variable composition, or biological origin substances) can contain numerous different individual chemical constituents with different properties and different levels of toxicity. In the workplace and as consumers and users of the substance throughout its life cycle, humans are generally exposed to the complex substance as a whole (though, for example, some constituents might tend to leach or volatilise preferentially). For most areas of the human health exposure assessment and risk characterisation, a DNEL based on the whole substance is appropriate.

One general exception is indirect exposure of humans via the environment. Exposure via the environment is so dependent on the properties of the constituent that it may be preferable to define separate DNELs for each 'hydrocarbon block'. This is particularly important if different constituents or 'blocks' have significantly different toxicological potency. The indirect exposure levels are estimated separately for each block by the environmental modelling, so that risk characterisation can be done in a focussed way where needed.

It may also be appropriate to set separate DNELs for any blocks that have very different exposure characteristics, for example a volatile gaseous impurity present in an otherwise low-volatility organic UVCB.

Some substances have hazardous properties that cannot be adequately controlled. These are the 'substances of very high concern', for example PBTs (persistent, bioaccumulative, and toxics) and CMRs (carcinogens, mutagens, and reproductive toxins). For such substances control is based on hazard rather than comparison of exposure with hazard thresholds and the approach is to keep exposure as low as possible.

11.3.2   How Can Risks Be Controlled Adequately?

The term used to describe the ways that risks are controlled is RMMs. RMM under REACH are concerned with reducing risk arising from normal use (as opposed to risk from accidental exposure, for example), from production and throughout the lifecycle of a chemical. RMMs are intended to prevent, reduce or limit exposure and often more than one tier of RMM is needed to control risk adequately. The types of RMM that may be needed to comply with REACH are similar to those needed to comply with other legislation including the IPPC (Integrated Pollution Prevention and Control) Directive.

RMMs needed depend on the type of exposure and the properties of the substance. A chemical with high vapour pressure is likely to require different RMMs than another chemical with low vapour pressure. Similarly a chemical with high toxicity severe effects or a low hazard threshold or both will need different RMMs from one with low toxicity.

Hazardous physico-chemical properties – characterisation of risk

Some substances have hazardous physico-chemical properties (flammable; explosive; oxidising). In the REACH CSA, risks must be assessed. A questionnaire process can be used to establish typical handling procedures in the workplace. Measures to manage risks should be included in the risk characterisation chapter and must be communicated through the eSDS (extended safety data sheet).

A simple means of managing risk is for there to be a barrier between the chemical and the organism at risk (receptor). Barriers include gloves and dust masks, which are used for protection of workers and consumers.

Risk can be reduced by removing the chemical from the local environment. This is achieved by various measures, such as ventilation of various types. For protection of the environment, RMM include treatment of waste water and waste gas treatment.

If the level of risk to workers cannot be adequately controlled by the use of barriers and removal of the chemical from the environment, it may be necessary to limit exposure by reducing the time that the worker is in contact with the chemical by patterns of working, changing shift length or limiting the time that may be spent on certain tasks or in certain areas of the manufacturing facility or formulation plant.

If risks are great because of the type of hazard and or a very low hazard threshold, other measures such as personal monitoring may be needed. The risks associated with some classes of chemicals are considered to be too great for such substances to be available for consumer use.

11.4   Where Safe Levels Cannot Be Established – CMRs and PBTs (and vPvBs)

CMRs very frequently have mechanisms of action which can elicit a toxic effect from even a single-event exposure; in theory, even one mutation or toxic response could give rise to an undesirable outcome. In reality the body has repair mechanisms but, nevertheless, the precautionary approach is that for most of such substances there is no dose-response relationship in which a low exposure can be considered to be acceptably safe. In this case it is hazard that has been assessed and there is no safe level. Therefore, the only logical response is to remove the substance from uses where exposure of individuals can occur. It is possible to apply for an authorised use; this topic is considered in depth in later chapters.

For the environment, the principle of assessment of persistence 'P', bioaccumulation 'B' and toxicity 'T' is a prominent part of REACH. A substance meeting the criteria P, B and T (as defined) simultaneously are considered to be of high concern. In addition, very persistent (vP) substances that are very bioaccumulative (vB) are also of high concern. It should be noted that there is a tiered system of assessment. The various criteria are defined in Annex XIII of the REACH Regulation. A revision of Annex XIII is being enacted which allows a less legalistic approach to be taken by consideration of weight of evidence. This will allow substances not meeting the definitions to still be considered as of 'equivalent concern' (e.g. known endocrine disrupting substances) and also, for others, mitigating factors can be accounted for. In legal terms, a substance only requires authorisation once it is listed on Annex XIV.

For substances such as these, even if a use is authorised, it is quite possible that customers will simply not want to buy one of these 'substances of very high concern'.

11.5   Responsibilities in the Supply Chain – Introduction

One of the fundamental objectives of REACH is to improve communication of hazard and risk in the supply chain. The responsibility to do this lies with the producers and users of a substance.

CHESAR – strengths and weaknesses

The CHESAR (CHEmical Safety Assessment and Reporting) software is a powerful and useful tool to simply and conveniently build exposure estimates using the standard algorithms and methods. Since it automatically extracts input data from the IUCLID (International Uniform Chemical Information Database) file, the convenience of the method is a significant advantage particularly for registrants inexperienced with exposure work. In some cases however, the tool has weaknesses, for example: when evaluating multi-constituent or UVCB substances, inorganics and ionisable or unstable substances.

Hazard is not a controversial or difficult topic; this has been part of supplier responsibilities for many years. Risk is quite another question, especially where the supply chain is long. In this section the legal requirements, and also the practical realities, are reviewed.

The system of codes to categorise a use in terms of sector of use (SU) and processes was intended to help the lines of communication. This is termed the Use Descriptor system and it may be summarised well in this extract from the introductory section of the ECHA (European Chemicals Agency) guidance document R.12 (ECHA, 2010):

> *Seven main groups of actors play a role during the life cycle of the substance: Manufacturers and importers of chemical substances (including metals and minerals), companies mixing and blending chemicals (formulators) to produce mixtures, distributors, industrial end-users, professional end-users, and consumers.*
>
> *The use descriptor system is based on five separate descriptor-lists which in combination with each other form a brief description of use or an exposure scenario title:*
>
> - *The sector of use category (SU) describes in which sector of the economy the substance is used. This includes mixing or re-packing of substances at formulator's level as well as industrial, professional, and consumer end-uses.*
> - *The chemical product category (PC) describes in which types of chemical products (= substances as such or in mixtures) the substance is finally contained when it is supplied to end-uses (by industrial, professional, or consumer users).*
> - *The process category (PROC) describes the application techniques or process types defined from the occupational perspective*
> - *The environmental release category (ERC) describes the broad conditions of use from the environmental perspective.*
> - *The article category (AC) describes the type of article into which the substance has eventually been processed. This also includes mixtures in their dried or cured form (e.g. dried printing ink in newspapers; dried coatings on various surfaces).*

Whilst this is admirably succinct, experience has shown that the detailed lists of SU and PC (product category) are often completely misinterpreted. In fact, DUs (downstream users) are frequently so confused that they tell their suppliers that they want all realistic combinations covered – which would be an impossible amount of work as well as being totally unnecessary.

11.6 Regulatory Requirements

The responsibility of manufactures and importers of a substance is to assess the hazards and exposure associated with the life cycle of the substance. This assessment will enable the RMMs required for adequate control of the risks to be decided. Importers and producers are also responsible for communicating with their customers, giving information on properties and hazards of the substance, and how it should be stored and handled for safe use. Communication is made via Safety Data Sheets (SDSs) (Section 11.8 in this chapter), which enable customers to make their own risk assessment and also explain emergency measures to be taken in case of accident.

There are responsibilities under REACH for DUs. DUs include formulators who receive the substance from the manufacturer/importer or from another formulator. Professional users are also DUs, as are industrial users, who may incorporate the substance into articles or use the substance or formulation as a process aid. DUs will receive SDS from their suppliers and must ensure that their use of a substance for which a hazard has been identified is covered by the exposure scenarios included in the SDS. If their use is covered then they must check whether they are implementing adequate RMMs to control risk. If a use is not covered there are several possible courses of action. The first possibility is to communicate the use up the supply chain so that it can be covered. If they do not want to do that, perhaps for commercial reasons, then they may be able to find another supplier who does cover their use, though this may not be possible especially for a novel use. When the use is not communicated up the supply chain, the DU is obliged to inform the ECHA of this use, and may need to conduct a CSA in order to comply with REACH. DUs would also need to communicate with ECHA if their classification of the substance differed from the suppliers.

DUs may discover that they are not implementing adequate RMMs, in which case they are obliged to put appropriate RMMs in place. This may be the case if the CSA conducted to comply with REACH has identified hazards that were not known about, even though the substance has been in use for some time. This may involve accepting that the hazard that has been revealed means that RMMs must now be used which were not considered necessary in the past.

DUs also need to comply with the classification and labelling requirements of CLP (classification, labelling and packaging) 2008. All hazardous substances must be labelled and packaged in accordance with the requirements of this legislation. Substances that have been labelled and packaged under the Dangerous Substances Directive (Directive 67/548/EEC) need to have their labelling and packaging changed to comply with CLP 2008. All hazardous substances had to be classified under CLP 2008 by December 2010. All hazardous mixtures (known as preparations under previous legislation, Directive 1999/45/EU) must be classified according to the Dangerous Substances Directive and CLP 2008 until 1 June 2015. After that date, only CLP classification applies.

If a substance is an intermediate only used under strictly controlled conditions (SCCs), DUs have to communicate their application of SCCs to the registrant.

There are other situations in which DUs are obliged to communicate up the supply chain. Firstly, if they have new information on hazard, that must be communicated to the supplier. Secondly, they are obliged to inform their suppliers of any information that appears to contradict the RMMs set out in the SDS.

DUs have responsibility for communicating the RMMs to their customers.

Consumers and distributors of substances may be considered as being at the other end of the supply chain from manufacturers and importers. Consumers do not have obligations under REACH but distributors are in a key position for the flow of information. The information distributors are obliged to communicate along the supply chain is described in the Guidance for DUs as including:

• Information related to the identification of uses, either from manufacturers/importers to DUs via questionnaires or from DUs to suppliers, for example via standard brief general descriptions of use.
• Specific requests for information from a DU who wants to make a DU CSR.
• SDS with and without exposure scenario.
• Information on, for example, authorisation of a substance.
• Distributors are also required to keep records.

11.7   Guidance

The ECHA has made available several pieces of official guidance on topics relating to the responsibilities of different players through the supply chain, and communication between them. ECHA guidance documents are free to access through ECHA's web site.

The ECHA's web site also offers a useful interactive navigator tool to assist users in identifying their obligations under REACH and CLP.

The guidance on information requirements and chemical safety assessment (IR and CSA) comprises a number of documents focussing on many separate aspects of preparation of registration dossiers, approaches and data needs.

Chapter 16 of this book provides more detail on the official sources of guidance.

Risk characterisation and sensitivity analysis – which properties can affect outcomes significantly?

Environmental assessment:

Octanol–water partition coefficient
Vapour pressure
Water solubility
Rate of degradation within waste-water treatment plants (WWTPs)
Effect concentrations
Consumption levels and/or release rate (kg/d) from the local site

Human health:

Physical state (at process temperature)
Vapour pressure
Toxicokinetics
Nature of handling
Effect concentrations and DNELs.

11.8   The Extended Safety Data Sheet

Since the adoption of the SDS Directive into Annex II of REACH Regulation, there have been a number of important changes to both the format and information requirements for this document. However, possibly the biggest challenge to both the chemical and allied industries has been the introduction of the eSDS. This is a SDS which contains an annex of relevant exposure scenarios of uses, for which 'safe use' can be demonstrated. The identified uses are listed in Section 1 of the SDS together with a list of any uses or activities where 'safe use' cannot be demonstrated.

The provision of an eSDS is only required for registered substances which are manufactured or imported in quantities greater than 10 tonnes/year and are classified as hazardous under Regulation (EC) No. 1272/2008 (CLP Regulation) or are assessed as being either PBT or vPvB (very persistent and very bioaccumulative). The exposure scenarios, which are annexed to the main SDS, describe, for each identified use, the conditions where a given chemical can be safely used without any risk of harm to man or the environment.

The first section of the exposure scenario provides information on the areas of use that is covered together with other activities that are linked to these areas of use, known as contributing scenarios. The subsequent sections address the recommended operational conditions of use (e.g. temperature, dilution factor of sewage treatment plant, etc.), together with the appropriate RMMs (such as gloves, ventilation etc.) that should be taken in order to achieve safe use for both man and the environment.

It is the responsibility of the DU to ensure that their own uses are covered by the eSDS and that their own RMMs/operating conditions for both human health and the environment are either in compliance with those specified, or equivalent in terms of providing the necessary level of protection to ensure safe use. Scaling may be applied where appropriate to determine whether the actual conditions of use by the DU are equivalent in terms of the required level of protection. The DU also needs to ensure that the relevant information contained within the eSDS is communicated to its customers in the supply chain.

11.8.1   Current issues surrounding the use of eSDS by DUs

At the time of writing there are a number of issues surrounding the use of eSDS by DUs:

1. In cases where there are a large number of identified uses in the supply chain for a given chemical, a large number of exposure scenarios would have to be generated for each of the identified uses; potentially resulting in a document which can be hundreds of pages long.

The recipient of the eSDS then has the problem of finding the information that is relevant to its identified uses as well as handling such a lengthy document, which even if sent electronically can problematic. Although the obvious solution would be to send what could effectively be considered as a 'tailored eSDS' for that particular DU, the practicality of doing this may not be so simple.

2. Another issue is related to the information that is provided in the exposure scenario and how to interpret it, particularly in cases where the operating conditions or RMMs are not exactly the same as those which are being used.
3. In addition, it is not often clear how the exposure scenario ties in with the main body of the SDS, meaning that important information that is contained in the 16 sections of the SDS could be overlooked in lieu of the information that is presented in the exposure scenario.

It is envisaged that with time these issues as well as others will be resolved the more familiar both authors and users of eSDS become with this new format.

11.9   When Communication Is Difficult

Table 11.1 sets out a quite common scenario which describes a set of commonly-experienced issues and some possible solutions.

Table 11.1 *Example problems with DU communication with possible solutions.*

Issues	Possible solutions
The producer has a broad idea of the supply chain but does not know all the details; the downstream user does not want to give away its intellectual property to a supplier who could become a competitor. Sometimes suppliers have unrealistic expectations of the DUs	It should be possible to use available information, including the descriptor code system, to establish a reasonable exposure scenario. Should a risk be identified then the DU will need, unavoidably, to try to refine the scenario. It is possible to refine scenarios by estimation of release levels without giving away commercially-sensitive information
The DU does not understand the technical details of REACH and so cannot assess what the supplier is doing	The only ways forward here are to discuss with the supplier or to acquire the expertise
The DU has a somewhat different use pattern to that described in the eSDS	The guidance sets out the principles of scaling very clearly, as long as the use pattern is reasonably similar
The DU is a formulator who has to assess a formulation based on information from several suppliers	Assessment of hazard and classification of mixtures (formulations) is well established in the CLP regulation. Assessment of combined risk is more difficult and in reality the DU will need access to significant expertise. The starting place is to make sure all risk characterisation ratios have been calculated on an equivalent basis, and then to sum the RCRs
The supplier has no information about the uses of the substance and the DU will not provide it	The supplier must clearly pass responsibility to the DU

11.10 Exposure Measurements in the Workplace – Occupational Hygiene

The British Occupational Hygiene Society (BOHS) defines that 'occupational hygiene is about the prevention of ill health from work, through recognising, evaluating and controlling the risks' (BOHS, nd). According to the BOHS, occupational hygiene is concerned with ... 'the assessment and control of risks to health from workplace exposure to hazards'. The hazards could be chemical, biological or even physical in nature. Occupational hygiene involves the anticipation and recognition of a potential or existing hazard in the workplace that could cause harm to health.

If a potential or existing hazard has been identified it is then necessary to evaluate whether exposure to the identified hazard is above or below a recognised standard; if it exists. In many cases there may not be a formal standard that can be used, which means that professional judgement is required. Finally, the hazard would need to be controlled in order to eliminate or reduce exposure. Under REACH the use of RMMs and/or changes in operating conditions are used.

In the context of chemicals and REACH, this can be taken to mean understanding and measuring sources of exposure in the workplace. It is, therefore, an experimental activity. REACH is often dealing with generic scenarios or exposures that are too low to be measured. However, there is a clear overlap.

Workplace measurements are often made to ensure that statutory workplace exposure limits (WELs) or occupational exposure limits (OELs) are being met. Such measurements may be very useful for REACH compliance. This can be for the specific substance or be applied to predict exposure for another substance for which no measurements have been made.

11.11   Control of Environmental Releases – Abatement Techniques

Chemical safety assessment for REACH purposes may reveal a need to control environmental releases in order to manage risks. In some cases this may be completely unforeseen. Many substances manufactured and used at high tonnage were simply not well characterised for hazardous effects prior to REACH. Furthermore, environmental properties and ecotoxicity are more likely to have been poorly understood than human health properties prior to REACH data requirements entering into force.

Any individual site in European Union at which chemicals are manufactured, handled or processed is likely to have duties and responsibilities under legislation other than REACH. Regulatory systems such as IPPC are wide reaching and cover installations undertaking a wide range of activities involving chemicals. IPPC is oriented to assessment of the site as a whole in the context of its local area. IPPC is administered by the relevant national competent authority of each country, for example in England and Wales, the Environment Agency.

11.11.1   Engineering Controls

Control of chemicals through use of engineering methods involves investment in technology within the installation.

Methods to reduce releases to the environment often include a degree of enclosure of the chemicals inside equipment within the plant. Examples include use of dedicated pipework and equipment; technology to fill packaging in a manner that avoids volatilisation and spillage. If spraying is necessary, enclosed chambers may be used. Enclosure reduces the possibility of volatilisation, spillage and incidental losses; it may also reduce waste.

In reality many processes involving handling and processing chemicals will mean multiple different substances or chemical constituents being handled together, possibly in the form of a blend or multi-constituent substance. The different chemical components or constituents may have very different chemical properties and this might mean different types of controls are more effective. For example, a volatile chemical present in a blend may tend to volatilise, whereas a chemical of high log K_{ow} present in the same blend may tend to adsorb to surfaces and predominantly be released after equipment cleaning processes.

11.11.2   Enclosure and Containment

Conversion to partially or fully closed processing is often the step with most significant impact on releases, regardless of chemical type and nature of hazards. Clearly, depending on the nature of the processing, enclosure may be difficult or inappropriate. Its greatest relevance is perhaps in manufacturing and formulation industries.

11.11.3   Bunding

A simple and effective step is to build bunds around designated areas where chemicals are loaded/unloaded, stored, handled, processed and filled into packaging. A bund is a lined enclosure built into the ground or floor to avoid uncontrolled spreading of liquids and solids around the installation. Incidental losses by spillage, leakage and waste fall within the bunded area and are retained there for collection and appropriate disposal.

11.11.4   Dedicated Equipment

The use of a set of equipment specific to one manufacturing process can have benefits such as reduction in wastage due to the possibility to always recycle substances from cleaning operations back into the next batch. When multipurpose equipment is used, cleaning wastes would be more likely to pass to drain or be sent for disposal. Therefore, in addition to reducing levels of polluting wastes, there can be advantages in terms of improving yields.

11.11.5   Investment and Scale of Use – Economic Viability

For a small chemical processing operation it may be questionable whether the benefits of installing significant new technology can outweigh the costs. Considerations may include what proportion of the business is associated with the affected substance and whether other substances in the same formulated product also require control measures.

11.11.6   Waste Stream Treatments

11.11.6.1   Scrubbing of Off-Gas Streams

Waste gas can include chemical contaminants and if vented directly to atmosphere can contribute to environmental concentrations. Off-gas scrubbing and filtration by measures such as activated charcoal cartridges can significantly limit the releases of substances to air. It should be noted that if washed, higher releases to waste water may be possible even while releases to air are significantly reduced.

11.11.6.2   Recovery and Recycling

Many companies use technological methods to recover valuable substances from residues or waste streams which are then recycled back into the process. Examples might be the recycling of rinsings from equipment cleaning in subsequent formulation batches and separation of oil from emulsified metalworking fluids for reuse. Effectively removing at least a proportion of the residual substance from waste streams would normally reduce the wastage, and hence the level of releases.

11.11.6.3   Oil/Water Separators and Interceptors

Sites processing oils can produce waste waters contaminated with high levels of oil. To avoid loading treatment plants and watercourses with high organic loadings, an interceptor unit may be built in to the waste stream. This allows the majority of the oil to separate from the waste water as a second phase, to be removed and disposed of separately.

11.11.7   WWTP Treatments

Waste waters containing organic chemicals can be treated alongside municipal waste waters. Larger sites may have facilities on-site for primary or in some cases secondary biological treatment. Types of treatment can often include physical (e.g. settling, filtration) and chemical methods (e.g. pH neutralisation).

11.11.8   Custom and Practise

Depending on the standards already in place, it may be the case that simple improvements to daily practise can make a significant difference to release levels.

11.11.9   Handling Standards

Typically, an environmental exposure assessment should work on an assumption of a low level of spillage during every day handling the substance in the workplace. While major accidents are outside the scope of REACH, low levels of unintentional releases during loading, transfer and packaging are within its scope. It is entirely normal to account for these.

Types of handling release would include, for example, small spillages of raw materials and finished products; minor leakage, settling of suspended dusts and condensation of off-gas and vapours.

11.11.10   Clean-Down Practises

Improvement to cleaning methods can have a significant impact, depending on the standards already in place. Improvements can include: collection and specialist disposal of contaminated cloths and clothing, choice of cleaning solvent (or water) and appropriate disposal of the used solvent/water.

Table 11.2 presents a quick reference summary of the typical efficacy of different measures to reduce emissions to the environment (more stars indicates higher impact on limiting the level of releases, and may be independent of control measures already in place).

The implementation of the IPPC Regulation is supported by best available techniques reference documents (BREFs), which provide a wealth of advice relevant to REACH (the following list is from the official source at http://eippcb.jrc.es/reference/):

Reference document
Cement, Lime and Magnesium Oxide Manufacturing Industries
Ceramic Manufacturing Industry
Common Waste Water and Waste Gas Treatment/Management Systems in the Chemical Sector
Economics and Cross-media Effects
Emissions from Storage
Energy Efficiency
Ferrous Metals Processing Industry
Food, Drink and Milk Industries
General Principles of Monitoring
Industrial Cooling Systems
Intensive Rearing of Poultry and Pigs
Iron and Steel Production
Large Combustion Plants
Large Volume Inorganic Chemicals – Ammonia, Acids and Fertilisers Industries
Large Volume Inorganic Chemicals – Solids and Others Industry
Large Volume Organic Chemical Industry
Management of Tailings and Waste Rock in Mining Activities
Manufacture of Glass
Manufacture of Organic Fine Chemicals
Non-ferrous Metals Industries
Production of Chlor-alkali
Production of Polymers
Production of Speciality Inorganic Chemicals
Pulp and Paper Industry
Refining of Mineral Oil and Gas
Slaughterhouses and Animals By-products Industries
Smitheries and Foundries Industry
Surface Treatment of Metals and Plastics
Surface Treatment Using Organic Solvents
Tanning of Hides and Skins
Textiles Industry
Waste Incineration
Waste Treatments Industries
Wood-based Panels Production
Wood and Wood Products Preservation with Chemicals

Table 11.2   *Summary of typical efficacies of possible abatement measures.*

	Inorganics	Volatile organics	Lipophilic organics	Moderately hydrophilic-lipophilic organics	Hydrophilic organics
Closed systems	****	****	****	****	****
Dedicated equipment	**	**	**	**	**
Off-gas scrubbing	–	***	*	*	*
Oil-water interceptors	–	*	***	**	*
WWTP	–	*	**	**	*
Improved handling	*	*	*	*	*
Improved cleaning	**	**	**	**	**

Measurement of concentrations in the workplace or in effluents may be used but they must be representative and statistically robust in terms of numbers of samples and their acquisition timing.

It may be appropriate and possible to conduct monitoring of releases leaving the site, for example monitoring of waste-water streams. Sites may conduct monitoring routinely in the context of other regulatory needs in the area of pollution control compliance (Chapter 3 of this book). In REACH such data may be useful to bypass some of the discussion on potential release fractions and control measures and use direct evidence from a monitoring programme to demonstrate the kilograms per day release or even the PEC (predicted environmental concentration). The data may also be useful to identify abatement techniques that could be useful in future.

For use in REACH, standards are high and registrants relying on monitoring data in their submissions must ensure it is shown to be representative, repeatable and well documented.

The ECHA has set out standards and considerations for collection of site environmental monitoring. The details can be found in part R.16 of the guidance on information requirements and chemical safety assessment (ECHA, 2012a). It sets out the relevant details to ensure that the monitoring programme:

- Is well designed in terms of sampling sites, timing and frequency and representativeness.
- Uses well-validated analytical methods against appropriate measurement objectives.
- Is correctly interpreted according to the context, for example making a conservative interpretation rather than applying averages.

11.12   Effectiveness of Risk Reduction – Risk Management Options

If the processes of exposure assessment and risk characterisation lead to the position that unacceptable risks exist for the substance in its present usage pattern, it is then necessary to take a course of action to reduce the risks. At the earliest stages of the process this course of action may still involve re-examining the estimations of exposure in case the predicted levels are too conservative and require refinement. The exposure associated with any particular type of handling and processing may vary considerably, depending on which of the available tools is used to model the exposure.

Before taking costly new action, the estimated exposures should be refined as far as possible (and, preferably, validated by obtaining reliable monitoring data). However, if the unacceptable risks remain, then the next step must be new measures to actively reduce the potential for exposure. This may involve application of new technology such as engineering controls, improved waste treatment methods, implementation of new personal protective equipment for workers and reformulation to reduce or even replace the substance of concern. These approaches are discussed in more detail in this section.

Should it become necessary for risk management to be enforced at a regulatory level, this would normally entail statutory or permit-driven controls on manufacture and use in order to limit the tonnage or type of end use.

This section discusses various options, their feasibility and typical efficacy to reduce exposure and control risks. Whatever method is selected, ultimately the aim is to reduce exposure to humans (as consumers or in the workplace) or to the environment, or both. Various aspects of strategy must be considered in order to plan to manage the risks effectively.

- The nature and severity of the hazard. In some cases, for example most carcinogens, PBT substances and so on, it is not normally possible to establish a safe threshold, which means that conventional risk characterisation is impossible.
- How large was the RCR found to be? An RCR that is above one (and hence unacceptable) but is still small in value, for example approximately 10 or less, means the exposure level will need to be reduced by only a small margin in order for the risks to be adequately controlled. Some types of RMMs offer only a small correction factor, so if the RCRs are found to be very large more drastic and widespread action is likely to be needed.
- How widespread is the problem? Unacceptable risks at only one or two specific industrial locations may be relatively straightforward to remedy, if the RCRs are not excessively large; whereas RCRs exceeding one for a large number of widespread dispersed downstream uses, particularly for consumers, can be extremely difficult to resolve. For the environment, if the RCR exceeds one at the regional level, this is particularly concerning.
- There may be complex business decisions for an individual supplier or DU depending on the additional costs of implementing the required RMMs (technology costs, staff training etc.).

In the case of any formulated products, it is necessary to consider risks from all components of a formulation. A very careful assessment should be made before expensive new measures are put into place.

11.13   Types of Risk Management – in the Workplace

This section concerns risk management where both an exposure concentration and a predicted DNEL for the human protection goals can be set, even if in some cases it is only a limit value. It does not discuss 'substances of very high concern'.

This section is concerned with the stage in risk characterisation when there is no further possibility to control risk under normal use patterns, and also describes some of the steps along that path. Risk management strategies depend to a considerable extent on the completeness of the property data set of the substance and the amount of information available about releases under normal conditions of use.

11.13.1   Options Overview

For human health, the amount of waiver based on 'exposure-based adaptation' is much less than for the environment. However, because the time scale for generation of long-term animal data can be significant, there will be times when a DNEL is not finalised pending arrival of new data – which could be less adverse or sometime more adverse.

11.13.2   Understanding Assumptions and Critical Issues

Step 1 is to get the best possible understanding of any modelled releases. What model has been used? Is it appropriate or not? Some models are time-based; others assume that the exposure concentration is achieved instantly.

Step 2 will be the level of releases themselves: how does the exposure model deal with duration, engineering controls such as ventilation or personal protective equipment such as gloves or respirators?

Step 3 may have to be elimination of the use of the particular substance. This is more complex in legal terms, and is described elsewhere in this book.

11.13.3   Risk Management Measures

To ensure that exposure to a hazardous chemical agent is either minimised or eliminated it is important to ensure that the correct RMMs are implemented. Ultimately, these are likely to be a combination of equipment and safe working practices together with appropriate training. However it should be noted that any RMMs that are taken should also be proportionate to the actual health risk.

The general approach as shown below is described in more detail in the ECHA Guidance document R.13 (ECHA, 2012b).

1. Elimination of the risk – this could be by using a suitable less hazardous alternative or by the use of totally enclosed systems.
2. Reduction of the risk – For example, altering the form in which a substance is used; sodium hydroxide supplied in the form of pellets rather than powder.
3. The use of ventilation such as local exhaust ventilation and other modifications to the processes that would reduce the risk of exposure.
4. The introduction of alternative work practices which reduce the number of employees who could potentially be exposed together with a reduction in the duration and level of exposure at any time.
5. The introduction of appropriate personal protective equipment.

11.14 Types of Risk Management – for the Environment

This section concerns risk management where both a PEC and a PNEC for every environmental compartment can be set, even if in some cases it is only a limit value. It does not discuss 'substances of very high concern'.

This section is concerned with the stage in risk characterisation when there is no further possibility to control risk under normal use patterns, and also describes some of the steps along that path. Risk management strategies depend to a considerable extent on the completeness of the property data set of the substance, and the amount of information available about releases under normal conditions of use.

11.14.1 Unacceptable Risk

Under REACH there are several environmental protection goals and an RCR value (PEC divided by PNEC) for any of these is unacceptable. However, the response to the problem may be influenced by which compartment is the cause of concern. Also, the magnitude of RCR relative to one should be considered, because it shows how severe the measures to solve the problem might need to be. It can also be the case that several protection goals have RCR values above one that could be 'fixed' by a single remedial action. It may also be the case that more than one such remedial action is needed to fix one or more concerns.

11.14.2 Options Overview

The options available to solve a particular RCR above are are described simply:

• Increase PNEC
• Reduce PEC
• Both of these.

It could be imagined that the PNEC is a 'given', which nothing can be done about it. That is rarely true of most substances. Even for the simplest case of toxicity to freshwater aquatic organisms, there is progression:

1. Acute studies
2. A standard set of three long-term studies (ECHA, 2012b)
3. Progressively more long-term studies allowing a lower assessment factor to be applied to the lowest NOEC (no observed effect concentration).

However, there will be a limit. Sometimes the limit is a scientific one – that no more studies can change the conclusion already reached. It can also be economic. Provided the mandatory REACH Annex requirements applicable for the tonnage band are met, then in some cases the optional studies might simply be omitted because they are too expensive for the registrant.

11.14.3 When a Data Set Is Not Complete

In the discussion above, the example of progression of aquatic toxicity testing was given. The same can also be true for studies on which terrestrial, sediment, and secondary poisoning PNEC values are based. The same type of principle applies to degradation testing, which is an important part of producing PEC values. Consider biodegradability in water; again there is a hierarchy of tests available:

1. Ready biodegradability
2. Inherent biodegradability
3. Laboratory simulations of waste-water treatment
4. Laboratory simulations of behaviour in the environment, in water, soil and sediment
5. Studies in the field.

Judgement is needed as to whether a proposed test might reduce the PEC by enough to solve the problem. Exposure models can be used to see outcomes from realistic scenarios. It certainly cannot be assumed that any study will improve the RCR; it could even make it worse.

11.14.4 When a Data Set Is as Complete as It Can Be

Consider the case when the data set has reached the point where the intrinsic properties are unlikely to be changed. If the RCR is greater than one then the only option is to reduce the predicted exposure (PEC) value or values. If the property data set is fixed, then the modelling of the environmental fate and behaviour of the substance is fixed, and the only area of discussion remaining is the size and scale of releases to the environment. Remember, reducing one area of release can fix several problems.

11.14.5 Understanding Assumptions and Critical Issues

Step 1 is to get the best possible understanding of any modelled releases.

- What is the spatial scale of the release to the environment? Is it widespread or to a limited number of point sources?
- Which releases have an influence on the modelled PEC? For example, PEC in soil is influenced by releases to waste water followed by spreading of sewage sludge; background concentrations; releases to air (with subsequent deposition onto soil from the air).
- Could measurements of the amount of substance being released help?
- What are the critical assumptions in the models that have been used?

Step 2 will be the level of releases themselves based on the use pattern and with use of more advanced ('higher Tier' models).

11.14.6 Strategies to Reduce the Amount of Substance Released to the Environment

11.14.6.1 General

Assume that all the above stages of gathering of property data and the understanding of models have been taken, and the RCR or RCRs are still stubbornly above one. The only option left is to try to find some agreed control on the level of release. For simplicity, this discussion assumes that the unacceptable RCRs are at the local scale, not regional. A regional RCR above one suggests a very intractable problem requiring widespread controls.

Consider some examples:

- Is a process change possible? For example, a change of temperature or by use of more enclosed engineering systems? The latter can include sealed systems, traps and the like.
- In a case where loss is an irreducible proportion of the amount of substance, can the quantities per day be reduced, or the number of sites of use increased? This would reduce the loss per day at a single location.
- In a case where loss is a fixed mass regardless of the amount of substance in the process, can the quantities per day be increased, or the number of sites of use decreased? This would result in a smaller fraction of wastage over time. This can be important in respect of losses in storage.
- Can waste streams be recycled back into the process itself?
- Can waste streams be subject to higher levels of purification, by use of extra treatment stages, such as settling tanks, oil–water separators, charcoal filters, biological treatment?
- Can waste streams be eliminated by such techniques as incineration?
- For a formulation stage, can the composition of the formulation be amended to reduce the amount of the substance of concern? Can the formulation be done by the chemical manufacturer in an integrated (enclosed) way, so as to reduce the potential for losses in handling and transfers?
- Is packaging managed optimally so as to reduce environmental exposure?
- Can the total tonnage be limited (without breaking competition law)?

11.14.6.2 Water and Sediment

The general points apply. Also, can means to dilute the waste be found (in respect of its final point of release)?

11.14.6.3 Soil

The general points apply. Also, is it possible to arrange for sewage sludge to not be spread onto soil, but to be burned?

11.14.6.4 Secondary Poisoning

See water or soil, depending on the secondary poisoning pathway under consideration.

11.14.6.5 Substitution

Step 3 may have to be elimination of the use of the particular substance. This is more complex in legal terms, and is described elsewhere in the book.

11.15 Consumer Protection

Consumers can, in principle, be exposed to a wide range of substances. Indeed, there are many regulations to protect consumers, including pharmaceuticals, pesticides, biocides, cosmetics, plastics additives, food additives and so on. Some of these are in part or wholly excluded from REACH.

Use by consumers must be taken into account in the CSA. In reality exposure models are not easy to amend when there is widespread use by consumers. However, amounts of substance are usually low. Should a risk be estimated then if the exposure scenario is realistic, then usually the only recourse is to propose maximum amounts of the substance.

References

BOHS (nd) The Chartered Society for Worker Health Protection, http://www.bohs.org/ (last accessed 25 July 2013).

CLP (2008) CLP-Regulation (EC) No 1272/2008.

ECHA (2010) Guidance on Information Requirements and Chemical Safety Assessment. Chapter R.12: Use Descriptor System, Version 2, March 2010.

ECHA (2012a) Guidance on Information Requirements and Chemical Safety Assessment, Chapter R.16: Environmental Exposure Estimation, Version: 2.1, October 2012.

ECHA (2012b) Guidance on Information Requirements and Chemical Safety Assessment. Chapter R.13: Risk management measures and operational conditions. Version 1.2, http://echa.europa.eu/documents/10162/13632/information_requirements_r13_en .pdf (last accessed 25 July 2013).

12

Avoiding the Use of Hazardous Substances: Substitution and Alternatives

Substances are typically used in the chemical industry either as intermediates in a synthesis of an 'effect chemical', or they are the end product itself. Intermediates in a synthetic process are not often a matter of choice, but their means of production may be. The particular substance may not be the only way to fulfil the need; indeed, there are almost always competitor products or methods. Competitors may be very similar substances or very different technologies. Consideration of the most acceptable method of meeting a need in respect of health, safety and environmental acceptability is becoming increasingly important.

This chapter expands on these principles, focusing on the end product. An example would be flame-retarded board used in the interior or buildings. There are a variety of polymers and inorganic fibres used to make board. For each of these broad types of product there may be different solutions concerning the required flame retardancy. Regulatory and commercial pressures combine as part of the decision about which type of board to use.

This chapter is important because it is a clear objective of the REACH regulation to reduce the use of hazardous substances.

When a substance possesses hazards, then good sense, as well as regulation, dictates that particular care needs to be taken in the use of the substance. Companies with a responsible attitude towards health and environmental safety will always prefer to use less hazardous substances, when an alternative exists. Beyond such sound principles of good practice, the regulatory-driven need to control use of a substance follows as a result of the identification of particularly severe hazards and risks. It is very clear also that there can be economic benefits to the use of less hazardous substances. These arise for several reasons:

- Users prefer them, so sales can be higher.
- The costs of risk reduction measures can be high.
- The costs of higher tier testing associated with hazardous substances can be high.

This chapter deals with some of the principles of both voluntary and imposed substitution of hazardous substances. Sections 12.6 and 12.7 of this chapter on 'Green Chemistry' and related sciences take a broader look at sustainability in substance and process improvement.

Substitution presents a number of challenges to affected industries. Where possible, substitution with a chemically similar product can have a significant advantage, as the changes to established formulations and technology may be minimal. However, a similar product may actually share the profile of hazardous characteristics of the restricted substance.

When considering if it is possible and appropriate to substitute an alternative substance as a direct 'drop-in' replacement, the business interests of individual suppliers of the restricted/withdrawn substance are likely to be a factor. Voluntary or early withdrawal of a hazardous substance may afford a competitive advantage due to positive marketing. However, investment costs in identifying and implementing a suitable substitute could be high. Whether the step is viable for a specific company may depend on whether it has interests in multiple hazardous substances, how amenable its products are to reformulation without significant new technology, and whether its portfolio already includes any potential substitutes. Chapters 13 and 14 build on this in respect of the formal regulatory processes defined in the REACH Regulation for risk management including socio-economic analysis.

It can be seen from the progress of REACH to date, that the topics covered in Chapters 12–14 are already becoming a focus of serious research effort at one end of the cycle, through to new regulatory controls.

The discussion in this chapter also draws on the experiences and practices of industries preceding REACH or exempt from it.

Does Regulation help or hinder innovation? In this book, Chapter 15 deals with the impacts of REACH. However, in the context of innovation, a recent report makes for interesting reading (Tunak, 2013). The report concluded that:

- There is consumer demand for safer chemicals.
- There are commitments from globally recognised brands promising to phase out certain hazardous chemicals.
- Following a regulatory intervention, there is an increase in the numbers of patents relevant to the possibility of substitution.
- Replacements must be thoroughly assessed.
- Restriction of the production of a hazardous chemical can cause its cost to increase, therefore making safer alternatives more economically competitive.
- Expenditure in R&D has actually increased in order to find suitable REACH-compliant substances (no data given).

12.1   Properties That Contribute to Hazard and Risk for Human Health and the Environment

Earlier chapters have defined and described hazards and risk in some detail. Although detailed exposure assessment is usually needed within REACH, it is often the case that substances which require substitution possess particularly adverse properties or combinations of properties, for which normal control measures are insufficient.

There are some potentially serious hazards that are quite manageable without exceptional measures. Even flammable solvents are handled by consumers, for example. More unusual hazards that can be controlled routinely (by engineering controls) include pyrophoricity and explosivity, in addition to flammability.

The application of engineering controls can also provide sufficient protection against human exposure to highly toxic substances such as carcinogens, or to sensitisers. REACH even allows for carcinogens to be used if controls are strict enough. However, some hazards are sufficiently concerning that authorisation of the use of the substance may be needed, because complete control cannot be achieved. Also, the imposition of controls may not always be realistic, practicable or achievable. This is discussed further below.

Properties that are indicators of particular problems in achieving safe use can include:

- For health:
  - Very toxic substances
  - Carcinogens, mutagens or reproductive toxins.
- For the environment:
  - Substances with severe ecotoxicological effects
  - Highly bioaccumulative substances
  - Highly persistent substances.

It is not inevitable that possession of such properties will lead to damage to human health and/or the environment. However, there may be business or regulatory reasons to avoid their use. These possibilities are discussed further in this chapter.

Which kinds of uses cannot be controlled completely? These include:

- Industrial uses that are not completely enclosed – which in reality means most uses outside the top end of the chemicals sector. It is frequently not realised that environmental exposure can occur over time due to volatilisation and dusts even from relatively involatile substances. Activities such as maintenance, storage, polymerisation, and formulation are often not completely enclosed.
- Substances or preparations used by professional users: this means users that are not within the chemical or other major industries, where the level of training found in industry is unlikely to have been taken by that user. Examples include the construction sector, trades, shops and suppliers, cleaning services, small garages.
- The general public, that is consumers' use of substances and preparations. Chemical substances can be found in many products, such as detergents and cleaners, inks, polishes, paint and paint products, textile treatments, solvents.
- Substances found in articles such as batteries or ink cartridges, furniture, toys, and electrical goods. Polymers usually contain additives, which will be lost from the article

over its life time, by processes such as volatilisation, leaching through contact with water and abrasion.

• Waste, for example electrical equipment, recycled paper and plastic goods at the end of their lives, when they are disposed of or recycled.

12.2   Assessment of Alternatives – Replacement of Use

This section considers what the alternatives are and how they can be assessed for consideration. Detailed socio-economic analysis may be part of the management of risks. However, this section is aimed at providing help and direction before such external regulatory actions are even needed at all.

Chemicals that are successful in the market do well because they have properties that are useful, at an acceptable price. Their hazards and risks, where known, need to be acceptable too. Sometimes risk assessment and socio-economic analysis may reveal that the best option is to substitute a particular chemical for another. Legislation such as the REACH Regulation and the Carcinogens Directive necessitate detailed consideration of the replacement of the most hazardous substances with more acceptable alternatives. Even if there are no regulatory pressures to substitute, there are economic benefits to the substitution of hazardous substances for 'green' alternatives.

However, this is not straightforward because 'off the shelf' substitution is rarely possible, and the new research and development needed for substitution is costly and time-consuming. The best way forward is dependent on the specific chemical, as shown in Figure 12.1.

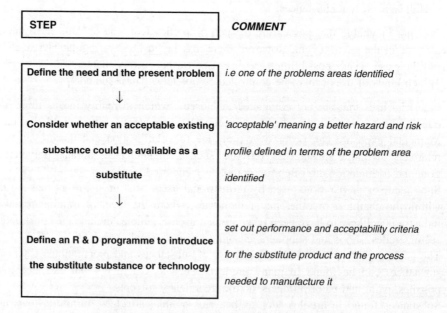

Figure 12.1 *Overview of Substance Substitution*

12.3 What Is an Alternative?

An alternative is not necessarily the replacement of one substance by another. Even in the simplest of applications, the *whole technology* or whole product is significant, not just the individual part. This is true in respect of both the technology and also acceptability. Therefore, when considering a replacement of a substance there will inevitably be consequences for the whole application. This fact is explored further below, and in the examples shown in Section 12.5 of this chapter.

Therefore, a *valid* alternative is one which represents an overall improvement. For example, there may be no benefit in replacing a hazardous substance with a less hazardous one should the second substance brings other types of concern. Introduction of a less toxic substance may be of little benefit if it requires a very dangerous process to manufacture it!

12.4 Analysis of Alternatives

Some general areas of consideration can be identified:

- Acceptability of a substance in respect of health and environmental safety.
- Substitution may necessitate a change in the whole formulation, usually for physico-chemical reasons.
- Substitution may require technology change, even for relatively simple cases; a change in physico-chemical properties, such as viscosity, density or vapour pressure, can cause a need for amended engineering methods and process conditions.
- A second substance is very unlikely to have exactly the same technical performance as the first, so new performance specifications will need to be researched and defined.

The analysis of alternatives requires consideration of the relative strengths and weaknesses of a replacement technology compared to another. Ultimately, socio-economic analysis might be needed. Considerations should include:

- Is there a regulatory need now, or will there be in future?
- Will there be a technology change?
- With a new technology, customers and/or suppliers will need to be involved. Will research and development work be needed?
- Will there be training needs?
- Will there be costs associated with establishing the regulatory acceptance of the new technology?

12.5 Substitution – Replacement with Substances of Reduced Hazard

Substitution of identified hazardous substances tends to be driven through one of three mechanisms.

1. *Voluntary activity*: there are various examples of industry groups taking action voluntarily to phase-out and substitute substances of concern. Processes such as the global ICCA/OECD (International Council of Chemical Associations/Organisation

for Economic Co-operation and Development) HPV (High Production Volume) assessment programme (and its equivalent within the USA) can help to highlight issues of concern for high-volume chemicals. In cases of voluntary action it is not unusual for the concerns to have come also to the attention of regulators, so that statutory restrictions may often follow after.

2. ***Statutory controls for general chemicals***: legislation setting out restrictions on manufacture and supply of substances identified as presenting unacceptable risks. Prior to REACH the major legislation for this purpose was the Marketing and Use Directive (M&UD) (76/769/EEC, as amended) (the controls established under this legislation were adopted as Annex XVII to the REACH Regulation). Legislation oriented towards substances presenting specific and severe hazards are also relevant in this regard, some examples being the Carcinogens and Mutagens Directive (1999/38/EC, as amended) and Persistent organic pollutants (Regulation (EC) No. 850/2004, as amended).

3. ***Application-specific controls***: legislation oriented towards certain industry sectors is also relevant to restricting named substances or setting general rules. Examples include the Restriction of hazardous substances in electrical and electronic equipment Directive (RoHS) (2011/65/EU[1]), the Pesticides Regulation (EC) No. 1107/2009 and the Biocides Regulation (EU) No. 528/2012.

A range of examples and case studies is presented here.

12.5.1   Examples of Voluntary Substitution

Perfluorinated alkyl chains have very unusual chemistry, being both water repellent and oil repellent, which makes them useful in a variety of applications involving surface treatment. ***Perfluorooctylsulfonates*** were used in high volumes worldwide but their accumulation in humans and wildlife was identified in the 1990s and their hazardous effects profile was also becoming clearer (carcinogenic, reprotoxic and harmful to breast-fed children, causes organ damage on repeated exposure, acutely toxic if swallowed or inhaled, and environmentally toxic). They came to the attention of regulators around the world but leading market suppliers acted voluntarily in 2000 to phase out PFOS (perfluorooctane sulfonate) and substitute structural analogues which lack the properties of concern. In 2009 PFOS was added to the list of persistent organic pollutants under the Stockholm Convention.

Sulfur compounds in ***engine oils and fuels*** are readily oxidised (particularly on burning of the fuels) causing sulfur dioxide to be emitted to the atmosphere. Among other issues sulfur dioxide emissions were directly linked to 'acid rain' and highly damaging, causing rapid weathering to masonry. In response to these issues, refineries introduced new technology to significantly reduce the content of sulfur compounds in refined petroleum streams.

High levels of polyaromatic constituents in ***petroleum-derived engine oils*** presented issues of toxicity. Synthetic engine oils, as well as having a much longer service life, lack these aromatics and are, therefore, considered to be a 'cleaner' technology. This is a perfect example of how a better technology has many benefits.

---

[1] Updating 2002/95/EC.

The ***cement industry*** uses large volumes of virgin raw materials which are extracted and require high energy inputs to manufacture the end products. Alternative raw materials inputs from other industrial processes, such as fly ash from power stations, furnace slag from steelworks and gypsum from desulfurisation in coal power plants, are now in use in cement manufacturing. A wide range of combustible waste is also used as fuels in kilns during cement manufacturing.

Biopolymers are increasingly being developed to replace polymers made from petrochemicals. However, the design of novel processes to address the high cost of producing biopolymers compared to polymers derived from petrochemicals is key to achieving industrial scale growth of the biopolymer sector. This touches on an area where replacement of a petrochemical product (e.g. diesel fuel from crude oil) with a natural material (biodiesel) may have advantages in respect of being less toxic but its production has wider impacts on the environment that are undesirable, such as planting crops instead of trees.

The ***manufacture of ammonia*** through the Haber process, using a magnetite catalyst, led to higher energy efficiency compared to the original electric arc process, operated at temperatures of over 3000 °C.

Further changes in fuel source from coal, oil and later gas led to improved ***energy efficiency***. Significant contributions to energy efficiency have also been made through improved steam distribution networks and flow converters with small-sized catalyst, without major changes to the chemistry of the process.

12.5.2   Regulation-Led Substitution – Case Studies

Hexavalent chromium (chromium in oxidation state VI) was formerly used extensively in the metals industries, particularly in anticorrosion coatings. It carries harmonised classifications for carcinogenicity, sensitisation and environmental toxicity. Its uses are restricted in the RoHS legislation (2011/65/EU) through the imposition of a maximum limit of 0.1% by weight in homogeneous materials. Additionally, chromium is often present in cement; in REACH (and formerly in the M&UD), levels of Cr(VI) in cement is limited to 0.0002% by weight.

Tetraethyl lead was used for several decades as the preferred additive in petroleum fuels as an antiknock agent. Lead has a number of toxic effects (nerve toxicity, reproductive/developmental effects, acute and repeated-dose hazards, environmental toxicity). The withdrawal of tetraethyl lead was partially driven by the need for petrol-engined road vehicles to comply with legislation on exhaust pollution, as tetraethyl lead was damaging to catalytic converters, though banning legislation with regard to lead pollution brought in the 1980–1990s was also instrumental. Technological changes in car engines were necessary, particularly changes to valve design.

Regulation (EC) No. 648/2004 brought together a number of prior legislative acts of the 1970s and 1980s to set minimum standards for European Union (EU) of ultimate biodegradability of ***detergents and surfactants***. Branched alkylbenzene sulfonates (ABS) are an example of a detergent commonly used in the 1950–1960s prior to control legislation, under which it was banned due to a lack of biodegradability causing accumulation in the environment. Readily biodegradable linear alkylbenzene sulfonates (LAS), which met the required standards, started to come into use in the 1960s and expanded in the market in response to the banning of ABS.

Bis(2-ethylhexyl) phthalate (DEHP) and various related phthalates were commonly used plasticisers particularly in PVC (polyvinyl chloride). It is a reproductive toxic. In MU&D and REACH Annex XVII, these phthalates are restricted when used in toys and childcare articles, to a maximum concentration of 0.1% by weight of plasticised material.

Tetrabromobisphenol-A (TBBP-A) is an additive and reactive flame retardant used in a wide range of resin plastics. It is hazardous to the environment. A risk reduction strategy to control the scale of releases at sites carrying out specific processes assessed in the risk assessment was set out as a result of the EU risk assessment process.

Nonylphenol and its isomers and their ethoxylates: most uses are subject to concentration restrictions, under Annex XVII, and some uses are banned. The EU risk assessment under the ESR (Existing Substances Regulation) (concluded 2002) concluded that risk management was needed throughout most of the life cycle for the aquatic and terrestrial compartments and for predators. Nonylphenol is classified for environmental toxicity, mammalian reproductive effects, acute toxicity and also corrosivity.

Musk xylene is a synthetic musk formerly used widely in fragranced products, including consumer personal care products. It has various intrinsic hazards, including ecotoxicity, physico-chemical hazards and carcinogenicity, and is very persistent, very bioaccumulative (vPvB) in the environment. Its use was substantially reduced in the 1980–1990s due to voluntary withdrawal by industry and since 2004 its use in personal care products has been formally controlled under the Cosmetics Directive (2004/88/EC, amending 76/768/EEC), which imposes concentration limits in various personal care products and bans its use in others. Musk xylene has largely been substituted by alternative synthetic musks that do not meet PBT (persistent, bioaccumulative and toxic) or vPvB criteria but are in some cases themselves highly toxic.

Polybrominated diphenyl ethers (such as the pentabromo congener) were widely used additive flame retardants in polyurethane foam among a wide range of types of substrate. They are persistent in the environment and in the body and are prone to bioaccumulate significantly. This range of substances is among the persistent organic pollutants listed in the Stockholm Convention and identified to have all manufacture ceased. A number of alternative flame retardants may be used in one or more of the same applications and the industry also researches changes to technology (such as methods to incorporate flame-retardant chemical functionality within the polymer substrate, greatly reducing the potential for leaching out of the finished product, rather than incorporating flame retardant additives later in the manufacture stage).

Trichloroethylphosphate (TCEP) had a large market in Europe as an additive flame retardant in polyurethane. The EU risk assessment for human health and environment (completed in 2000) identified risks associated with carcinogenicity, reproductive toxicity and other chronic health effects, concluding that risk management was required. Two structurally analogous chlorinated phosphonate flame retardants were particularly identified as candidates as 'drop-in' replacements of TCEP in different applications, these being trichloropropylphosphate (TCPP) and tris(dichloropropyl)phosphate (TDCP). These were also subject to in-depth risk assessment under ESR. For TCPP the conclusion of the RCR (risk characterisation ratio) was a need to manage risks to fertility and reproduction from manufacturing, for TDCP the same effects and additionally carcinogenicity are at issue, with several conclusions on hold pending further information. TDCP has subsequently been identified by various regulators as presenting concern in its own right.

12.5.2.1 *SUBSPORT – Web-Based Chemical Substitution Database Project*

There have been a number of nationally led initiatives to promote and support substitution of hazardous chemicals. SUBSPORT (substitution support portal) is a project that intends to bring a number of such projects together in one place and offer examples and advice on the replacement of hazardous substances (SUBSPORT, nd).

The SUBSPORT project is funded by the EU Life programme,[2] the German Federal Institute for Occupational Safety and Health (BAuA) and the Austrian Federal Ministry of Agriculture, Forestry, Environment, and Water Management. The project ran for three years from January 2010 to March 2013.

The project partners are: The Danish engineering consultants Grontmij; ISTAS, the Spanish Institute of Work, Environment and Health; the Sweden-based NGO (non-governmental organisation) ChemSec (the International Chemical Secretariat); and the German occupational safety, health and environmental protection consultants Koopera-tionsstelle Hamburg IFE GmbH.

The goal of the SUBSPORT project is to develop an internet portal that constitutes a 'state-of-the-art' resource on safer alternatives to the use of hazardous chemicals. It is intended to be a source of information on alternative substances and technologies and also of tools and guidance for substance evaluation and substitution management.

SUBSPORT is aimed at supporting companies in fulfilling substitution requirements of EU legislation, such as those specified under the REACH authorisation procedure, the Water Framework Directive or the Chemical Agents Directive. It is assumed that other stakeholders, such as authorities, environmental and consumer organisations, as well as scientific institutions, will benefit from the portal.

The project aims to create a network of experts and stakeholders who are active in substitution. The network aims to assist in content development and promotion of the portal as well as ensuring sustainable updates and maintenance. It is hoped that this will contribute to the project's goal of raising awareness and promoting safer alternatives. In addition, it is stated that training on substitution methodology and alternatives assessment will be provided.

SUBSPORT partners offer the information on the internet portal and make it publicly available (in four languages):

- A structured presentation of legal information on substitution throughout the European Union and, in part, on an international and national level.
- A database of hazardous substances that are legally or voluntarily restricted or subjects of public debates.
- A compilation of prevalent criteria for the identification of hazardous substances.
- A description of existing substitution tools to compare and assess alternative substances and technologies.
- A database comprising case studies from companies and literature with general information on alternatives to the use of hazardous substances.

---

[2] LIFE is the EU's financial instrument supporting environmental and nature conservation projects throughout the EU, as well as in some candidate, acceding and neighbouring countries. Since 1992, LIFE has co-financed some 3708 projects, contributing approximately €2.8 billion to the protection of the environment.

- A database containing detailed and evaluated case studies that document practical experiences in the substitution of 10 selected substances of very high concern in various essential applications.
- Concepts and materials for substitution training programmes.
- Interactive elements for discussion, networking, exchange of information and experience as well as for portal updates.

Of particular assistance is SUBSPORT's overview of the methods and tools that may be helpful in the substitution process. The web site allows links/access to the original method or tool presentations, which are intended to allow the user to acquire basic information on the available methods and tools, identify which of them are appropriate for their needs and proceed to further investigations. For each tool SUBSPORT comments on the reliability, applicability, user friendliness, limitations and availability of the tool. The 'tools' described are detailed in the following sections.

Column Model for Chemical Substitutes Assessment. The German Institute for Occupational Safety, Health and Social Accident Insurance developed the 'Column Model' to provide industry with a practical tool for identification of alternative substances. This is a simplified method to make a preliminary comparison between the risks of the different substances and products and offers a quick judgement on the convenience of substitution.

COSHH Essentials. COSHH (Control of Substances Hazardous to Health) Essentials is an assessment method developed by the UK Health and Safety Executive to help firms comply with the COSHH Regulations.

The method is used to determine the appropriate control measures for a given task and not specifically to assess risk levels. However, it can be used to compare alternatives by determining hazard levels for different substances and products.

Technical Rules for Hazardous Substances (TRGS) 600. Developed by the German Committee on Hazardous Substances, the aim of the Substitution Technical Rules for Hazardous Substances (TRGS) is to support employers:

- In avoiding activities involving hazardous substances.
- Replacing hazardous substances by substances, preparations or processes which are not hazardous or less so under the relevant conditions of use.
- Replacing hazardous processes by less hazardous ones.

TRGS 600 includes a framework for deciding on substitution that considers criteria for assessing technical suitability and health and physicochemical risk of alternatives.

Substitution Green Screen for Safer Chemicals. This tool was developed by Clean Production Action (CPA), a consultancy based in the USA and Canada.

The Green Screen for Safer Chemicals is a hazard-based screening method that is designed to inform decision makers in businesses and governments as well as individuals concerned with the risks posed by chemicals and to advance the development of green chemistry. The Green Screen defines four 'benchmarks' on the 'path' to safer chemicals, with each benchmark defining a progressively safer chemical:

- Benchmark 1: Avoid. Chemicals of high concern.
- Benchmark 2: Use but search for safer substitutes.

- Benchmark 3: Use but still opportunity for improvement.
- Benchmark 4: Safe chemical.

Each benchmark includes a set of hazard criteria – including persistence, bioaccumulation, ecotoxicity, carcinogenicity and reproductive toxicity – that a chemical, along with its known and predicted breakdown products and metabolites, must pass.

Green Screen assesses chemicals on the basis of intrinsic hazards determined by their potential to cause acute or chronic human and environmental effects and on certain physical and chemical characteristics of interest for human health.

Determination and Work with Code Numbered Products (MAL Code).   The   National Working Environment Authority, Denmark, developed a code number system to provide users with a practical tool for choosing the less harmful product and decide working routines and prevention measures for products with different code numbers.

Once a code number is designated to a product it is easy for a user to compare products. The higher number – the more hazardous.

Pollution Prevention Options Analysis System (P2OASys).   This was developed by the Toxic Use Reduction Institute (TURI) – Massachusetts USA – to provide companies with a framework for complete and systematic evaluation of potential hazards of processes and products in use and of alternatives.

The tool can be used in two different ways:

1. To systematically examine the potential environmental and worker impacts of toxic use reduction (TUR) options in a comprehensive manner, examining the total impacts of process changes, rather than simply those of chemical changes.
2. To compare TUR options with the company's current process based on quantitative and qualitative factors.

Priority-Setting Guide (PRIO).   Developed by the Swedish Chemicals Inspectorate (KEMI), this is part of a wider web-based tool called PRIO – a tool for risk reduction of chemicals to facilitate the assessment of health and environmental risks of chemicals, so that people who work as environmental managers, purchasers, and product developers can identify the need for risk reduction. PRIO also provides a source of knowledge for environmental and health inspectors, environmental auditors, risk analysts, and those who in some other way can influence the use and handling of chemicals.

PRIO includes:

- A database of 4472 substances with properties hazardous to the environment and health that should be prioritised in risk-reduction work.
- A priority setting guide.
- Guidance on how to reduce chemical risks in practical use.

PRIO allows users to:

- Search for substances and obtain information on properties hazardous to the environment and health.
- Obtain information on prioritised health and environmental properties.

- Identify substances contained in chemically characterised substance groups and product types.
- Obtain help in developing routines for purchasing, product development and risk management.

Quick Scan.   The Dutch Ministry of Housing, Spatial Planning, and Environment developed Quick Scan as part of a new chemicals policy to ensure that the potential risks and hazards associated with the use of substances in each stage of their life cycle are sufficiently controlled so as to remove, or to reduce to a negligible level, any harmful effects caused by substances on man or the environment.

Quick Scan describes measures to be taken for each chemical depending on their intrinsic hazard and potential exposure.

Stockholm Convention Alternatives Guidance.   The Persistent Organic Pollutants Review Committee of the Stockholm Convention on Persistent Organic Pollutants, 2009 guidance provides a general description of the issues to be considered in identifying and evaluating alternatives to listed persistent organic pollutants and candidate chemicals included in the Stockholm Convention on Persistent Organic Pollutants (UNEP, 2005). It is intended for use by the Persistent Organic Pollutants Review Committee and by Parties when considering the listing of new persistent organic pollutants. It may also be useful for manufacturers or users of listed persistent organic pollutants and candidate chemicals in identifying and deploying alternatives.

Stoffenmanager.   Developed by the Health and Safety at Work Department of the Ministry of Social Affairs and Employment of The Netherlands, Stoffenmanager is a tool for prioritising worker health risks to dangerous substances. Stoffenmanager was developed as a tool allowing small and medium-size enterprises to prioritise health risks to dangerous substances and to determine effective control measures. The tool combines hazard information of a substance or product with an inhalation and/or dermal worker exposure assessment to calculate a risk score. When risks are presumed, effects of control measures can be examined. An action plan shows an overview of the risk assessments with control measures. Stoffenmanager can also be used as a quantitative inhalation exposure tool and as a REACH Tier 1 quantitative inhalation exposure tool.

12.6   Sustainability and Green Chemistry

Increasing legislation in the United Kingdom, EU and North America (e.g. REACH, US Pollution Prevention Act, IPPC (Integrated Pollution Potential and Control): please refer to Regulations Chapter 3), as well as NGO-led public pressure, are forcing the chemical industry to minimise the impacts of its products and processes on the environment. These are among the strongest drivers towards a more sustainable and 'green' approach. Cost savings are also an important incentive.

The phrase 'green chemistry' has come to represent a particular stream of activity, largely led by chemists in industry, US regulatory bodies and academics all around the world.

Green chemistry, in broad terms, involves the application of chemistry knowledge to design products and processes in order to reduce or eliminate the use and generation of hazardous substances.

In practice over the last few years, the emphasis of most green chemistry activity has largely been on improving large scale chemical industry processes in respect of the types of raw materials used, the quantity of releases and, ultimately their efficiency. As is so often the case, measuring performance against these criteria has resulted in massive cost savings reported by many major companies. Reduction in emissions, avoidance of the use of expensive processes (in energy or chemical terms) and increase in recycling have made big impacts.

The 12 core principles of green chemistry were set out by Anatas and Warner (1998). They have been listed in many subsequent publications without any significant change (EPA, 2011). The twelve core principles spell out objectives, which together form a highly effective approach for pollution control, as this approach applies innovative scientific solutions to real-world environmental situations. They comprise targets for all industrial chemical processes.

1. It is better to prevent waste than to treat or clean up waste after it is formed.
2. Synthetic methods should be designed to maximise the incorporation of all materials used in the process into the final product.
3. Wherever practicable, synthetic methodologies should be designed to use and generate substances that possess little or no toxicity to human health and the environment.
4. Chemical products should be designed to preserve efficiency of function while reducing toxicity.
5. The use of auxiliary substances (e.g. solvents, separation agents etc.) should be made unnecessary wherever possible and innocuous when used.
6. Energy requirements should be recognised for their environmental and economic impacts and should be minimised. Synthetic methods should be conducted at ambient temperature and pressure.
7. A raw material of feedstock should be renewable rather than depleting wherever technically and economically practicable.
8. Unnecessary derivatisation (blocking group, protection/deprotection, temporary modification of physical/chemical processes) should be avoided whenever possible.
9. Catalytic reagents (as selective as possible) are superior to stoichiometric reagents.
10. Chemical products should be designed so that at the end of their function they do not persist in the environment and break down into innocuous degradation products.
11. Analytical methodologies need to be further developed to allow for real-time, in-process monitoring and control prior to the formation of hazardous substances.
12. Substances and the form of a substance used in a chemical process should be chosen so as to minimise the potential for chemical accidents, including releases, explosions and fires.

12.7   What Is Green Chemistry in Practice? Principles and Concepts

At the heart of green chemistry is the aim to 'design out' the health and environmental hazards posed by chemical products and reduce pollution from chemical processes. This focuses on the lifecycle of chemicals from cradle to grave, rather than the more traditional 'end of pipe' control.

This section looks at how far green chemistry has gone in achieving this aim and how green chemistry and sustainability concepts are helping to change the ways chemical products are designed, processes are run and products used. While some of the contributions of green chemistry are driven by regulation; others are voluntary initiatives by the chemical industry or driven by competitive innovation and market forces.

12.7.1   Why Is Green Chemistry Important?

Green chemistry focuses on the principals of sustainability, health and the environment. Society faces several issues in these areas, such as:

- Decreasing stocks and dependence on finite resources such as fossil fuels and metal ores.
- Increasing energy demand coupled with decreasing supplies of coal and fossil fuels.
- Human impact on the rate of climate change from the production of greenhouse gases.
- Industrial production and use of hazardous chemicals that can lead to accidents/spills and industrial production of large quantities of waste and pollution.
- Overuse of water resulting in decreasing supply of fresh water.
- Developing better methods of sustainable waste management and reducing waste.

It is thought that green chemistry could be extremely useful in helping find solutions to these problems, which makes it a vital area of scientific study.

12.7.2   Research in Green Chemistry

Whilst it is not a direct explanation of importance, the fact that a lot of active research is being carried out indicates that the scientists carrying out the research deem it important. The Royal Society of Chemistry (RSC) even has a 'Green Chemistry' journal dedicated to publishing peer-reviewed papers specifically relating to green chemistry. One of the key ideas being carried out in the current research is the idea of greener synthesis, which involves trying to make chemical processes as efficient, sustainable and as safe as possible.

12.7.3   Substance Design

Sometimes, a suitable substitute is not available, and a new chemical substance may be the only option.

In order to design substitutes, it is necessary to optimise performance for the desired function at the same time as health, safety and environmental acceptability.

Substance design plays an integral part in following the core principles of green chemistry, but also R&D in general. The design stage allows researchers to incorporate green chemistry at the earliest possible stage and allows avoiding environmental and commercial acceptability issues to develop at a later stage.

12.7.3.1   Property Prediction

Prediction of the properties of a substance encompasses both health and safety and the environment, and also desired effects.

In respect of health, safety and environmental properties, the following endpoints and tools can be considered:

Endpoints	Types of tools
Physiochemical properties	Fragment methods
Degradation	Property correlation
Ecotoxicity	Molecular modelling
Toxicity	
Mutagenicity	

The statistical techniques to use the tools need to be well understood.

- Fragment methods describe chemical structures in terms of the atoms or functional groups present.
- Property correlation is straightforward, for example ecotoxicology often relates to simple physico-chemical properties.
- Molecular modelling is more complex, often involving quantum theory to describe structure.

In REACH, as described in Chapter 5, if a predictive method is used to fill a data point rather than measurement, there are strict validation criteria. In R&D the process is less formal.

Prevention of Toxicity. Bioisostere design is predominantly used in the drug design process but its use for green chemistry could be applied more widely. Bioisosteres are functional groups with similar physical and chemical properties producing broadly similar biological responses. Bioisosteres can be used to reduce toxicity by replacing an unfavourable and toxic group with a safer alternative.

Oxygen and sulfur atoms exhibit similar binding interactions in the same molecular environment. Carbonyl ($C=O$) and thiocarbonyl ($C=S$) as well as ether (-C-O-C-) and thioether (-C-S-C-) are two such bioisosteric pairs. Drug design nearly universally will prefer the oxygen containing functionality.

Other classical bioisosteres are monovalent atoms replacing each other like hydroxy (OH), amine ($-NH_2$), methyl ($-CH_3$), hydrogen (-H) or halogen (-F, -Cl, -Br, -I). In this case it is often less clear which bioisostere represents the greenest solution. The molecular environment and the intended purpose of the substitution dictate the best overall solution. It is likely that synthesis and subsequent testing of the individual compounds is necessary to verify the best alternative.

Avoiding Persistence in the Environment. Substance design often needs to balance favourable properties like durability and interaction with other materials with the fate of the substance at the end of their functional live.

What kind of features help biodegradability, and which hinder?

- Presence of labile functions such as esters, alcohols or activated double bonds is beneficial.

- Use of linear alkyl groups and avoidance of branching/fused rings can help primary degradation start.
- Limiting use of halogens, especially multiple substitution (some halogenation is not necessarily a problem) is usually a good strategy.
- Limiting molecular weight and size helps bioavailability.
- Avoidance of some types of cyclic systems and of certain functional groups (e.g. nitro) on aromatic rings can be necessary.
- Use of functions founds in natural compounds, for example acids and amines, can help degradation, although there can be toxicity risks in mimicking biotic substances too closely.

12.7.4   Process Design

One of the key strategies of sustainable development is to use innovations to create value. For the chemical industry, this is achieved not only by making improvements at the molecular and product level but also, and perhaps more importantly, at the process and system level. It is, therefore, not surprising that improving industrial sustainability through technology is one of green engineering's main focuses.

Green engineering goes beyond consideration of baseline engineering quality and safety specification in process development. It incorporates the 12 principles of green chemistry and engineering by considering environmental, economic and social factors in process design. This can be narrowed down, in very simplistic terms, to input consumed and output generated.

12.7.4.1   Process Input Demands

Design for energy efficiency and the use of renewable raw materials or raw materials with low environmental are two key aspects of sustainable chemical production. The raw material and energy demands of a chemical process can be said to be a 'litmus test' of how green and sustainable the process is.

Biotechnology is helping chemical industry make use of more sustainable raw materials in their production processes. Bioprocesses are now being developed to replace traditional chemical processes for the manufacturing of many organic and other chemicals. This incorporates the concept of industrial ecology where unused material from one industry can be used as a feedstock for another. The result is the intersection of materials and energy within the industry, rather than sourcing for virgin raw materials and energy input.

Chemical industries also consider developing processes with improved yield but with less energy consumption. Process adaption with reduced energy considerations include:

- Control and optimisation of process condition.
- Integrated and multifunctional processes, which includes recycling of by products, including waste or process heat, within the process.
- Energy efficient separation technologies such as membrane separation.
- Reactive separation techniques.
- Limiting the number of drying steps within a process.

Prevention of waste and inclusion of all raw materials into the synthesis starts with the design of a new substance. Certain chemical structure motives are more difficult and, therefore, often more wasteful to synthesise then other. Ideally, the newly designed substance has a carbon skeleton that is readily available from renewable feedstock. This reduces the number of chemical steps needed in producing the final substance.

Alternatively, readily available feedstocks can be modified or fine tuned in a way that delivers a new substance that can replace a less green alternative.

12.7.4.2   *Overall Conclusions Concerning Substitution Driven by Acceptability Criteria*

- Since the first data call in 2010, a number of substances are on, or will be on, the Annex XIV list of substances of very high concern. It is too early to say what the impact of such listing is.
- REACH encourages new chemistry developments but at the significant cost of registration on top of R&D costs. It will only be in future that assessment of the effects on green innovation will be quantifiable (see also Section 15.12.3, which includes an assessment of whether regulation is affecting innovation).
- The main driver for innovation is likely to be the financial gain from the introduction of a totally new technology, when R&D scientists solve an issue in its entirety.
- The need for a totally new technology can be regulator driven or user driven.
- Substitute substances, that is replace A with B, can be profitable but do not seem to occur easily, and are quite often only short-term fixes.

References

Anastas, P.T. and Warner, J.C. (1998) *Green Chemistry: Theory and Practice*, Oxford University Press, New York, p. 30.

EPA (United States Environmental Protection Agency) (2011) Twelve Principles of Green Chemistry. http://www.epa.gov/sciencematters/june2011/principles.htm (last accessed 26 July 2013).

SUBSPORT (Substitution Support Portal) (nd) Moving towards safer alternatives. http://www.subsport.eu/ (last accessed 26 July 2013).

Tuncak, B. (2013) Driving Innovation, The Center for International Environmental Law.

UNEP (United Nations Environment Programme) (2005) Ridding the World of POPS: A Guide to the Stockholm Convention on Persistent Organic Pollutants. http://www.pops.int/documents/guidance/beg_guide.pdf (last accessed 5 August 2013).

13

Hazards, Risks and Impacts – The Development and Application of Frameworks for the Assessment of Risk

This chapter focuses on how risk is considered at a regulatory and policy level and the scientific concepts that underpin the application of policy and regulation for the control of risk. It introduces and explains the processes for the progressive withdrawal from the market of the most hazardous substances and limitation of risks that are use-specific, namely, the processes of *Restriction* and of *Authorisation* within the REACH Regulation. Also considered are the decision making and scientific frameworks that underpin appraisal of different options for the management of risk from exposure to substances on their own and in mixtures. The final section introduces the application of socio-economic analyses (SEAs) and the assessment of the costs and benefits of regulatory action on chemicals. This is explained and explored in more detail in Chapter 14 of this book.

The concept of risk is at the heart of this book and both this book and, of course, REACH itself are concerned with the assessment and control of the risks of chemical substances in order to establish chemical safety. The control of risk or the action of making something that has inherent hazard or danger 'safe', comes at a cost. It can also be assumed that the reason why a hazardous substance was used in the first place, despite the inherent cost of hazard containment, is that the users of the substance consider it to be the most cost effective in its application. REACH introduces an additional level of assurance that the potential risks are under control and that also carries a cost. It takes time and resources to take any action to make a hazard safer. This action may be in the form of some kind of containment, instruction/training or equipment, but in any, or perhaps even all cases, an action being taken must have a cost, even if those costs

Chemical Risk Assessment: A Manual for REACH, First Edition. Peter Fisk Associates Ltd.
© 2014 John Wiley & Sons, Ltd. Published 2014 by John Wiley & Sons, Ltd.

cannot be directly or immediately measured in terms of money. The concept, then, is quite simple: If a risk needs to be controlled, the application of control will require some kind of action and that action will require time and/or resource which will have a cost. The question for the risk manager is: How does control/management of the risk compare to the risk; that is, what is the cost of the action versus the benefit that it brings? This rather simple concept applies at the level of wearing rubber gloves, safety goggles or other protective equipment, right through to whether the introduction of new legislation on chemicals will bring about the desired level of increased environmental and health safety in society at European or even global scale.

The controls that are imposed, called 'risk management measures' (RMMs), are required because of the inherent hazards that chemicals have. However, it is the likelihood that living things will come into contact with those hazards that determine the risk and, therefore, it is the control of risk that is the main focus of legislative and policy regimes for the control of chemicals. In the end, the objective of such legislation is the prevention of actual harm or damage to living organisms and systems. The actual *harm* that is done to living things as a result of being exposed to the hazards from chemicals is the *impact*. Imposing restrictions or controls on activities that lead to impacts would mean that the damage would have to have been done in order to measure the impact, and thus the actions to control impact would be retrospective, for example cleaning up after an oil spill.

The *prediction* of possible impacts allows the severity of harm to be estimated *in advance* and, thus, the proportionality of preventative actions to be evaluated accordingly. This is explored in Section 13.2 in this chapter.

13.1   Policy Context – Risk, Hazard and the Precautionary Principle

The difficulty for the risk assessor, or at least the risk assessment framework, is providing meaningful measures that allow the risk manager to take appropriate and proportionate action to control the risk and prevent unacceptable harm to receptors that matter.

The aim of legislation and policy in context of the control of chemicals, such as REACH and other legislation, is assessing substance safety. This is so that uses of substances placed on the market are shown to be safe. The objective, therefore, is to improve, maintain and sustain human health and environmental quality by prevention of the harm cause to these *receptors*, that is living systems. Policy is about protecting 'what matters' to us as a society. For human health, this seems obvious – health and well-being should not be impacted by the production and use of substances. However, for the environment, 'what matters' is more than the health of a single species, but the integrity of the environment itself (Box 13.1). This usually means the normal functioning of the ecosystem, which itself is made up of populations of many species and communities of organisms functioning at different levels within the ecosystem (from plants producing food from sunlight, water and carbon dioxide – 'primary producers', right through to the predators at the top of the food chain). This can be very difficult to measure, as is discussed in Section 13.3 in this chapter and also in Chapter 14 of this book.

Box 13.1   Harm to things that matter – receptors

The concept of a *receptor* is not something that is explicitly mentioned in REACH. It is, however, central to other risk and impact assessment frameworks such as environmental impact assessment (EIA), the assessment and remediation of contaminated land, and the prevention of major accident hazards (Seveso II – Council Directive 96/82/EC on the control of major-accident hazards). The receptor is a useful concept in the context of understanding hazard, risk, and impacts. Put simply, the *receptor* is the thing that gets damaged or harmed. In most risk assessment frameworks this is a living system, whether a single species, such as humans, or parts/compartments of the environment – as in REACH. It can also be non-living things, such as important geological or geographical features or historic or important buildings, for example as considered in Seveso II or EIA assessments. The risk assessment frameworks are set up to allow assessment of the risk to receptors that are representative of the things that matter to us – that is the *protection goals* for the legislation. For human beings that means protecting health, life-span, and life itself; for the environment it is a wider span and more complex as it is the functions and services that the ecosystem provides, including resources, processing (carbon and nitrogen cycles, for example), as well as the existence of wildlife.

Protection of health should, and does, take into account different sections of the population, for example the old, infirm, the young or pregnant; it requires understanding of the sensitivities in human populations. This can be difficult to understand even when only one species, that is humans, is the focus. For the environment, however, the complex interactions between species and the functioning of parts of and whole ecosystem, makes it very difficult to decide upon what the protection goal for the environment should be. In addition, all systems have a range of response within which normal functioning can be maintained or brought back within a normal range. Therefore, exposure may cause a perturbation to a living system, that is within the range of normal functioning or from which it can recover (Box 13.2).

Box 13.2   Effects – response and recovery

Living systems have a range of responses to being exposed to stresses. In the context of chemical risk assessment this is about humans, as well as plants and animals in the environment, coming into contact with man-made chemicals that are released through their manufacture and use.

In toxicity testing – the controlled exposure of animal and plant species to a range of concentrations of a given substance in order to measure a specific set of effects – the aim is to determine the dose or concentration that presents a specific level of harm (for example, lethality to half of the tested individual organisms). Ultimately, the level of toxicity can be used to find which dose or concentration is considered to be safe by determining the dose or concentration at which no adverse

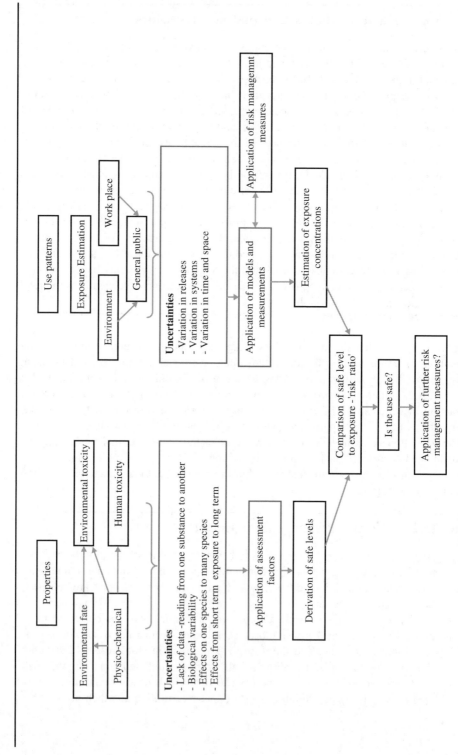

Figure 13.1 A Simplified Risk Assessment Framework, with Areas of Uncertainty

effects are measured or expected. Laboratory testing is controlled to reduce experimental variability as much as possible in order to get as precise an estimate of the toxicity of the test substance as possible.

Toxicity testing is an experimental model that is used to understand the *relative potency* of the substance's toxic effects. However, in the world outside the laboratory, humans and wildlife show wide variability in response to stress, in this case being exposed to man-made chemicals. This variability within and between species is accounted for to some extent by the application of assessment factors to toxicity results to derive safe levels for assessment (Figure 13.1), but the ability for both recovery and capacity to absorb stress is arguably poorly accounted for in the current system of risk assessment of chemicals. This applies perhaps more for the environment because ecosystems can (i) often continue to function during exposure to chemicals and (ii) recover from that exposure. In the first case it must be accepted that there may be some change in the number of species or structure of the ecological community (but the function of the ecosystem remains) and, in the second case, the recovery is only possible as long as the exposure stops or is reduced at some point (i.e. it is not continuous). For humans it is clearly not acceptable to view society in the same way as ecosystems and accept some harm to some individuals as long they will recover or that society will continue to function!

How does policy set out the protection goals and how are the assessment frameworks set up to allow measurement and prediction of harm, so that those goals can be achieved?

Different policies/legislation can have differing specific protection goals, depending on the main objective(s) of the policy and the receptors that are to be protected. Nevertheless, all of them have an objective: the protection of receptors from *harm*, that is to humans and/or the environment and also, in some cases, non-living receptors, such as geographical, geological or architectural features.

In some legislative regimes the definition of harm is relatively clear and measured (for example, Box 13.3) but in others the definition of harm is deliberately vague or undefined, leaving the regulator to assess, define and enforce the legislation accordingly. The protection goals may be based on preserving the integrity of the receptor, such as maintaining good environmental quality (such as in the Water Framework Directive – Box 13.3), or based on safety and thus control of exposure at or below a level which is considered to be safe, as in REACH.

Box 13.3   The Water Framework Directive and achieving protection goals for good ecological status of water

The Water Framework Directive (Directive 2000/60/EC of the European Parliament and of the Council establishing a framework for the Community action in the field of water policy) has an overall objective of maintaining all inland, coastal and ground waters at a state of 'good status'. As part of that, stressors on the ecological quality are defined and controlled, one of which is the release of chemical substances into the water body.

The European Commission's Environment Directorate web site declares (EC, 2012):

> *Good chemical status is defined in terms of compliance with all the quality standards established for chemical substances at European level. The Directive also provides a mechanism for renewing these standards and establishing new ones by means of a prioritisation mechanism for hazardous chemicals.*

The establishment of safe levels for different compartments of the aquatic environment called Environmental Quality Standards (EQS) (under Annex X, and later the EQS Directive 2008/105/EC) sets out concentrations limits (in surface waters, sediment and biota) for certain hazardous substances which must be adhered to by member state authorities. This is as well as a commitment to control or phase out releases and emissions of these substances (depending on their status as *priority* or *priority hazardous* substances). Adherence to EQS values can be achieved by ensuring that allowed levels in environmental permits, for example under the Industrial Emissions Directive (Directive 2010/75/EU), do not lead to concentrations in the environment in excess of the EQS, as well as through compliance with REACH, by ensuring concentrations in environmental compartments are below the safe level determined for those compartments.

Here the *safe level*, that is the EQS, already integrates hazard, exposure, and levels found in the environment. This is somewhat like a predicted no-effect concentration (PNEC) as defined in the REACH risk assessment process, but one that accounts for what is already in the environment and also considers the ecological functioning of the environment the is exposed to the chemical, to some extent.

It is worth noting the comparative situation for derived no-effect levels (DNELs) in REACH and the application of Occupational Exposure Limits (OELs) for worker protection. Whilst the DNEL derived in REACH is a level based on toxicity and the application of assessment factors, OELs integrate some level of operational achievability.

In REACH, the assessment for the registration of substances for demonstration of safe use/adequate control is based on a risk assessment (Section 1.2.8 of this book introduces the topic). This is a process in which exposure is compared to a notional safe level and the derivation of a simple ratio of the exposure to the safe level allows the use pattern to be judged as safe or not. The assessment frameworks set out the method for the assessment and how to present the information to the regulator. Importantly, they also set out how uncertainty should be handled.

13.1.1   Assessment Frameworks – Hazard and Risk and Impacts

The assessment framework set out in REACH is described in detail in Chapter 1 of this book; this section is about how and why the assessment framework is set up as it is and how hazard, risk and impact are dealt with. The diagram in Figure 13.1 shows a simplification of the risk assessment process in REACH, highlighting the areas of uncertainty and how these are dealt with in the framework (grey outline boxes).

From Figure 13.1 it can be seen that there are essentially two sides to the risk assessment framework – the hazard side that deals with the inherent properties of substances and the exposure side – how humans and the environment come into contact with the chemical and at what concentrations. On both sides there can be considerable uncertainty due to lack of data. For the environment a key uncertainty on the hazard side is that the data used for the assessment are from single specific tests done in the laboratory. Whilst this gives a level of precision due to the controlled conditions, it does not give a good picture of what effects might be caused in the 'real world' (Box 13.2). There are uncertainties in applying effects observed in one species to another, or to a number of other species and 'reading across' data from one chemical to another (for example using data from one substance that has a similar molecular structure to the substance of interest). On the exposure side the uncertainties come from variation in how chemicals are used and how they are released into the workplace and the environment (including how they may enter the food chain and, eventually, get into the food we eat and the water we drink).

To counter these uncertainties within the risk assessment framework, for estimation hazard *assessment factors* are applied to allow for the limited data that is used. In addition, the application of assessment factors is necessarily quite conservative in order to 'reward' the provision of better data by the assessor. For example, as explained in detail in Chapter 7 of this book, for the environmental hazard, the derivation of a PNEC is required. To do this, basic data on the short-term toxicity of the substance to aquatic organisms is required (i.e. fish, invertebrates and algae – a so-called base set of aquatic toxicity data). The PNEC is derived by taking the toxicity estimate from the most sensitive of the three test organisms and applying a factor of 1000 to account for the variation (i.e. within and between species, from short-term to long-term exposure and from laboratory to field and from measuring actual effects to a level where there would be no effects). This means that if the lowest toxicity estimate from the base set was 10 units (for such testing this will be a concentration such as milligrams per litre or parts per billion), then the PNEC would be 0.01 units. However, if the data set included measures of long term toxicity for the most sensitive species and/or tests on a number of species then the assessment factor can be reduced to ten or even five, possibly giving a much higher PNEC (depending on the estimate of toxicity from these better data) and thus a higher concentration before the notional safe threshold is crossed. This would potentially allow more release and/or the need for fewer controls on releases, because concentration of the safe level would be higher. This scheme of rewarding more data allows data to be gathered in a tiered and proportionate way, because if toxicity was very low or absent in the base set, there would be no point in doing further testing in an attempt to reduce the assessment factor and possibly the PNEC.

On the exposure side, the inherent uncertainties are associated with the variability in the releases of the substance. Although it can be assumed that the way a chemical comes into contact with humans or the environment is constant, this is not in fact the reality, as emissions vary widely in both time and space and are dependent on the physical and chemical conditions into which they are released; for example, for humans the internal conditions in the air passageways and lungs are quite different from the stomach, and for the environment releases into water will be quite different from releases into the air.

Models of varying degrees of sophistication are used to understand and estimate the exposure to humans and of the environment. These models take into account way that a chemical behaves in the environment into which it is released. In addition, the output from these models can be compared to actual measurements, for example from monitoring in the workplace or indeed from biological monitoring from workers themselves in order to verify and correct the models accordingly.[1] Both the application of assessment factors for hazard and the use of models for exposure estimation are explained in more detail in Chapters 8 and 10 of this book.

What is described earlier and illustrated in Figure 13.1 is a 'deterministic' risk assessment and although effort is made to account for uncertainty by the use of assessment factors and models for hazard and exposure, respectively, it ends up with a single figure that tells the risk assessor if the use under assessment is safe or not. Although this is very helpful for assessor and regulator because it gives a yes/no answer, it does not give the risk assessor or regulator an idea of the variability. Variability in risk assessment can be dealt with by the application of methodologies that explicitly account for uncertainty by applying a 'probabilistic' approach using statistical methods to generate estimates of the variability around point estimates both of hazard and exposure and showing these explicitly in the risk estimate. These approaches and their relative advantages and drawbacks are discussed in Section 15.8 of this book.

The risk framework described here relies on being able to measure or estimate hazards and exposure, and for hazards it is dependent on being able to determine a level below which there are no effects. However, for the specific properties of some substances it is considered not to be possible to derive a safe level due largely to the uncertainties associated with the effects that may happen as a result of exposure to these properties or because the very properties of the substances enable them to accumulate over time and/or in the food chain to levels that may become toxic. This is discussed further in the next section.

13.1.2   Precaution – Where No Safe Level Can Be Established

As described in detail in Chapter 9 of this book, substances that have carcinogenic effects (i.e. that may cause cancer) and effects on the genetic material of living organisms (so called mutagens), can in theory effects at any level of exposure, that is there is not a 'no-effect concentration'. These are called non-threshold carcinogens and mutagens, because no safe threshold can be determined. This can also be the case for some effects on the reproduction of humans and other mammals, such as effects on the reproductive organs or the developing foetus. Similarly, for the environment, substances that do not degrade in the environment accumulate in food chains and are particularly toxic to wildlife can be considered not to have a threshold for effects. While for the human health non-threshold carcinogen, mutagen, and reproductive toxins (so called CMRs) the scientific justification for not being able to derive a safe level is more clear, it is less clear for the persistent, bioaccumulative and toxic (so called PBT) substances. For PBT substances, it is possible

---

[1] Actual data can help validate the models and, conversely, discrepancies between model predictions and real life data can help identify model flaws. By an iterative process the models can be improved until predictions can be made with a reasonable level of accuracy.

to derive a safe level (i.e. a PNEC) based on the toxicity, but because the substance can be present in the environment for so long and also accumulate in food chains there is thepossibility for long-term effects to occur that may not be captured by the current risk assessment framework (Figure 13.1). In addition, substances that may not have toxic effects but are very persistent and very bioaccumulative (vPvB), that is those which stay and accumulate in the environment, are also considered in the same way because of the uncertainty associated with potentially unknown effects. This is a policy-led addition to the risk assessment framework within REACH that was arguably prompted by some high profile cases in which there was perceived to be a lack of timely regulatory action on substances. It was inserted in the Regulation in order to capture unforeseen problems or those that may have fallen though the regulatory net of previous legislative regimes. Examples of such substances are shown in Box 13.4.

Box 13.4   Examples of persistent and bioaccumulative substances that have been subject to bans

- The polychlorinated biphenyls (PCBs). PCBs are chemically stable substances that were used as coolants in electrical transformers. Although they were banned in the European Union (EU) some 20–30 years ago, concentrations are still found in organisms far from where they were produced and used.
- Organostannic compounds (hydrocarbons containing the metal tin), used as biocides, notably as antifoulants on marine vessels, that can cause specific effects on the reproduction of shell fish at concentrations so low that it is difficult to measure.
- Some fluorinated compounds, such as perfluorooctane sulfonate (PFOS), used in various applications, including as surfactants in fire-fighting foams and mist suppressants in the metal plating, that stay in the environment for a very long time and are found to cause effects on reproduction.

The approach to CMRs and PBT/vPvB is precautionary in that the conventional risk assessment framework, from which a safe level is derived, is not applied and these substances are regulated within REACH on the basis of their hazard alone. The assessment is based on reduction of 'risk' or rather the greatest level of control and the residual risk or possible impact is compared to the benefit that the use of the substance brings to society as whole. This is the basis for the Authorisation process in REACH and is discussed in more detail later.

This precautionary approach to the most hazardous chemicals is one that has been explicitly applied by government and regulator policies in the last 15–20 years. A notable example is the application of this approach to substances that may cause effects on the hormonal (endocrine) systems of wild life. The link between the effects of these endocrine disrupting chemicals on individual species and what that means for populations and the functioning of ecosystems is still not well understood. Nevertheless, because the effects, in particular those that affect the reproductive hormones,[2] *could* have effects on wildlife populations, policy makers deemed it prudent to take a precautionary approach to the regulation of substances that cause such effects. An extra layer of assessment is applied to these substances in order to prevent the release of such substances, which could cause long-term effects. This is particularly the case for substances that have endocrine disrupting effects that are also persistent. The action is one of preventing the opening of the 'Pandora's Box' in the first place, until more is understood about the long-term effects of these substances.

13.1.3   Application of Assessment Frameworks to Human Health and Environmental Protection

The assessment framework applied in REACH as summarised and simplified in Figure 13.1 is similar in principle to frameworks used in other legislative regimes. It is most similar to regimes and frameworks that are applied to substances that are regulated on the basis of use, for example biocides and plant protection products (pesticides). For pesticides and biocides the difference between these uses and the uses covered by REACH is that they are confined to those which usually have the objective of killing pest organisms, so are intentionally toxic. In addition, there is part of the assessment that considers efficacy, that is the effectiveness of the product – Does the product work? – in contrast with REACH, which does not assess, for example, if a substance used as a process aid or as a cleaning product is actually effective in that use.

Other risk frameworks are aimed at protecting a specific receptor/s or specific situations. For example, there is an ecological assessment framework for the protection of land and soil quality, and the Water Framework Directive is aimed specifically at the protection of water bodies and coastal waters (Box 13.3). The Seveso II Directive (Council Directive 96/82/EC on the control of major accident hazards) sets out a framework for assessing the safety of installations that hold significant quantities of very hazardous substances (e.g. highly flammable, explosive or highly toxic) and the prevention of accidents that could have a major impact on human life and the environment.

The assessment framework for the Seveso II Directive includes a process for identification of risks based on inherent hazard and the likelihood of exposure to those hazards, and considers the possible impacts that would happen if those risks were realised. The severity of the risks is managed by the application of mitigation measures to reduce the risk to acceptable levels (similar to the application of risk management measures in REACH). An important difference between Seveso II and REACH is that Seveso II requires the consideration of *impacts* (i.e. the actual harm that is cause to humans and the environment) in terms of a quantitative and qualitative description of the harm. The EIA Directive (85/337/EEC) is concerned with the assessment of the possible impacts from the building of new industrial installations and large commercial building and infrastructure. In addition to assessing the possible severity of environmental impacts both specially and temporally to wildlife and to humans, including noise/vibration and air quality, other more subtle impacts, such as visual impact, are assessed (for example, currently of high profile in the assessment of on-shore wind turbines).

---

[2] Mostly because the substances 'mimic' the molecular structure of the hormones and thus interfere with the reproductive process.

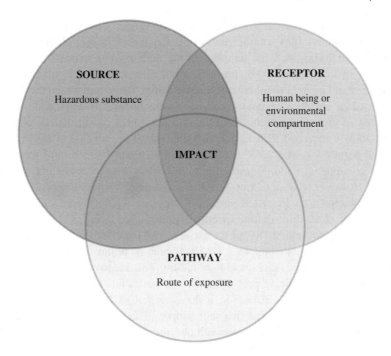

Figure 13.2   *A Simple Illustration of Source, Pathway, Receptor and Impact*

A key concept in these assessment frameworks is that of source–pathway–receptor. This is a deceptively simple concept that states that for an impact to happen there must be a source of hazard – this could be a chemical, or a physical hazard such as noise/vibration or electromagnetic radiation (light, UV light, microwaves etc.); a way in which that hazard can come into contact with systems or objects that it may damage (e.g. living organisms), that is a pathway; and, finally, there must be the presence of systems or objects that can be damaged/harmed – that is receptors. All three must be in place for an impact to happen; remove one and the risk is removed (Figure 13.2).

Whilst this is explicit in the Seveso II and EIA Directives, because the concept helps the assessor to do the assessment, it is implicit in REACH. REACH seeks to establish that the substance under assessment is used safely, that is that exposure is below the safe level, not what would happen if those levels were exceeded.

13.2   From Hazards to Risks to Impacts – Understanding the Implications of Exposure to Dangerous Chemicals

This section is about how safety is established, the level of evidence for establishing the risk of harm (and thus safety) and the level of uncertainty (and therefore precaution) applied to that.

The easiest measure of potential harm to understand is hazard – that is, something is intrinsically poisonous or dangerous. Controlling substances on the basis of their hazards, however, is the most precautionary; this is because, as discussed in Chapter 11 of this book, it is the opportunity for that hazardous property to come into contact with a 'receptor' that determines whether harm will happen. The control of substances on the basis of their hazard is what happens in the authorisation process under REACH (Section 13.3 in this chapter). For the estimation of risk more information is needed because understanding of both inherent dangerous properties – hazard and the opportunity for the hazard to act on a receptor – exposure – are needed to understand risk in the simplest type of risk assessment. Assessment and control of substances on the basis of risk are widely regarded to be a scientifically robust approach and are defended by industry and regulators alike. Risk, however, informs about estimation of the likelihood that harm will happen. In addition, usually it establishes thresholds beyond which safety is assumed. However, risk does not establish what actually might happen if exposure to the hazard happens or indeed if sustained exposure to the hazard happens. This is because laboratory tests done on one specific organism used to establish a safe level for a particular effect cannot reliably tell us the incidence of the effect in the whole population or whole ecosystem. As an example, a DNEL (which quantifies the hazard) is set based on a toxicological endpoint, such as kidney failure. This means that if exposure exceeds the DNEL, there is a significant probability that at least a few humans will suffer kidney failure in a large population. However, it does not tell us how many will suffer from kidney failure or when and where that will happen. The environment is more complex because in that case we are dealing not with one species, but very many species, as well as exposure in different environmental compartments.

The need to assess risk is about the need to *predict* what might happen and to make an a priori assessment based on laboratory studies and physico-chemical properties, but understanding what really happens if receptors are exposed to risk is usually about understanding what has happened and assessing a posteriori the impacts. Nevertheless, the assessment of impacts may also have to be predictive. The impacts are what it is actually necessary to control and prevent, but it is most difficult to be predictive about because it relies on the damage having happened in order to be measured. It is difficult to quantify impacts and understand how they will be distributed in time and space. For example, for the substances in Box 13.5, given the understanding of the fate and effects at the time these substances were first marketed (and the risk assessment frameworks in place), it was not possible to have predicted where these substances would end up or some of the specific effects they have on living organisms. For the first time in the regulation of chemicals in normal use, REACH, in the authorisation process, demands the assessment or consideration of *possible* impacts from the use of substances. Nevertheless, as mentioned previously, the predictive assessment of impacts is already a feature of other legislative regimes such as the EIA regulations (for the assessment of the environmental impacts of new buildings and factories, etc.) and the Seveso II Directive (assessing the control of major hazards from accidents from the accidental release of specific very hazardous chemicals).

13.2.1   Introduction: The Need for a Culture of Safety

A manager of a chemical production or formulation site has the challenging task of balancing manufacture and use of chemicals at the lowest cost possible and ensuring safe operations. Legislation in most countries make him/her (and or the directors of the firm/she works for) personally responsible for the health and safety of employees, that is if workers suffer health effects due to exposure to hazardous chemicals, or to the environment, and for people living in the neighbourhood. His/her management expects the manager to maintain the value of the investments, to help increase returns by continuing to lower costs and to help the business retain a competitive edge. Especially in the larger companies, he/she is not alone. He/she will employ staff directly reporting to him/her dedicated to implementing health, environmental and safety programmes and to help maintain a constructive relationship with local authorities as well as with the senior management of the company to which the plant belongs.

There have been a number of high profile incidents that illustrate the accidental release of substances and the harm caused. However, this book and REACH itself are concerned with the control of risk from use of the substances under circumstances of normal manufacture and use. Both accidental releases and releases from the normal manufacture and use of substances and products have led to an industry initiative to set standards and procedures for control of risks, taking the form of a coordinated product stewardship programme. The Responsible Care® programme, in addition to ensuring compliance to the regulations enacted following incidents, are the industry's response to the need for coordinated product stewardship. Responsible Care ultimately aims at creating a culture of safety, emphasising prevention rather than cure.

In addition, statistics of fatalities due to exposure to chemicals, such as cancers and respiratory, while not all directly attributable to the chemical industry, attract the attention of non-governmental organisations (NGOs) with concerns about the perceived increasing level of harmful substances workers are exposed to, and released to and accumulating in the environment.

Table 13.1 is a summary of some notable environmental impacts connected to releases of chemical substances and to exposure via the environment.

A number of other chemical substances found in the environment are suspected of having an impact, largely based on laboratory data. It is extremely difficult to demonstrate in a convincing way that some of them have an effect on the environment, that is impacts on populations of plants and animals, since there are so many mechanisms that interact in the environment, compared to laboratory conditions. However, because of the uncertainties and difficulties in identifying effects in the environment and attributing specific effects to specific substances, the authorities are taking an increasingly precautionary approach to avoid the potential long-term consequences of releases of man-made chemicals to the environment. Within REACH this precautionary approach is applied by basing the selection of substances of very high concern (SVHC) on specific hazardous properties of substances that cause long-term effects on humans and the environment.

The application of the precautionary approach may be raising fears that over-regulating the chemical industry results in less innovation, loss of competitiveness (relative to under-regulated producing regions) and, therefore, loss of jobs. It is the view of the authorities that the precautionary approach within REACH promotes innovation as the progressive withdrawal of hazardous substances from the market will open the market for safer alternatives. A track record of safe manufacturing, transport and use of chemicals is expected to rebuild public trust in the longer term.

Table 13.1   *Environmental impact of chemical contaminants.*

Chemical substance	Main use/source (primarily historical)	Impact on environment
Chlorofluorocarbons and methyl bromide	Respectively refrigerants and soil fumigants	Lowers ozone levels in stratosphere
Mercury	Chlorine production, thermometers, coal-fed power plants, gold mining	Converts to methyl mercury in the environment, biomagnifies through the food chain and affects human reproduction and mental health
Sulfur dioxide	Flue gases, internal combustion engines	Oxidizes to sulfuric acid in the atmosphere and increases acidity of rain, which can affect vegetation growth; promotes respiratory diseases
Tributyltin derivatives	Antifouling paint	Toxic to shellfish and causes disruption to reproduction in marine snails
Tetraethyl lead	Petrol additive ('antiknock' agent)	Biomagnifies through the food chain and affects mental health of children

In summary, the chemical industry intends rebuilding trust in the safety of industrial chemicals by a combination of:

- Awareness of the inherent hazards and risks related to exposure, including effective programmes to identify them before problems arise.
- Management principles that clearly define chains of accountability and of responsibility.
- Supervision/auditing by independent organisations (contractors or regulators), including dissuading fines or criminal charges.
- Creation and nurturing a culture of safety including the right attitudes from top to bottom in all hierarchies.

13.2.2   Responsible Care®

Responsible Care® was developed by the Chemicals Industry Association of Canada (CIAC) in 1985. Since then, Responsible Care® has been made into the condition of membership of chemical producers of the main international chemical industry associations. The International Council of Chemical Associations (ICCA) actively promotes Responsible Care®; as of 2012, 52 countries accounting for nearly 90% of global chemical production have implemented the Codes of Practice. In view of the importance of Responsible Care® for the chemical industry, it is worth quoting in full its Ethic and Principles for Sustainability (CIAC, nd; ICCA, nd):

We are committed to do the right thing, and be seen to do the right thing.
We dedicate ourselves, our technology and our business practices to sustainability – the bet-
terment of society, the environment and the economy. The principles of Responsible Care®
are key to our business success, and compel us to:

- *Work for the improvement of people's lives and the environment, while striving to do no harm.*
- *Be accountable and responsive to the public, especially our local communities, who have the right to understand the risks and benefits of what we do.*
- *Take preventative action to protect health and the environment.*
- *Innovate for safer products and processes that conserve resources and provide enhanced value.*
- *Engage with our business partners to ensure the stewardship and security of our products, services and raw materials throughout their life-cycles.*
- *Understand and meet expectations for social responsibility.*
- *Work with all stakeholders for public policy and standards that enhance sustainability, act to advance legal requirements and meet or exceed their letter and spirit.*
- *Promote awareness of Responsible Care, and inspire others to commit to these principles.*

Taking into account practical experience, Responsible Care® has evolved from six to three Codes of Practice:

- The Operations Code (manufacturing and transportation).
- The Accountability Code (engagement with other stakeholders, both locally and broadly).
- The Stewardship Code (customers and distributors).

Responsible Care is much broader in scope than any single regulation.

Since compliance to the Codes of Practice of Responsible Care® is voluntary, it is necessary to implement a system of verification, otherwise it can be accused of having 'no teeth'. Verification is essential to ensure continuous improvement of the companies involved, of the industry and also to improve credibility and trust. Verification takes place every three years by teams of industry experts, local citizens and by public advocates; verification reports are made public.

13.2.3 Standards and Management Tools

In addition to the example of the Responsible Care programme, there are other incentives to improve the overall performance of industry, which all have a direct or an indirect impact on operations and eventually on the profitability of firms; examples are:

- Environmental standards, such as ISO (International Organization for Standardization) 14001, covering environmental management.
- Quality standards, such as the ISO 9000, 9001, and 9004 family, concerning, respectively, the principles of quality management, their requirements, and their sustained success.

In many cases evidence of compliance to these standards is a condition of doing business, that is it is a condition of supplying to customers.

Implementation of quality standards such as ISO 9000 is particularly important in that it provides, in particular, for the documentation and review of operating procedures and for the application of the Deming Cycle, which was initially developed in the 1950s (Arveson, 1998). To quote Arveson:

> *Deming proposed that business processes should be analysed and measured to identify sources of variations that cause products to deviate from customer requirements. He recommended that business processes be placed in a continuous feedback loop so that managers can identify and change the parts of the process that need improvements. As a teacher, Deming created a (rather oversimplified) diagram to illustrate this continuous process, commonly known as the PDCA cycle:*
>
> • *PLAN: Design or revise business process components to improve results*
> • *DO: Implement the plan and measure its performance*
> • *CHECK: Assess the measurements and report the results to decision makers*
> • *ACT: Decide on changes needed to improve the process.*

In practice, no reportable incident takes place without managerial review, identification of root causes, development and implementation of corrective actions and reporting, whether occurring in manufacturing, distribution or in general management.

Since prevention is better than cure, potentially hazardous processes are identified and regularly reviewed to ensure that they are carried out at the highest practical level of safety. Several methods have been developed, such as the fault-tree analysis, to identify and then address any unacceptable risks.

The boards and the leadership teams of several manufacturers begin with review of the safety performance of the operating units, instilling a culture of safety at the highest levels, so that no business decision is made while ignoring possible implications.

Some manufacturers have taken these tools even further. For example, DuPont™ in particular is promoting its STOP™ programme (Safety Training Observation Programme), which is based on the following principles (DuPont, 2013):

• All injuries can be prevented.
• Employee involvement is essential.
• Management is responsible for preventing injuries.
• All operating exposures can be safeguarded.
• Training employees to work safely is essential.
• Working safely is a condition of employment.
• Management audits are a must.
• All deficiencies must be corrected promptly.
• Off-the-job safety will be emphasized.

While the term 'safety' primarily relates to the reduction of loss time injuries and protection of assets, the same principles are equally applicable to environmental protection and regulatory compliance.

Compliance to these standards and management tools has greatly facilitated implementation of regulations such as REACH and of product stewardship programmes such as Responsible Care®.

What is particularly interesting is that the implementation of these processes has had a profound effect on the chemical industry, simultaneously making it more competitive and safer in operation.

Finally, in response to consumer demand, the voluntary ecolabels or environmental standards have proliferated in the EU with a European scheme being run alongside other national schemes. They have been met with varying success from one country to another, affecting consumer behaviour in preference for labelled products. Some manufacturers have responded by changing their processes to other technologies that were viewed as less damaging to the environment. For example, the association of the use of halogenated substances in a number of industrial processes and the possible formation of dioxins has resulted in the substitution of chlorine bleach in the papermaking industry such that in 2007 it was reported that about 75% of world bleached Kraft pulp production is elemental chlorine-free (ECF) (Roberts *et al.*, 2007).

13.2.4   Risk Control and Management

In the preceding section a combination of regulatory compliance and self-regulation were discussed as the 'drivers' behind environmental and health safety. National and European-wide regulations also serve an essential role by creating an even playing field for manufacturers and importers. Therefore, REACH and other chemical safety regulations are tightly integrated in the daily operations of chemical companies and downstream users. Even consumers, that is the general public, should be provided with sufficient information to understand the risks they are potentially exposing themselves to when handling any chemicals in products. This is particularly true of chemicals present in most homes, such as cleaning or disinfecting products, construction materials, medicines, paints, pesticides and biocides.

The obligatory use of danger symbols on consumer products when required by the hazard properties of ingredients, especially of mixtures (classification and labelling requirements – Appendix A), is playing an important role in educating consumers, in ensuring they are used safely and that they can allow a choice in certain cases.

13.2.5   Risk Control and Management in REACH

REACH is about the normal manufacture and use of chemicals and deals with the assessment of safe use based on risk assessment, while other legislation such as the Seveso II Directive deals with conditions that are outside of normal operation, that is in major accident situations. A Seveso II Directive safety report requires an estimation of impact, while REACH does not require assessment of impact, except within authorisation and restriction (Section 13.3 in this chapter). Given the large number of legislative instruments already at the disposal of the Regulatory Authorities, one may wonder, perhaps provocatively, what gap REACH was intended to address when it was enacted?

The main arguments were put forward during the political discussions that took place when the legislation was being developed are summarised in Table 13.2.

Great importance was placed on severe impacts having occurred in the past and the need to put in place instruments designed to prevent them from happening again.

In REACH we try to understand the inherent hazards of substances and describe them, we look at how use leads to exposure and that allows us to define risk. This procedure normally stops there if the registrant of the substance can show that the risks are controlled but for SVHCs, for example, an additional layer of control is imposed (based on precaution) and for authorisation there is the need to describe the type of

Table 13.2   *How REACH was implemented.*

Main REACH drivers	How REACH was implemented
The potential hazards posed by chemicals placed on the market before 1981 were not always clearly understood due to lack or unavailability of data	Requirement to provide base set of data proportional to tonnage and dependent on main use category
There is no use inventory of chemical substances so, therefore, no way to ascertain whether they are safe	Requirement to inventory each use of chemicals and exposure scenario, and to undertake a risk assessment defining conditions of safe use
Previous legislation had failed controlling hazardous chemicals in timely fashion	Manufacturers and importers are required to provide dossiers instead of regulators
Persistent, bioaccumulative and toxic (PBT) or very persistent and very bioaccumulative (vPvB) substances are unregulated	CMRs and PBTs/vPvBs are considered substances of very high concern (SVHCs) and subject to authorisation
Substitution by less hazardous chemicals or functional substitution is feasible when economic incentives are put on the same level than health and environmental protection	Substitution of SVHCs is required under authorisation

impact that the use of the substance may have. Therefore, it is the control of risk that we are concerned about, and REACH is essentially about staying below threshold (safe) levels, that is DNEL and PNECs, except under the special circumstances of authorisation when a further layer of assessment is needed and then it is necessary to get further data on the magnitude of risk (i.e. how many workers would potentially get ill, how much of the environment will be harmed etc.).

It may be useful to consider a specific example of a substance within the regulatory process. In Box 13.5 an example of a substance under regulatory scrutiny is considered, including the action of other non-REACH legislation.

Box 13.5   Example of legislative action on a chemical – decabromodiphenyl ether – a case study

The first version of the RoHS (Restrictions of Hazardous Substances) Directive 2002/95/EC was adopted in February 2003. The main provision of the Directive limits the use of the flame retardants polybrominated diphenylethers (PBDEs), along with five other substances, to a maximum concentration of 0.1% by weight of homogenous material (understood as material that can potentially be separated mechanically). Since flame retardants are only effective above a minimum concentration by weight (typically in the range of 5–30%) the measure, which is intended to reduce the toxicity of waste electrical and electronic equipment (WEEE), implies a ban on the use of PBDEs.

The decision affecting the PBDEs was taken irrespective of other EU instruments, such as the Marketing and Use Directive, which itself was superseded in 2006 by REACH. In principle, the Marketing and Use Directive in force at the time the RoHS were adopted; it was based on a risk assessment of the use of substances intended to be controlled.

This case is a good example of how different EU instruments can conflict with each other, often for reasons of political expediency.

PBDEs are a class of potentially 209 individual substances, but only three of them are commercial as the pentabromo, octabromo and decabromo diphenylethers. The commercial pentabromo and octabromo BDEs are, in fact, mixtures where the penta and octabromo derivatives are the major component. Commercial decabromodiphenylether ('Deca') is the nearly pure substance. The three commercial PBDEs 'Penta-BDE', 'Octa-BDE' and 'Deca-BDE' are mainly used in polyurethane foam, ABS plastic and polystyrene, respectively. This explains why Deca-BDE was the major brominated flame retardant used in television set and computer cathode ray tube (CRT) monitors. Deca-BDE is also used in the formulation of flame retardant barrier coatings in upholstered furniture. Therefore, the main impact of the RoHS was on Deca-BDE in electronics and electrical equipment.

Deca-BDE and Octa-BDE were on the first EU list of Priority Substances for Risk Assessment under the Existing Substances Regulation 793/93/EEC, published as Commission Regulation (EC) No 1179/94 of 25 May 1994. They were joined by Penta-BDE on the second priority list (Commission Regulation (EC) No 2268/95 of 27 September 1995). The Anglo-Welsh (UK) Environment Agency was asked to report on its conclusions regarding the risks posed by the uses of these three substances in the environment and France was asked to do the same on its health aspects.

In 2005 the Commission published a Decision (2005/717/EC) through comitology which stated explicitly 'since the risk assessment of Deca-BDE, under Council Regulation (EEC) No. 793/93 of 23 March 1993 on the evaluation and control of the risks of existing substances, has concluded that there is at present no need for measures to reduce the risks for consumers beyond those which are being applied already, but additional studies are required under the risk assessment, Deca-BDE can be exempted until further notice from the requirements of Article 4(1) of Directive 2002/95/EC'. This decision was then challenged by the European Parliament and Denmark and subsequently reversed in the European Court of Justice. In its judgement (published on 1 April 2008), the court found for procedural reasons that the Commission abused its administrative powers by exempting Deca-BDE through comitology as opposed to co-decision.

The European Commission then published a Communication on the results of the risk evaluation of Deca-BDE (2008/C 131/04), confirming the conclusions already published in its decision (subsequently reversed) to exempt it from the RoHS. In summary, for workers and humans exposed through the environment there was a need for further testing (developmental neurotoxicity, biomonitoring) and for consumers there was no need for further information and/or testing or for risk reduction measures beyond those already applied. Further monitoring was required to follow environmental trends over a period of 10 years. No risks were identified for the atmosphere or for sewage plants microorganisms. Therefore, retrospectively the Deca ban in the RoHS hardly seemed justified, since no significant risks were identified and no restrictions were recommended.

The RoHS were subsequently revised in 2011 (RoHS 2, Directive 2011/65/EU) but despite high hopes by industry based on the relatively favourable outcome of Deca's risk assessment, the PBDE ban was maintained.

In July 2012, the UK Environment Agency submitted a REACH Annex XV dossier identifying Deca-BDE as a SVHC due to its potential to degrade in the environment to PBT substances. This would be the first step towards a possible Annex XIV listing for Authorisation or Restriction.

The inclusion of Deca-BDE in the several versions of the RoHS could, therefore, be seen as pre-empting risk reduction measures in REACH, which have not yet been decided.

13.3   Risk Management Options – REACH Processes for Control of Hazardous and Risky Substances

The first section of this chapter considered why hazards, risks and impacts are controlled and the policy and legislative instruments that allow hazard and risk to be assessed and controlled. It also considered the different frameworks that are applied to the assessment of hazards, risk and impacts. Section 13.2 in this chapter looked at how risks are managed with a specific focus on REACH and also considered voluntary industry initiatives and product stewardship as a driving force for risk control in chemical production and use. This last section considers what happens when risks cannot be controlled, or rather when the risk assessment framework, specifically REACH, does not allow the demonstration of safe use. In particular, this section looks at risk management with REACH in the processes of Authorisation and Restriction.

13.3.1   Restrictions and Authorisations in REACH

In Section 13.1 in this chapter how hazards and risk are defined and assessed in REACH were considered. As explained in Section 1.2.4 of this book, the processes of Authorisation and Restriction in REACH are two sides to the same coin. Where Restriction is applying limits or prohibition on specific uses of a substance and in principle allowing all other uses to continue (so long as they can be shown to be safe), Authorisation places a prohibition on all uses of a substance, unless it can be shown for a specific use or uses that there are no viable alternatives and the benefits to society outweigh the risks. In that case only those uses will be permitted and all others banned, but permission for time limited continued use must be applied for by the manufacture or user and subsequently granted by the European Commission. Another distinction can be made between Restriction and Authorisation; that is the Restriction is about the control of *risk*, but risk at the scale of the European Union, while Authorisation is about *hazard* and the need to drive the most dangerous substance off the market in order to promote the use of safer alternatives.

Restriction is perhaps better understood by industry and regulators because it is very similar to the system that was in place in Europe before REACH. Within the existing substances regulation (ESR) there was a framework for the assessment of risk, much like that set out in Figure 13.1. As set out in Chapter 1 of this book the assessment led to one of three conclusions and where a risk was not controlled the risk management process was initiated. A risk reduction strategy, carried out by the same Member State rapporteur that completed the risk assessment, had the objective of further investigating the risks identified in the risk assessment to ascertain if the risks were likely or if, in reality, they were controlled and, if not, to propose the best measure(s) to control the risk(s). The measures would take into account the effectiveness, 'monitorability', enforceability, and proportionality of the proposed measure(s). The costs and benefits of proposed measures were also taken into account.

It could be that if the risk(s) were confined to a specific country then national level action would be the best measure to control the risk. Alternatively, if the risk was confined to a specific part of the environment that was well controlled by other legislation, it could be that further measures within that legislation should be proposed (for example an occupational exposure level (OEL) for workers or an EQS for the aquatic environment). The ultimate sanction was a limitation on use that could ban the use of the substance for a particular application. Examples are:

• Limitation of the use of cadmium for certain plastics.
• Ban on the use of hexavalent chromium in cement.
• Ban on the use of perfluorooctane sulfonate as a surfactant in electroplating baths.

There were, of course, 'derogations' to these limitations, whereby some specific sub-uses were not included in the ban; for example, cadmium was permitted in some plastics on safety grounds, as it gave the bright yellow colour in gas pipes and identification of gas pipes is very important from a safety point of view (EC, 2011). This process is familiar to the regulators – to the Member State competent authorities and also to the European Commission and the restriction process is rather similar. Indeed, existing restriction from the Limitations Directive (also known as the Marketing and Use Directive) were directly carried over to REACH as listed in Annex XVII. Restriction leaves the assessment of community wide risk in the hands of the Members States, as it was before REACH, but the level of information that should be provided on the costs and benefits of a restriction of a specific use of a substance is arguably greater than it was prior to REACH.

Authorisation, in contrast, puts all of the assessment of whether a substance should remain on the market for a specific use in the hands of the industry. That is, if a use is required beyond the date that the substance will be phased out for all uses, it is those that want to continue to use the substance that must gather and present the evidence (to the ECHA and the European Commission) in order to prove that continued use is justified. Whilst REACH puts the risk assessment of all substances to be placed on the market in the hands of those wishing to do so (though the registration process), the assessment of substances that would otherwise be banned is in the hands of industry and, in some cases, in the hands of those that are users rather than manufacturers of substances. This is a process that is new to regulators and alien to all those in the supply chain apart from manufacturers. The authorisation of substances is arguably the greatest challenge for industry, especially downstream users, because it requires a level of argument and understanding that was until now the preserve of a few specialists in regulatory agencies. The processes of restriction and authorisation are explored below in terms of the process itself and the consequences and developments to date.

The distinction between the application of authorisation and restriction within REACH might seem clear. Substances that meet the criteria for being SVHC are placed on the candidate list and those that are prioritised for placing on Annex XIV will require authorisation; here hazard is driving the process. For restriction, uses of substances that pose a risk(s) at EU level that is not otherwise controlled, require restriction; here risk is driving the process.

As ever, the boundaries between the two processes of authorisation and restriction are blurred and different Member States interpret the application in different ways.

The decision point, for Member State Competent Authorities and the ECHA/European Commission on how to best control substances, is the risk management options process (RMOs). In fact, the RMO process is discussed in most depth in guidance and communications related to the restriction process, for consideration of different possible options to control the residual risk (as in the risk reduction strategy process described earlier). However, some Member States take the view that the RMO process is one which also includes consideration of the best regulatory option within REACH, including considering whether authorisation or restriction could be the best option to control the substance.[3] This is not a view shared by all. Nevertheless, the reality is that member states and regulators are using the RMO process to consider if restriction or authorisation represents the best regulatory path. An additional consideration here is also that in restriction the time and resources come from the regulator, whilst in authorisation the application is compiled by the industry. It should also be noted that Member States are still in the process of handling regulator action of substances that they 'inherited' from the pre-REACH legislation; for example, unfinished risk assessment and risk reduction strategies that are being completed in the transition between ESR and REACH. However, the community rolling action plan (CoRAP) process based on Member State evaluation of substances coming through from registration in REACH will now be the focus for Member States, rather than substance that are still 'on their books' from the pre-REACH regimes.

13.3.2   Restrictions

Restriction as set out in 'Title VIII' of the REACH regulation is focussed on control of risks of substances at a EU-wide level that would not otherwise be captured by the risk assessment framework within REACH. For example, it could be that in the REACH registration process there are a number of registrations for a substance that show a particular use is safe. However, each of the registrants, although they should be aware of each other through the substance information exchange forum (SIEF) process,[4] may

---

[3] At the time of writing, the ECHA has communicated on how to address chemicals of concern in 'roadmap 2020 25 March 2013' indicating the application of the RMO process and its role in the selection for substances a candidates for SVHCs.

[4] SIEF – in which all registrants of a substance are in contact with each other in order that they may share/purchase information, principally in order to avoid unnecessary duplication of animal toxicity testing but also to give them access to the owners of dossiers to enable the right of access to dossiers in order to also register.

make specific assessments of uses that are safe, perhaps because the assessment is based on their own tonnage and measure that they apply to control risk, but the cumulative tonnage of all registrants may lead to a concern over the whole of the EU. In addition there may be concerns about how risk management measures are applied.

Chapter 11 of this book explained in detail how risks are managed and the supply of information down the supply chain on safe use via the extended safety data sheet (eSDS), which is a task initiated by the manufacturers and importers of the substances. It is, of course, the users of the substances that put these risk reduction measures and occupational conditions (OCs) in place. Therefore, Member States or the ECHA or indeed the European Commission may have concerns about the application of risk management measures in reality and the level of control achieved. This is of course an enforcement issue at national/Member State level, but it may be that European-wide action is needed in order to control the risks. As mentioned in Section 1.2.4 of this book the intention of the restriction process is as a 'safety net' for the registration process in order to capture and control the risks of substances not relevant to Authorisation because they do not fulfil the criteria for SVHCs, but still present a possible risk.

It might have been anticipated that the application of a *safety net* would be limited and a relatively rare event, but the activity by Member States and the ECHA itself to bring forward restriction proposals suggests that the process is being treated in reality like the risk management phase of the ESR regime, which was in place prior to REACH. This is not intended as a cynical statement but an observation that is also connected to the way that REACH is interpreted by different Member States as well as the application of the RMO process, as discussed above.

The restriction process is about assessing a particular use or uses of a substance that present an EU-wide risk. Similar to risk reduction strategies of ESR, the process is one of understanding the *reality* of the risks. For example, in the first instance, it is finding out if the users actually control the risks through the application of abatement/control measures and/or through adherence to other legislation. Once the need for risk control is established, a process of considering the 'advantages and drawbacks' of different control options though a RMOs process is initiated. Should a restriction – some limit on use or even a specific prohibition on certain uses – be decided upon, a comparison of the risks versus the benefits in a SEA can be presented. This would normally take the form of a cost-benefit analysis (CBA) in which two scenarios are compared: (i) the *status quo/business as usual/baseline* – that is a situation without the restriction – and (ii) the *non-use scenario* – in which the ban is in place. The net costs and benefits, that is the difference between the two scenarios, that have been quantified and monetised as far as possible, are presented in an SEA report to support the restriction proposal, although this is not mandatory. The proposal is presented to the ECHA by a Member State in the form of an Annex XV dossier. As discussed below for authorisation, and in more detail in Chapter 14 of this book, the SEA needs to compare the risks with the benefits. For the 'risks' this means understanding what the impacts – to human health and or to the environment – are in order to be able to understand the costs and benefits.

13.3.3 Authorisations

The process of Authorisation in REACH was described and illustrated in Section 1.2.4 of this book. As indicated above, the authorisation process is one that applies exclusively to SVHCs. Title VII of the REACH legal text indicates that the one of objectives of Authorisation is to:

> ... *ensure the good functioning of the internal market while assuring that the risks from substances of very high concern are properly controlled and that these substances are progressively replaced by suitable alternative substances or technologies where these are economically and technically viable...*

In simple terms this means banning SVHCs, unless it can be shown that there are not alternatives and the risks of continued use are outweighed by the benefits.

Again, as explained in Chapter 1 of this book, there are two ways in which the applicant (the company or person seeking authorisation) can get an application granted and that depends on the status of the substance as a threshold or non-threshold CMR (Section 13.1.2 in this chapter), namely via the adequate control or socio-economic routes, respectively. It should be noted that all PBT and vPvB substances are considered to be non-threshold in REACH. Demonstrating adequate control for substances that have a threshold for the specific CMR toxicity (that it is listed on the candidate list for) and showing that there are no alternatives or presenting a substitution plan to phase out the SVHC and replace it with a safer alternative, means that an authorisation *shall* be granted. In the adequate control route to authorisation the hazard of the SVHC is considered in terms of risk in what is essentially a conventional risk assessment. All that is different from the registration risk assessment is presenting additional evidence for an analysis of alternatives and a substitution plan, if relevant. This may need to be supported by a SEA but it is likely that analysis would be supporting the arguments about the economic viability of alternatives or the actions and timings of the substitution plan, rather than fully analysing the risks versus the benefits of continued use, since the risks are already shown to be adequately controlled.

The SEA route to authorisation is somewhat more complicated and relates closely to the earlier discussion in this chapter of hazard, risk and impacts. The SEA route to authorisation applies to non-threshold CMRs and to PBT/vPvB substances – that is for which, within the rules of REACH, it is not possible for demonstrate adequate control. The applicant must in this case show that, for the use(s) authorisation is being applied for, that there are no alternatives *and* that the benefits outweigh the risks. Note that it is risks (not hazard) and note also that this requires a comparison of risks and benefits. To make such an analysis requires a reduction of the things to be compared to the same units, otherwise the comparison is entirely subjective – 'comparing apples and oranges'. For economists, this means quantifying risk in terms of impact – that is what would happen if the risks were realised – and applying units of money to those impacts, so that they can be compared to the benefits – which can also be monetised.

In this part of authorisation in particular, we see substances identified for stricter control on the basis of hazard, the need to show that risks are reduced to the lowest levels and in addition the implicit need to convert those risks to estimates of actual impact in order that a comparison of all the impacts, including health, environment, social and economic, of the situation in which the substance is banned to the situation that specific use(s) is continued. It should be noted that a granted authorisation would always be time-limited and subject to review by the regulator (the ECHA).

The practicalities of SEA are discussed in more detail in the next chapter.

References

Arveson, P. (1998) The Deming Cycle. Balanced Scorecard Institute (BSC), http://www
.balancedscorecard.org/TheDemingCycle/tabid/112/Default.aspx (last accessed 29 July
2013).

CIAC (Chemistry Industry Association of Canada) (nd) Responsible Care Ethic &
Principles. http://www.canadianchemistry.ca/ResponsibleCareHome/
ResponsibleCareBREthicPrinciplesBR.aspx (last accessed 29 July 2013).

DuPont (2013) Welcome to DuPont STOP (Safety Training Observation Program).
http://www.training.dupont.com/dupont-stop (last accessed 29 July 2013).

EC (European Commission) (2011) Commission Regulation (EU) No 494/2011 of 20
May 2011 amending Regulation (EC) No 1907/2006 of the European Parliament
and of the Council on the Registration, Evaluation, Authorisation and Restriction of
Chemicals (REACH) as regards Annex XVII (Cadmium).

EC (European Commission) (2012) Introduction to the new EU Water Framework
Directive. Last updated 21 September 2012. http://ec.europa.eu/environment/water/
water-framework/info/intro_en.htm (last accessed 28 July 2013).

ICCA (International Council of Chemical Associations) (nd) Responsible Care. http://
www.icca-chem.org/en/Home/Responsible-care/ (last accessed 29 July 2013).

Roberts, J. (ed.) (2007) The State of the Paper Industry. Monitoring the Indicators of
Environmental Performance. Steering Committee of the Environmental Paper Network, Asheville, NC.

14

Socio-Economic Analysis in REACH

As mentioned throughout this book, there is ample guidance for many of the issues covered by this book and technical guidance on how to do a socio-economic analysis (SEA) for both an Authorisation application and a Restriction proposal is no exception. It is not the intention of this chapter (or indeed this book) to go over or reiterate the detailed technical guidance published by the European Chemicals Agency (ECHA). The purpose of this chapter is to explain, hopefully in a relatively simple way, what an SEA is, why it needs to be done, what the difficulties are and the main features of doing an SEA.

This chapter discusses the SEA, which plays an important role in both the authorisation and restriction process that were introduced in Section 1.2.4 of this book. The need for the SEA originally developed from the risk management process in the – now defunct – Existing Substances Regulation (ESR), which was discussed in Chapter 13 of this book and is elaborated further in Section 14.1 in this chapter.

A key component of an SEA is a desire (an objective) to express risks in physical terms (e.g. life years lost, tonnes being released to waste water, number of workers affected, number of hospital visits etc.). Therefore, this chapter leads nicely on from some of the discussions previously in Section 1.2.4 of this book, in which the connections between the assessment of hazard, risk and impacts and what is required within authorisation and restriction within REACH were considered.

In the present chapter, SEA is discussed in the context of authorisation and restriction in REACH; however, the SEA process has been, and continues to be, applied in other legislation (e.g. Biocidal Products Directive/Regulation) and, in particular, in policy analysis and options appraisal. Indeed, the imposition of a ban and 'derogations' – that is exceptions, from that ban, which is what authorisation and restriction are, it is essentially a policy options appraisal process.

The process of applying SEA in a chemicals legislation framework is not new; neither is carrying out an economics-based analysis in order to assess the most effective course of regulatory action. An SEA is a formalised way of weighing up the pros and cons or advantages and drawbacks of a course of regulatory action. Such assessments form the

Chemical Risk Assessment: A Manual for REACH, First Edition. Peter Fisk Associates Ltd.
© 2014 John Wiley & Sons, Ltd. Published 2014 by John Wiley & Sons, Ltd.

basis of 'impact assessments' (formally called regulatory impact assessments), in which the costs and benefits of proposed new legislation or a change in legislation or policy is assessed. The simplest and perhaps most straight forward analysis of this type is a 'cost–benefit analysis' (CBA). Put simply, this is an assessment in which the 'costs' and 'benefits' of the current situation (without the change, sometimes called 'business as usual') are compared with the costs and benefits of the situation in which the change is imposed.

Comparing the difference between these two situations – so called *scenarios* – is, in principle, simple. However, with SEA 'the devil is in the detail'. The aim of this chapter is to help the reader better understand issues such as;

- What are 'costs' and what are 'benefits' and how are they measured? What represents a cost and what a benefit depends on who you are; for example, a loss of sales for one manufacturer could be beneficial in terms of increased sales of another manufacturer.
- How does one compare benefits that may be measured, for example in terms of well-being and safety against the costs to industry to implement specific risk management measures?
- How can improvements in water or air quality be compared with the loss of a product on the market?
- How does one value the benefit to society from having, for example antifungal paints, an MP3 player, smart-phone or having plastics that can withstand extreme temperatures in comparison to improvements to the environment but may result in less desirable product characteristics like durability, ease of use, effectiveness and aesthetics?

All these issues do not involve a simple comparison of 'like with like' or more of one product versus less of another. SEA is, therefore, a useful tool within government policy, as it is used to address trade-offs by helping to compare risks (costs) with the benefits.

This trade-off analysis is at the heart of economic analysis, which, put simply, is about how best to allocate resources that may be scarce and maximise the theoretically infinite requirements of society now, that is our desire for goods and services, and scarce resources available for future generations.

The purpose of this chapter, as with this book, is not to give detailed instructions on how to *do* an SEA – there is plenty of technical guidance for that already – but rather to give some understanding and explanation of the main features of the process and how it is applied. In addition, it is hoped to perhaps give some insights into the practicalities and problems of doing such work. We do not go into any detail about the application of specific economic analysis techniques or how impacts are assessed – the reader is referred to the technical guidance as well as other sources, as each assessment and the techniques applied should be judged on their own merits.

The structure of this chapter is as follows:

- Section 14.1 – Considers briefly the background for the need for SEA – this has already been touched upon in the previous chapter.
- Section 14.2 – Builds on the introductory text and sets out what an SEA is, and why it is applied in the context of REACH.
- Section 14.3 – Briefly examines the purpose of the SEA, with explanation of how to do an SEA in REACH.

- Section 14.4 – Considers the difficulties in getting and the right kind of existing information of in order to assess impacts of substances on human health and the environment in an SEA.
- Section 14.5 – Looks at the process and who makes the decisions on an SEA in REACH.
- Section 14.6 – Takes a wider perspective and looks at how an SEA can be used in the chemicals risk management process, that is in a product stewardship context.
- Finally, in Section 14.7 some observations are made on SEA to date and some thoughts are offered on future developments.

14.1 Background – the Need for and Development of Socio-Economic Analysis in the Regulation of Chemicals

In the context of legislation on chemicals an impact assessment or SEA is concerned with a measure/regulatory action/policy that puts a limit in some way on the production, import or use(s) of a chemical and compares what happens when that change is imposed with the current situation, that is to answer the question: 'What benefit does this action bring'? There are, therefore, two 'scenarios' being compared:

1. Without the action
2. With the action imposed.

If the regulatory action is a ban or limit on the use or uses of a substance then the scenarios becomes:

1. The current situation or 'baseline'
2. The non-use of the chemical.

In the second scenario above, this may be quite specific to particular chemicals and their particular uses.

As indicated in the previous chapter, SEA is about impacts and assessing the environmental, health, economic and social impacts (both positive and negative) of a course of action.

SEA under REACH has its background in the need for consideration of the social and economic impacts of risk reduction measures under the ESR and also the application of more detailed regulatory impact assessments by policy makers at national and at European Union level.

Nowhere in REACH is it stated that an SEA is a *mandatory* part of the process, that is it does not *have* to be part of an Authorisation application from industry or an Annex XV restriction proposal from a Member State or the ECHA. The REACH legislation makes clear, however, that for an authorisation to be granted in the SEA route, that is for use or uses of a substance for which a safe level *cannot* be established, it must be shown that the benefits of the use(s) outweigh the risks. This strongly indicates that some kind of SEA is needed and, indeed, there is detailed and comprehensive technical guidance from the ECHA on SEA in support of Authorisation applications and Restriction proposals.

As noted publicly by ECHA officials, an SEA does not have to be a long and detailed analysis but should be proportionate to the assessment of the impacts being assessed.

From this it is understood that in a CBA if, whilst being sufficiently robust, it can quickly and clearly be shown that the benefits outweigh the costs by a significant margin, or vice versa, then further detailed analysis is not required. However, it will be in the 'grey areas' where the differences are not so clear. This, for example, could be where some aspects are monetised and others not, or where some impacts are more uncertain that others, so that one assessor's view might be that costs outweigh benefits whilst another's view, based on similar data, would make the opposite conclusion. It is in these cases that perhaps more detailed arguments and analysis are needed. However, the ECHA advises applicants to submit their 'best shot', because the Authorisation applicant will not be able to resubmit a revised and improved SEA should the first version fail to satisfy regulators. It is, therefore, wise for applicants to give more evidence than perhaps they consider absolutely necessary, knowing that the reviewers are bound to take a more conservative view on the level of evidence required to support their conclusions.

The difficulty with giving any definitive judgements on the detail of an SEA is, at the time of writing, that no Authorisation application has yet been evaluated by the ECHA committees and, thus, apart from guidance, seminars and statements from the ECHA and some of the more proactive Member States with particular interest in the field, there is nothing to indicate the level of evidence that should be presented or how the ECHA will view certain analyses or the application of particular techniques. However, it should be noted that there is experience of restrictions proposals with accompanying SEAs.

Another difficulty, or perhaps an interesting development at least, is that the SEA process brings a new discipline, and thus a new body of practitioners into the area of chemicals assessment and regulation for the first time – the economists. Whilst risk assessors and risk managers are used to dealing with specialist scientists for specific information on hazards, exposure and risks (indeed most, if not all, risk assessors involved in chemicals regulation have a background in the physical, chemical or biological sciences), they are not used to dealing with the techniques, paradigms and language of economists. SEA requires that impacts are compared and this means that risk assessors, scientists and economists must work together to produce such assessments, but there is not always a direct connect between what the risk assessor and what the economist understand. As the demand for SEAs continues, however, these multidisciplinary teams will no doubt become more established on both sides of the regulatory fence.

14.2   What Is SEA and Why Is It Needed and Applied in REACH?

14.2.1   What Is SEA within REACH?

An SEA is the analysis and an SEA report is the reporting of the findings. The study itself is usually an iterative one whereby the 'boundaries' of the study are defined (i.e. what is to be considered and what is not relevant), the use and non-use (of the chemical) scenarios defined and the main impacts indicated. As the study develops there may be a need to gather more data on particular areas and refine or strengthen certain parts of the analysis. This first stage of defining the scope of the analysis is very important because it allows these main aspects to be focused on. It may be that it is clear from that point (i.e. results of a 'scoping' study) that further analysis is not required because it is clear that the costs outweigh the benefits or vice versa.

At the heart of the SEA is the comparison between two scenarios – the use and non-use scenarios. The 'baseline' or *status quo* situation is continued use (but for the uses that are being applied for only), the other scenario is the *non-use* or situation in which no use of the substance is permitted.

The SEA is concerned with assessing the impacts – that is the costs and benefits in both scenarios and comparing them. It is the difference in costs and benefits between the two scenarios that matter, that is in the end it is the *net* cost and benefits that are presented.

The ECHA technical guidance on SEA usefully sets out a five stage process for doing SEA for a restriction or authorisation. The five stages are:

Stage 1 is the setting of the ***aims*** of the SEA. This is to answer the question: Why is the SEA or input to one being developed?

Stage 2 is the ***scoping phase***, to define what economic and other responses and changes will occur as a result of either (i) the proposed restriction – a key question to answer is: How will actors in the relevant supply chain react if they are subjected to the proposed restriction? – or (ii) the Authorisation application – What are the 'applied for use' and the 'non-use' scenarios and what are the supply chains involved?

Stage 3 involves the ***identification and assessment of impacts***. The aim is to answer the question: What are the impacts of the 'non-use/restriction' compared to the 'baseline' scenario?

Stage 4 focusses on interpreting the impacts identified and assessed in stages 2 and 3. It is about ***bringing the information on different impacts together to determine the net impact*** and undertaking an ***uncertainty analysis*** to test the robustness of the SEA. Based on the assessment and the uncertainty analysis, it can be decided to either conclude the SEA or undertake more analysis by reverting back to stage 2 or 3.

Stage 5 is the final stage. In this stage the ***main findings and results of the analysis are summarised***. It is important to present all data in a systematic and transparent way in order to aid the decision making process.

Uncertainties can and will arise during the SEA process and they need to be:

- Considered throughout the process.
- Minimised where possible.
- Assessed for their importance with respect to the outcome of the SEA.
- Documented, so that they can be well understood alongside any SEA outcomes.

14.3   Role, Purpose and Performing an SEA in REACH

14.3.1   Role and Purpose of an SEA in REACH

An SEA in support of an Authorisation application and that in support of a Restriction proposal can be rather similar, although they have opposite objectives. The Authorisation SEA is seeking to demonstrate that the benefits of continued use/s of a substance outweigh the risks; while the Restriction SEA is seeking to show that the benefits of banning specific uses outweigh the risks of keeping those uses on the market. As mentioned earlier, the SEA in both authorisation and restriction may also have different supporting roles. An Authorisation SEA may be supporting the timing and actions in a substitution plan, for example, or demonstrating that a potential alternative (functional or substance) is not economically feasible in a reasonable timeframe. A Restriction SEA may be considering the costs and benefits of a set of risk management options (RMOs) in order to decide upon the most beneficial way to control the risk.

An important role of an SEA not yet mentioned is that of supporting a 'third party', that is not the applicant and not the Member State Competent Authority (MSCA), but another interested organisation or individual that wishes to present an analysis on an authorisation application or restriction proposal. For example, it may be an NGO wanting to demonstrate the benefits of introducing an alternative, or perhaps the manufacturer of that alternative, or it could be an industrial trade organisation putting forward its case for the impacts of a proposed Restriction. Within the processes the ECHA's Socio-Economic Analysis Committee (SEAC) takes account of these third party SEA submissions within the consultation period, but third parties are necessarily limited by the information that they have access to. The availability of information may also be limited by its commercially sensitive nature. Competitors may not be willing to submit data that might reveal their commercial plans. Manufacturers and/or end-users may be hindered by the requirement to comply with competition law.

14.3.2   Doing an SEA in REACH

As mentioned at the beginning of this chapter, it is not the intention to reiterate guidance that is already available. This section sets out the main features of doing an SEA and directs the reader to guidance and other documents that may help, but it is assumed that practitioners who are actually compiling an SEA will not need this chapter and will already be familiar with the ECHA guidance on this subject. There are three main 'actors' who can submit and SEA as part of the REACH regulatory processes within authorisation and restriction. Table 14.1 summarises the actors and roles for the SEA.

The scoping phase of the SEA is critical to understanding the full study and should be the process in which the main impacts are identified; equally, it is the process in which impacts that are not relevant to the study are eliminated from further consideration. Unlike with some other legalisation (for example environmental impact assessment), scoping is not a formal part of the application process, so that reasons for 'scoping out' irrelevant impacts need to be retained in the full study documentation. An initial iteration or 'scoping' SEA could and should be undertaken relatively quickly (e.g. one to three month's work). It should focus primarily on the authorisation applicant and their likely response to a refused authorisation application and any associated impacts.

Figure 14.1 illustrates the process for an authorisation SEA; the final step, which is not illustrated, is the documentation of the SEA in an SEA report.

14.4   The Difficulties of Moving from Risks to Impacts

As discussed in Chapter 13 of this book, a key difficulty in this process is the estimation of *impact* and the apparent disconnect between the outputs of the risk assessment that is required for the chemicals safety assessment in REACH and the data on impacts that

Table 14.1 *Actors and roles for the SEA.*

	Authorisation		Restriction
	Adequate control route	SEA route	
SVHC manufacturer/ importer/ user	SEA is not required, but can be: (i) Supporting substitution plan if alternatives or (ii) Supporting review period if no alternatives	SEA demonstrating the benefits outweigh the risks	N/A (although it is possible to engage in the RMO process and offer SEA-type information)
ECHA/MSCA	N/A		SEA supporting a restriction proposal
Third Party[a]	SEA in support of/against the authorisation	SEA in support of/against the authorisation	N/A (although it is possible to engage in the public consultation process and offer SEA-type information)

[a]The ECHA Technical guidance defines a third party as – Any organisation, individual, authority or company other than the applicant or the Agency/Commission with a potential interest in submitting information on alternatives or other information, for example on socio-economic benefits arising from use of the Annex XIV substance and socio-economic implications of a refusal to authorise.
SVHC – substance of very high concern.

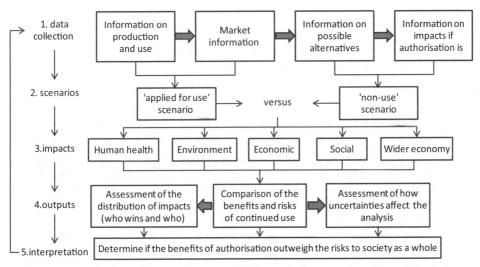

Figure 14.1 *Illustration of the Main Processes and Steps in an Authorisation SEA (adapted from Nickel Institute, 2012)*

are needed for an SEA. Conceptually, the assessment is a simple one: assess the costs (C) and assess the benefits (B) in the two scenarios and compare them; if B > C then the authorisation or restriction is justified.

For economists, the outcomes of a risk assessment currently are insufficient for valuing possible human health and/or environmental impacts. A risk assessment is concerned with determining whether the exposure exceeds the no-effect level, and thereby resulting in possible adverse effects (i.e. that there is a risk). The general problem for economists with this approach is that it does not describe or quantify what types of effects might occur as a result (e.g. how will fish stocks will change, how many people might get cancer, how many tonnes end up in waste water, what is the loss in land availability/productivity, etc.). Essentially, economists (and indeed policy makers) want to better understand; what this means in actual physical terms, why any release/risk is a problem and how the environment or humans might be affected. Until these questions can be determined from the risk assessment, it is very difficult to value benefits in the same terms as economic costs.

In Box 14.1, the output from risk assessment and the needs of the risk management process, that is to quantity impacts in order to do a CBA, is explored further with reference to a key paper that considered this issue by Valery Forbes and Peter Calow of the University of Nebraska, USA (Forbes and Calow, 2012).

Box 14.1   Risk assessment, risk management, impacts and cost benefit analysis

From a policy maker or risk manager's point of view there is the need to weigh the benefits of an action, such as a restriction on a specific use (as in Restriction) or indeed continuation of a specific use (as in Authorisation) against the costs to society of that action.

Benefits

In the context of a CBA this can be expressed – Forbes and Calow (2012) set this out succinctly for benefits and costs – for benefit:

$$\Delta B = f(\Delta A \times V)$$

This is explained as a change in benefit (B) from a policy action that is a function of the impact avoided (A) and the value (V) placed on what is protected (V/A). Forbes and Calow (2012) refer to human and ecological impacts but this could also include other impacts, such as social or economic ones (e.g. increased sales and employment from competing products (alternatives) that will benefit from a Restriction/refused Authorisation).

The difficulty of course for SEAs for REACH is obtaining (i) an estimation of impact and thus impact avoided (i.e. and estimation of 'A'), in particular for the environment, and (ii) obtaining values (V) for those impacts. However, if it is possible to measure A, then it is possible that economists, by using various studies that reveal the value of the impact and thus 'A', to assign a value 'V' to it.

For impacts on human beings this is arguably more straightforward, since there are known values for impacts, such as medical treatment of disease and disability as well as standardised values for lost quality of life and even loss of life-years. The quantification is done by estimating/predicting the number of individuals affected. However, these approaches underestimate of the true impact, as they do not capture aspects not reflected currently in monetary terms (called 'non-market' impacts), such as fear, emotional and physical pain as well as suffering/well-being. Economists have tried to encompass all these aspects, for example through non-market valuation tools such as the willingness-to-pay (WTP) values (i.e. the value that individuals would attribute to avoid getting the disease/illness) or willingness-to-accept (WTA) values to reflect the true value required to compensate someone from getting that disease/illness. These values are not widely used or available due to a lack of availability of the underling research to create such values across a number of types of health impacts (as they are case/risk specific) as well as some negative perceptions of these values. However, in policy terms these values are likely to become increasingly more important in order to be able to better compare health impacts against more easily quantifiable financial costs such as equipment costs or loss of sales.

For the environment, valuing benefits is even trickier, since the risk assessment does not give information on exactly how different parts of the environment will be affected or what the distribution of effects will be in time and space. The risk assessment simply indicates that above a certain concentration (the predicted no-effect concentration, PNEC) that there will be a risk of harm, without any indication of the impacts and its severity. Environmental economics as a discipline has significantly grown in the last 10–15 years, with increasing valuations of various types of environmental and health benefits, which can be seen, for example, in the Environmental Valuation Reference Inventory (EVRI). Nevertheless, more work is required to bridge the evident gap between risk assessment outputs and the quantification of impacts needed for an SEA.

Costs

The costs (C) of the action are a function of the loses that result from losing the use of a quantity (Q) of the chemical, which can be expressed as $-\Delta Q$, so that cost is expressed as $\Delta C = f(-\Delta Q)$. The costs should also include the wider costs of the loss of the use of the substance, such as impacts on the supply chain, loss of product quality (for example due perhaps to using a substance that does the same job but not quite as well), change in prices and durability incurred by consumers from using competing products that may be on the market, and so on.

As Forbes and Calow (2012) rightly point out – in a CBA B > C, in order for the action (such as a ban, or conversely continuing a use) to be justified – that is the benefits must outweigh the costs.

However, the important point is that the risk assessment must give outputs that are 'As' – that is impact on the things that of value that need to be protected and can be expressed as **dose-responses** in terms of A and Q – that is a relationship between exposure and impact is established. Otherwise, if there are only endpoints that are not related to the impacts, then subjective judgements have to be made on their importance and seriousness. Therefore, the values will not necessarily reflect those of the affected groups. This happens when a precautionary approach is taken which may inflate the relationships between the endpoint and the impact avoided (A) or vice versa. Currently, in REACH, because the risk assessments do not produce the relevant information on impacts, there exists a situation whereby other information on impacts has to be sourced or generated in order to estimate the avoided impacts.

Some regulatory agencies have attempted to derive impacts for SEA studies to show how this might be done 'in anger' for SEA's in support of restrictions and authorisations. These studies have been based on substances that already have good data from existing studies and or substances that were likely to end up on Annex XIV as case studies. At the time of writing, it also seems to be the case that the ECHA itself is commissioning studies on some substances that it suspects will be the subject of Authorisation applications in anticipation of the application, (presumably as an exercise in informing themselves and the SEA Committee of the likely magnitude of costs and benefits that they may be presented with).

The Dutch Government Agency 'the National Institute for Public Health and the Environment' (RIVM) attempted to directly address the difficulties with assessing environmental impacts in its study report entitled: *From risk assessment to environmental impact assessment of chemical substances. Methodology development to be used in socio-economic analysis for REACH* (Verhoeven *et al.*, 2012).

In the study RIVM proposed a stepwise approach, which has much in common with the ECHA technical guidance on SEA. The approach is based on six steps:

1. Scope and scenario definition
2. Exposure and hazard estimation
3. Determination of endpoints and assessment method
4. Environmental impact assessment
5. Dealing with uncertainties
6. Comparison of the scenarios and providing comparable information on environmental impacts.

In step 4 the assessment of environmental impacts of compounds involves the possibility of a ranking of persistent, bioaccumulative and toxic (PBT) characteristics of substances and impact assessment based on a deterministic or probabilistic approach.

A number of possible approaches are presented by RIVM. These are essentially surrogates for not being able to really understand environmental impacts. However, in reality these may be the only ways to get around the problem in the absence of data that properly link concentration to impact.

The RIVM report suggests:

The RIVM study considered a number of well-studied substances in order to test the proposed framework, including (1) the historic replacement of nonylphenol (a surfactant) in detergents by alcohol ethoxylates, (2) the replacement of zinc gutters by PVC gutters, and

(3) the replacement of the fire retardant hexabromocyclododecane in insulating building material by two alternative flame retardants.

In another study the European Commission Environment Directorate General (DG Environment) commissioned UK-based consultants Risk and Policy Analysts in association with the University of Stuttgart and Imperial College London, to produce a study entitled 'Assessing the Health and Environmental Impacts in the Context of Socio-economic Analysis Under REACH' (EC, 2011). The DG Environment report presented a 'logic framework', again not dissimilar from that set out in the ECHA technical guidance. The DG Environment report is focussed on health impacts but offers some case studies that also consider environmental impacts. While the assessment suggested is certainly thorough, it depends on gathering quite a lot of additional data to that which would be in a chemical safety assessment, in a stepwise process as follows:

1. Hazard characterisation
2. Exposure characterisation
3. Qualitative description of potential impacts
4. Benchmarking of environmental hazard
5. Assessment of the potential for quantification of impacts.

Both of these reports show creative and robust attempts to solve the problems of gathering impacts data and are to be commended. However, it is clear that estimation of impacts especially for the environment is difficult and subject to much uncertainty and perhaps also a subjective assessment.

It has been suggested by a ECHA study (ECHA, 2011) that it may be possible to adopt an approach based on getting an idea of the likely costs and benefits by focussing on the costs side. For example, it is possible to construct a 'cost curve' in which measures to control releases to the environment of a PBT or vPvB (very persistent and very bioaccumulative) substance can be ranked in order of the cost per unit mass that they remove. If a desired reduction is known then this then allows a relatively quick and easy assessment of the cost of achieving that reduction, so long as most of the available measures are included and the costs of those measures is reasonably accurate. In principle this (cost effectiveness) approach could be useful within a restriction context by comparing the cost effectiveness of various restriction options potentially against other risk management options, to help determine which restriction option (and whether restriction itself) is the most appropriate risk management option (e.g. compared to emission limit value, occupational exposure limit, permit/tax scheme etc.).

It should be stressed that this *does not* replace the assessment of impacts and gives no estimation of the value of the benefits gained from the reductions that may potentially be achieved. Therefore, despite recommendations for its use (due to ease of data), cost effectiveness (or a cost-curve approach) has limited benefits in determining if $B > C$ unless the benefits can assessed separately.

To date, most attempts to value risks have been approached from the risks assessors' perspective as shown in Section 14.4 in this chapter. However, it might be interesting to look at the same challenge from an economists' viewpoint. Box 14.2 sets out a possible approach to valuing the environmental risks and thus environmental impacts from the perspective of an economist.

Box 14.2   Valuing the environment – an economists perspective

A study for the UK government (Defra), *Valuing Environmental Impacts: Practical Guidelines for the Use of Value Transfer in Policy and Project Appraisal* (eftec, 2009), provides guidance for valuing environmental impacts via a technique known as value transfer (i.e. the use of existing valuations in literature).

Whilst in this book it is not proposed to repeat the practical guidance on 'how', which is quite extensive, the discussion presented below provides an insight into how a similar approach can be applied in the context of REACH SEA and chemicals management. It can be summarised in four steps:

1. Define the 'goods' (e.g. parts of the environment) and population being affected.
2. Define and quantify the change in provision of the policy 'good'.
3. Identify and select monetary valuation evidence/methods available.
4. Estimate the value of the policy 'good'.

As mentioned in Section 14.4 in this chapter, the outputs from the risk assessment can help identify if there is a problem, but not the magnitude of the risk or the likelihood actual impacts. In this first step, it is important to describe what the risk might be (qualitatively or quantitatively) and how it might affect the environment, taking into consideration the services the environment provides.

For example, if a chemical potentially is being released into the aquatic environment, the first steps are to understand how it is getting into the environment, in what concentration/volume and why this might be an issue. This is unlikely to be very different to how a risk assessor might approach the problem.

Where differences may occur is that economists are likely to have a preference for this analysis to be firstly described qualitatively in terms of how the release/use of the chemical affects total economic value (TEV), which is made up of various components, as listed below and illustrated in Figure 14.2:

- Use value:
  - *Direct uses* obtained from the environment, for example food, timber, drinking water, navigation routes, recreational activities, visual amenities and so on.
  - *Indirect uses* obtained from the environment without directly using it, for example carbon sequestration, water regulation, pollution filtering, soil retention and so on.
  - *Option value* – the value that people place on having the option to use a resource in the future even if they are not current users, for example being able to make use a particularly landscape or recreational activity or having the option to use certain natural resources in the future.
- Non-use value: (simply knowing that the natural environment is being maintained)
  - *Bequest value* where individuals attach value from the fact that the ecosystem resource will be passed on to future generations.
  - *Altruistic value* where individuals attach values to the availability of the ecosystem resource to others in the current generation.
  - *Existence value* derived from the existence of an ecosystem resource, even though an individual has no actual or planned use of it.

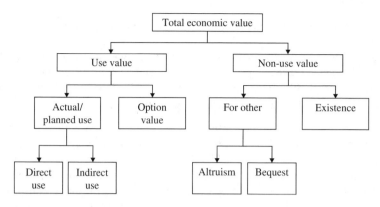

Figure 14.2   *Total Economic Value*

Essentially, it is necessary to understand how the substance helps/inhibits the services that the environment provides, examining both the use and non-uses that the environment provides. For PBT and vPvB substances, it is more difficult, but it is still important to think carefully about how the substance affects the various components of TEV, otherwise if any link cannot be found it is unlikely that any valuation will be possible.

It is recommended that this is done qualitatively at this stage as the next (second) step, is then to critically assess how any policy action (Authorisation or Restriction) will change any of the components affected. If the policy option makes no difference on certain components, then these components can be ignored initially, to focus on the most significant changes in the environment and the services that the environment provides. These first two steps require economics input, but the majority of the work should be performed by those with expertise in risk assessments/chemicals/environmental science.

The third step is to identify how these changes in services can be valued. In order to aid in identifying which type of method to use, it is worth following this valuation hierarchy:

1. Market prices
2. Shadow prices
3. Revealed prices
4. Stated preference.

Market prices should be used where possible, since these are values that people can best relate to with greater confidence. These are particular relevant for direct uses such as the value of fish stock, raw materials, timber and so on. However, the ability to use market prices for environment services can be limited. When market prices are not available, shadow prices should be used. The best way to describe them is through the use of an example, where shadow prices have been determined for carbon emissions. These values have been widely adopted and accepted in policy making to value the benefits or costs associated with marginal changes in carbon emissions.

If market and shadow prices are not available, changes in environmental impacts could be valued using revealed preference/prices. There are many techniques available and again for simplicity a few examples are set out to illustrate how these revealed prices can be used rather than explaining each technique:

- The market price for equipment used to filter water or even bottled mineral water could be an indication of the value people attribute to an improvement in drinking water.
- The difference in house prices from a house with quiet surroundings compared to an identical house nearby which is subject to heavy road traffic could provide an indication of the value of externalities associated with road transport/vehicle noise and air emissions.
- The amount of time or travel cost people are willing to spend to visit a particular site, such as a beach or national park (e.g. for cultural, visual amenity or recreational purposes), could provide an indication of the value society places that use of land/environment service even though, for example, people are not required to pay directly to use the park or beach.
- An improvement in water quality or removal of certain chemicals may result in avoided costs to users and providers of water, such as improving the lifetime of their equipment, reducing cleaning costs, reduced use of energy, improving fish stock which makes fishing easier/more sustainable, improves water recreational activities and so on.
- Some chemicals may lead to depletions in the ozone layer, resulting in consumers adapting their behaviour to avert any negative impacts. The value of peoples' purchases in goods and services to avert these negative effects, such as use of sun creams, hats and sunglasses, could be used to reveal the environmental costs associated with ozone depletion.

If none of the above techniques is suitable, then a final method available is stated preference, where people's willing to pay or accept compensation for change in environment risk are elicited through the use of hypothetical markets. There are a number of stated preference methods available using various survey techniques in order to establish a willingness-to-pay or willingness-to-accept value. These studies can take time and should be carried out by environmental economists. To avoid associated costs of developing new valuations, it is possible to use existing valuations (e.g. see the EVRI database) under certain conditions. This technique is called value transfer (eftec, 2009).

The final step is to make use of the most appropriate techniques available based on the type of environment impact occurring and what data are available to value the changes in the environment. It is recommended that SEA practitioners follow the valuation hierarchy using market and shadow prices where possible and then move to revealed prices and stated preference thereafter. Regardless of the technique used, it is important to be transparent in what and how values are used, to carry out sensitivity analysis on key variables used to appropriately reflect any uncertainties that might be apparent and to ensure consistency in how values are reported (e.g. consistent use of currency, unit of value, per year or total costs over a given time period etc.).

It is stressed that it is only possible to value (and monetise) changes in the environment (changes in risk) if it is possible to qualitatively and quantify how risks affect the environment and the services the environment provides.

14.5   Regulatory Processes – Who Are the Decision Makers and What Are Their Roles?

In Section 1.2.4 of this book a schematic diagram illustrates the process for both Authorisation and Restriction. For the SEA it is the ECHA's SEAC that assesses the SEA report. The SEAC comprises nominated experts from Member States, who although nominated and funded to attend by their respective Members States are supposed to represent themselves and not necessarily the views of the national regulator bodies they come from. By background (and therefore not taking into consideration their experience with doing SEA) the SEAC currently comprises about one third economists and social scientists, with the majority having engineering or scientific (mostly chemistry) backgrounds. The SEAC Membership with curricula vitae and declaration of interests can be viewed from the ECHA web site (ECHA, nd).

The fundamental issue that needs to be addressed by an SEA is whether society as a whole is better off with or without the continued specific uses of the substance in question. The SEA report should, therefore, seek to justify to the SEAC in these terms whether an Authorisation or Restriction should be granted or not. By definition, this takes a broad and societal view of potential impacts that extends beyond the actual use of the substance.

The SEAC can request further information on an SEA submitted. The SEAC forms an opinion on an SEA and makes a recommendation to the ECHA, which in turn makes its recommendation to the European Commission. This is done in tandem with the recommendation from the risk assessment committee (RAC) that considers risk related parts of the authorisation application or restriction proposal.

As an example a timeline of actions for an authorisation application including an SEA might be as illustrated in Figure 14.3. The timeline is illustrative only and intended to give an indication of actions that have to be taken and also the likely length of time needed to prepare the relevant material including an SEA.

14.6   The Wider Benefits of Performing an SEA

Although the focus of this chapter has been on carrying out an SEA in support of restriction and authorisation in REACH, there is also a more subtle purpose for SEA within the context of chemicals assessment. This is not in a direct regulatory function but one more aligned with product stewardship. For example, a CBA allows the demonstration

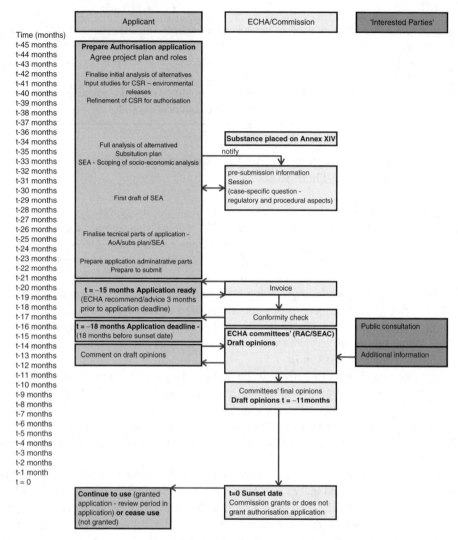

Figure 14.3   *Illustration of a Possible Authorisation Timeline with Actions for Applicant, Regulator and Interested/Third Party*

to customers that the benefits for specific uses are justified, by the control of the risks and the benefit that the uses bring. An SEA or, more specifically, a CBA, might show that early and proportionate action could allow an industry body to be proactive and set the agenda on its product, based on good science and careful analysis. One standpoint is that an SEA is actually a business justification. For example, even an SEA supporting an Authorisation application should be treated as a *business case*, whereby applicants are demonstrating the value of their use of the substance. In reality, of course, these decisions can be difficult and subjective, for example when considering examples like cosmetics: How can a value be put on the contribution to individual and social well-being if a given beauty product must be abandoned? The SEA will, therefore, determine if society is better or worse off if authorisation is granted. The same can be applied to a CBA of any substance as a justification that the product brings real value/benefit to society as a product stewardship task and demonstration of product sustainability, whether or not the substance is a current risk from regulatory action.

14.7   Developments and the Future

There are no hard and fast rules about how an SEA must be done, although it might be unwise to depart too far from the ECHA guidance, at least until if or when it is revised. What is important, however, is that the analysis is proportionate and transparent, so that it is clear where data are from and that there are no 'black boxes' in the analysis – that it can easily be seen how impacts were derived and how the analysis was done.

Although life cycle analysis (LCA) methodologies potentially may bring much to SEA, it is the derivation of subjective scores for impacts that are not a good fit with what is required for REACH. To date, all indications are that a CBA approach is favoured because it potentially delivers what is required for SEA, as long as all the main impacts are covered, uncertainties explained and justified and sensitivities explored (e.g. in worst case and reasonable case scenarios).

It is clear that a stumbling block for SEA within REACH is actually the methodologies around assessing impacts on human health and, in particular, the environment. It may be that, as indicated in Chapter 13 of this book, other assessment frameworks that already consider impact in a quantitative or semi-quantitative way may offer the way forward, or it may be that the risk assessment outputs from the chemicals safety assessment process will have to be altered in order to provide the type of data that is needed to do SEAs for REACH. Future reviews of the implementation of REACH and its effectiveness will surely take this into consideration. In the meantime, it is urged that risks to human health and the environment are well described in terms of possible impacts, so that those assessing the SEA can at least subjectively decide how these compare to tangible costs, rather than just focusing on a cost effectiveness/cost curve approach, which provides no fruitful assessment of risks.

References

EC (European Commission, Directorate-General Environment) (2011) Assessing the health and environmental impacts in the context of socio-economic analysis under REACH. ENV.D.1/SER/2009/0085r. Part 1: Literature Review and Recommendations (http://ec.europa.eu/environment/chemicals/reach/pdf/REACH%20SEA%20Part%201 _Final%20publ.pdf; last accessed 5 August 2013); Part 2: The Proposed Logic Framework and Supporting Case Studies (http://ec.europa.eu/environment/chemicals/reach /pdf/REACH%20SEA%20Part%202%20LogicFrame%20Final%20publ.pdf; last accessed 5 August 2013).
ECHA (nd) Committee for Socio-Economic Analysis. http://echa.europa.eu/web/guest /about-us/who-we-are/committee-for-socio-economic-analysis (last accessed 28 July 2013).

ECHA (2011) Direct Contract: Abatement Costs of Certain Hazardous Chemicals. ECHA/2011/140 (last accessed 5 August 2013).

eftec (2009) Valuing Environmental Impacts: Practical Guidelines for the Use of Value Transfer in Policy and Project Appraisal. http://archive.defra.gov.uk/environment /policy/natural-environ/using/valuation/documents/vt-guidelines.pdf (last accessed 28 July 2013).

Forbes, V.E. and Calow, P. (2012) Promises and problems for the new paradigm for risk assessment and alternative approache involving predictive systems models. *Environmental Toxicology and Chemistry*, **31**, 2663–2671.

Nickel Institute (2012) AUTHORISATION SEA 'ROADMAP' Practical Guidance For Undertaking Socio-Economic Analysis, edn 1. http://nickelconsortia.eu/assets/files /consortia/SEA_ROADMAP_12_09.pdf (last accessed 5 August 2013).

Verhoeven, J.K., Bakker, J., Bruinen de Bruin, Y. *et al.* (2012) From Risk Assessment to Environmental Impact Assessment of Chemical Substances. Methodology Development to be Used in Socio-Economic Analysis for REACH. RIVM Report 601353002/2012. Dutch National Institute for Public Health and the Environment (RIVM). Ministry of Health, Welfare and Sport.

15

REACH: How It Is Working and May Develop

15.1 Introduction

This chapter gives a somewhat subjective view of the development of REACH to date, drawing upon the authors' experiences and some published sources. Some of the issues are still in their relative infancy; it should be remembered that REACH does not come to an end when all the phase-in substances are registered.

At the time of writing (early 2013), what observations can be made about REACH? There has also been some official analysis; this is drawn upon in this chapter. From there, it can be speculated about how REACH will operate over the next 10 years. By 2023 many dossiers will have been evaluated, test plans made and completed, and a new 'steady state' may have been achieved.

For now, the impacts of REACH are manageable for large companies. However, in a market where speciality substances are vital for profitability, it is presenting huge challenges to smaller business units and small and medium-sized enterprises (SMEs). In respect of the promotion of health, safety, and protection of the environment, REACH has been very effective overall.

15.2 Experiences and Observations

The regulation itself and its enforcement:

- 'New' and 'existing' substances have been brought into a unified system and the previous 'new substances' regulation was, as expected, a good guide to how REACH would work.
- Thousands of registrations of existing substances of high tonnage or priority have been submitted, and also some in lower tonnage bands; the quality of these submissions is highly variable.

Chemical Risk Assessment: A Manual for REACH, First Edition. Peter Fisk Associates Ltd.
© 2014 John Wiley & Sons, Ltd. Published 2014 by John Wiley & Sons, Ltd.

- Preparations are well advanced for the 100–1000 tonnes/year band.
- Many new reports of important properties have come to light and been shared.
- Substances of very low hazard profile still seem to attract too much attention from the regulators. Classifications (and labels) have been reviewed and substance of very high concern (SVHC) candidates identified.
- The importance of adequate analytical characterisation has been stressed
- Authorisation work is starting.
- Some enforcement actions have occurred due to non-compliances.
- Read-across, category and (Q)SAR (quantitative) structure-activity relationshipmethods have been very demanding of time if the strict letter of the Guidance is observed.
- Test plans have been reviewed, often with the European Chemicals Agency (ECHA) demanding more animal tests than registrants expected.
- Contrary to expectations REACH-IT has not failed, despite frequent updates to IUCLID International Uniform Chemical Information Database.
- Guidance and requirement changes continue to pour out from the ECHA, stretching the capacity of the industry to adjust, against unrealistic time scales and resource capacity. The amount of guidance is overwhelming for the inexperienced, or for a small group.
- Suppliers and downstream users have not had the capacity to communicate to the level that the Regulation intends.
- The property parts of dossiers are now available on the ECHA web site, which will help development of further read-across and (quantitative) structure-activity relationship development. Turning to the authorities: Member State involvement with review and with CoRAP (community rolling action plan) is of varying quality and commitment, with a wide range of interpretations of information.
- The ECHA has been unswervingly competent (and very formal) in enforcement of a 'level playing field'; the emphasis has been on property data, with relatively little feedback on exposure issues for the most part; the exception has been where registrants have sought to justify data waivers with insufficient discussion of exposure.
- The European Commission believes that risk characterisation ratios (RCRs) are decreasing (Eurostat, 2012).

15.2.1   Observations

- The overlaps with other legislation have tended to confuse some registrants, particularly given that the Regulation makes no cross-references to some highly relevant existing measures.
- Interaction of registrants and regulators with stakeholders outside the European Union (EU) has been somewhat minimal; that is expected to change as worldwide harmonisation looks increasingly likely.
- The exposure scenarios have been highly contentious, with the defaults being impossibly cautious, placing huge pressure on industry to understand exposures better.
- The level of comment on dossiers from the ECHA has ranged from highly insightful to naïve; such variability confuses registrants.
- Animal welfare groups are highly concerned that not enough is being done to minimise test requirements; industry shares some of those concerns.

- Other non-governmental organisations feel that the ECHA is too secret about its inter-actions with industry.
- A number of unintended consequences have arisen, particularly concerning recycling of waste and by-products, which although probably harmless require to be registered.
- Importers of articles have had challenges to ensure that the goods do not contain SVHCs above the statutory limit (0.1%).

Some key themes are now developed further.

15.3   Basics of Successful Submission

The process of registration and of upgrading 'new substance' dossiers has been much less eventful than some feared. It was always clear that REACH would never be 'softer' than the old 'New Substances Regulations', which some people had (somewhat optimistically) hoped.

The IUCLID technical compliance check does provide a crude quality check, but behind that, many different attitudes of registrants (and other stakeholders) can be found, including some of the following:

- Do the minimum and wait to see what happens.
- Do the most complete and fully developed registration you can, because it will help your customers and keep ECHA off your case.
- Never propose testing where other options exist.
- Use prediction or read-across as much as possible OR do not use prediction or read-across because it is risky.
- Be prepared to challenge the ECHA, in court if necessary – but do as much as possible to avoid that.
- Estimation of realistic exposure levels is difficult but consolidated effort can often be successful (and very necessary).

For the most part registrants need support from consultants, in addition to staff time, and the cost of writing dossiers and interacting with downstream users is significant. For the 2010 deadline there were some huge consortia, so costs were also wrapped up in man-agement, which sometimes got out of hand. This had to be balanced against the uncertain costs associated with being found to be non-compliant. The quality of a dossier certainly needs to be high when it contains large numbers of data waivers or read across proposals.

15.4   Testing, Prediction and Read-Across

Discussions with regulators, registrants and consultants reveal a wide variety of attitudes and interpretations of the regulation and guidance. The interpretation of the guidance is made difficult by the sheer quantity of it, also because relevant topics are spread across different documents. Judgement on this can turn into debate about the science and does depend on clarity of communication in reports.

With the application of (quantitative) structure-activity relationship, registrants should ensure to follow the requirements concerning validation. The most common general error concerning read-across is insufficient justification in respect of the target and source substances. In respect of toxicology, there is often insufficient consideration of the mode of action of the substance and its absorption, distribution and metabolism properties.

To perform good read-across work, it is essential to have good knowledge of the scientific literature. It is to be hoped that the quality of science in submitted dossiers will improve.

15.5   The Community Rolling Action Plan

This part of the evaluation process is underway, with members state rapporteurs producing reports on substances as nominated. The different member states have different emphases and it can be confusing when it seems that substances are nominated on the basis of mistaken interpretations of data. This can only be seen as a part of evaluation.

Despite the clear position of the ECHA, some registrants still seem to believe that a successful registration, and even agreement about testing, implies that 'our dossier is good'. In reality, the door on further comment is never closed.

15.6   EU and National Responsibilities

A review of the enforcement of existing restrictions under REACH has drawn out some useful insights (EC, 2013a). As with other areas of legislation, one of the main issues facing the relevant authorities in the implementation and enforcement of restrictions is a lack of resources. Given the current economic climate, this is not likely to improve in the near future and, therefore, it will be more important than ever for authorities to set priorities for enforcement action.

Some Member States have described difficulties in enforcing specific restrictions because of the lack of in-country facilities and testing equipment for analysis of samples. However, even where access to laboratories is available, there are a number of exceptions to this general conclusion, as issues have arisen in relation to certain restrictions regarding the reliability of the test method and specifically in relation to the cost of carrying out testing.

Actions are being taken by Member States to improve the exchange of information among enforcement authorities.

15.7   Risk-Based Regulation and the Precautionary Approach

Chemical safety assessment (CSA) includes characterisation of hazard and risk. There is real tension between all stakeholders on this. Many endpoints have to be examined in order to assess hazard according to REACH. Exposure considerations (and hence risk) are not part of this for some endpoints (particularly in mammalian toxicology). Only in the case of proving 'strictly controlled conditions' can some studies by excluded. However, proof that such conditions are met is very costly to obtain. In other words,

REACH chemical safety assessment is a highly precautionary system. The Regulation would need to be revised for that to change.

15.8   Higher Tiers of Assessment

Risk characterisation is required for substances that possess a definite hazard or when exposure-based adaptation is applied validly. The quantification of exposure is tiered, moving broadly from default methods to higher level models and then to comprehensive determination of exposure by measurement. Various kinds of error can be seen in dossiers. Typically, simple models, such as ECETOC (European Centre for Ecotoxicology and Toxicology of Chemicals) TRA (Targeted Risk Assessment) for human exposure, are seen as definitive, when they are not. At the other extreme, registrants often produce too few measured data points to try to over-ride models. Many registrants find that the best compromise is to use a higher Tier model right from the start.

It is easy for users to consider that the regulatory method is scientific best practice. That is not the case necessarily, because it is, in fact, the regulation and guidance that set out a realistic framework for assessment. Different regulatory regimes around the world apply different methods! The validity of a characterisation of risk is highest when the RCR is very much above one or very much below one; near one it must be understood that there is uncertainty in the outcome.

One specific topic is now considered, as an example of how more science can always be done: can we use deterministic or probabilistic risk assessment for the environment? Is this important?

Various uncertainties are inherent in the process of extrapolating estimated exposures from property inputs and industry data, and also in extrapolating a safe exposure concentration from laboratory (eco)toxicity data.

The uncertainties arise from such sources as:

- Confidence intervals on property data values.
- Fluctuations in releases data.
- Varied interspecies and intraspecies sensitivity to (eco)toxic effects.
- Local environmental parameters and weather conditions.

Any reported individual measurement (or valid estimate) of a property data value actually lies within a range associated with the confidence interval. Statistically, were further measurements to be made (e.g. at different laboratories or using different but valid methods or species) an overall probability distribution would become apparent. To meet the requirements of REACH, it may be the case that only one reliable value is available. It will not then be apparent where, within the distribution, this value falls. Speaking more broadly, the same issues apply for the predicted exposure level (dose or PEC (predicted environmental concentration)) and for safe level (DNEL (derived no-effect level) or PNEC (predicted no-effect concentration)).

In REACH the standard approaches to both environmental and human exposure and risk are deterministic. The requirement on registrants to conduct exposure and risk assessment in a uniformly fair and achievable manner means a standard method for practical purposes can only use a deterministic method. This way the inputs need minimal assessment; the exposure level and no-effect level (DNEL or PNEC) are derived as single values, and the overall risk characterisation conclusion is simply based on the ratio of two figures. The conclusion of risk 'acceptable' or 'not acceptable' is based very simplistically on the value of the ratio.

A deterministic approach means that a single value is accepted as representative for the purpose of the assessment and its implications are directly followed through the equations. Normally, the uncertainty in individual inputs that are relatively weak can be readily assessed. It is appropriate to make such an assessment in order to be confident that the conclusions of the assessment are adequately robust to the uncertainties present.

A probabilistic assessment takes into account more specifically the uncertainty distribution of inputs, outcomes or both. A stochastic approach (for example a Monte Carlo model) assesses this by following through a large number of assessments in which different combinations of values are each played out.

The result of a probabilistic assessment is then a distribution, from which a percentile interval may be selected as the basis of the overall conclusion. In a probabilistic approach it is normally important that the uncertainty and data distributions for all inputs are well understood from the outset.

For the environment in particular, it is unusual to have sufficient data available to permit probabilistic risk assessment on a large scale, although there are some exceptions, such as toxic heavy metals. It is not unusual for organisations with a special interest on understanding exposure in specific local environments to use such methods to model them. Having said that, probabilistic methods have been factored into the approaches. For example, the assessment factors used to derive PNEC from EC_{50}s/NOECs (no observed effect concentrations) and to derive DNELs from NOAELs (no observed adverse effect levels), draw on real data sets to ensure a statistically relevant level of protection is conferred on the conclusion drawn. Real evidence on the environment and waste treatment facilities, though statistical assessment, are used as the basis of values used. Many environmental emission scenario documents also use this approach.

Other topics have their own hierarchies of better science. For workplace exposure, chemical monitoring is common and several probabilistic models have been developed based on the availability of such information. Examples are the Advanced REACH Tool (ART) and Stoffenmanager tools. Human exposure via the environment modelled by ACC-HUMAN is another example.

It can be anticipated that more use of such methods will grow.

15.9   REACH Developments

15.9.1   Methods of Operation and Constant Change

REACH compliance has been subject to almost constant revisions of the IT system and new versions of guidance come out frequently. This was occurring even up to the 2010 deadline. However, dossiers then valid, if being revised for voluntary updates, have to be updated to comply with the latest software versions. This has placed significant extra demands on registrants and seems to be a process that industry cannot influence or even question. For the 2013 deadline there was a suspension of updates.

It is to be hoped that a more structured approach to changes can be agreed. It will also be beneficial if serious breaches of compliance are punished quickly and clearly.

15.9.2 Improved Efficiency of Operation

There seems little doubt that the ECHA is running a well-organised system. There are, however, areas where rationalisation could help. The amount of guidance is disproportionate to need and generally impenetrable except for the most experienced. This has allowed for some 'advisors' to come into consultancy with minimal track records, with inevitable consequences.

More structured approaches could help laboratories and consultancies plan for realistic demand levels, which at the moment are not at all predictable with reliability.

15.9.3 Increased Scope

15.9.3.1 Regulation and Guidance

What changes can be anticipated in the future? (EC, 2013b)

Both Regulation and guidance are subject to review, and will be changed as a result of experience and scientific advances.

Changes may be made to the scope of the REACH Regulation. Reviewing the scope of the legislation is required by the Regulation (Article 138). Under consideration at present is a review of the information requirements and exemption from obligation to perform a chemical safety assessment for substances manufactured or imported at levels below 10 tonnes/year. No conclusion had been reached when the Commission General Report on REACH became available (EC, 2013b). Work is continuing and proposals will be made by 1 January 2015, if it is considered appropriate to do so.

This review is also assessing polymers, to put together information that can lead to an informed decision on whether REACH should be extended to include polymers. The study is considering if information from monomers already registered can be used to determine the hazards associated with polymers. It is attempting to discover if polymers pose risks, how such risks compare with other substances and whether or not REACH manages the risks from polymers. The preliminary findings presented in the General Report are that more information is needed, and that if changes are considered necessary, proposals will be made by January 2015.

Reviews have been undertaken to look at the implementation of REACH restrictions and the information requirements of REACH and Classification, Labelling and Packaging (CLP). The effectiveness of REACH in achieving its objectives has also been considered, to answer such questions as:

- Has REACH reduced the nominal risk of chemicals?
- Has REACH benefited the environment and human health?
- What has been the impact of REACH on innovation in the chemicals industry?
- How has REACH contributed to the development of emerging technology?

Reports have been published for most of the 2012 reviews but it is not yet known if any of them will lead to changes in REACH (EC, 2013c).

It is anticipated that at some point in the future evaluation of potential for endocrine disruption will be included in the REACH Regulation when methods for assessment of this property have been sufficiently validated. The scope of authorisation may be extended to cover substances with endocrine disrupting properties; further guidance has been issued during 2013.

The ECHA has a process for reviewing and updating guidance. The process includes a consultation with the EC, Member States' Competent Authorities and industry organisations. There are no current consultations at the time of writing; the ECHA web site has details of past and present consultations.

One of the changes to the guidance that has already taken place is the inclusion of guidance on the assessment of nanomaterials to Parts R07 a, b and c. Nanomaterials are substances defined in REACH, so they are subject to the Registration requirements that apply to other chemicals. Information on nanomaterials in REACH and CLP is available on the EU web site (EC, 2013d).

Enshrined in the legislation is the principle that test methods are to reviewed. It is likely that there will be changes to testing for skin irritation; when the *in vitro* methods are fully validated, testing *in vivo* will no longer be needed in any circumstances. Other changes may occur, for example when the Organisation for Economic Co-operation and Development (OECD) guidelines change or new guidelines are introduced.

15.9.4   Policy Development on the Control of Chemicals – EU and Global Perspectives

At a worldwide level, OECD countries continue with the high production volume (HPV) programme of reporting data sets to OECD and then to the United Nations Environment Programme (UNEP). The value of the hazard assessments is questionable for substances already registered in REACH at above 1000 tonnes/year.

Regulators around the world have different methods for risk characterisation and assessment, whereas classification and labelling is now more consistent.

Bringing in non-OECD countries into these systems would be very beneficial.

15.10   Rationalising Overlap with Other Legislation

An area that caused concern for industry even as REACH was being framed was the apparent lack of integration with other areas of regulation. The exemptions are clear enough but the overlap with other regulations has not been explored in any guidance. Companies have information that is relevant to REACH compliance from:

- Major accident regulations (Seveso II, or Control of Major Accident Hazards (COMAH) in the United Kingdom): this requires site-specific information.
- Integrated pollution prevention and control regulation (IPPC, enacted in different ways in various member states); again, consents for site-specific practices contain much useful information. The IPPC best available techniques reference documents (BREFs) are very well-researched sources.

- National health and safety legislation, for example Control of Substances Hazardous to Health (COSHH) in the United Kingdom.
- Agricultural chemicals – in respect of pesticide formulations there are areas of overlap.
- Cosmetics: although the definitions are clear, companies should use information about use pattern for REACH too.

The following is largely derived from an EC publication (EC, 2013e).

REACH is a broad Regulation and was intended bring together some of the previous legislation that applied to chemicals. Analysis has been carried out looking at where and how REACH overlaps with other EU chemicals legislation that is still in force (Milieu, 2012). The study has identified areas where REACH does not contribute anything, because adequate legislation exists (referred to as double regulation). However, some areas of overlap between REACH and other chemical legislation act together to the same end, that is protection of the environment and of human health while at the same time promoting the internal market for chemicals. Gaps in the legislation have also been identified. Recommendations have been made to deal with double legislation and with legislative gaps.

Opportunities have been identified where REACH can interact with another act in a way that makes the EU legislation more coherent. In some instances this is because it was intended that the legislation work together. Other interactions that were not expressly intended are already in existence or could take place. For example, risk management measures or risk assessments from other legislation are used under REACH. Data gathered for REACH purposes could be used to identify substances that need to be controlled under sectoral legislation. It has been recommended that links between REACH and sectoral legislation be strengthened, so to reduce costs (for regulators and those trying to comply with the Regulations) and give better protection to health and the environment. This could be done by amending REACH to make more specific reference to other EU acts in Annex I 0.5, as well amendment of other legislation to require that information from the chemical safety assessment be taken into account in risk assessment and risk management measures.

Double regulation can be very problematic because not only are parallel restrictions on some substances or uses but there are also contradictions, where the restrictions in REACH Annex XVII are different from those in another act for the same substance or use. Examples of double regulation exist in the Toys Directive, the Packaging Directive and the PCB/PCTs (polychlorinated biphenyls/polychlorinated terphenyls) Directive. Two recommendations have been made to resolve the issues of double regulation.

- Preparation of a database detailing all the restrictions in place for certain substances.
- Amend specific legislation and/or REACH Annex XVII so that duplicates are removed or the restrictions are consistent.

Gaps in the legislation exist where classes of substances are exempt from REACH, or considered as registered, but the risk assessment that has been carried out for them does not cover some stages of their life cycle, that is manufacturing, formulation and waste management. This applies to cosmetics and pharmaceuticals, which are exempt from REACH, and to plant protection products and biocidal products, which are considered to be registered. There is legislation in place such as the IPPC/Industrial Emissions

Directives, the Seveso II Directive and workers' protection legislation, REACH requires a more detailed risk assessment than these. One possible way of overcoming this would be an amendment of REACH to cover the stages of the life cycle that are missing. There would be disadvantages to doing this, so the second option may be the one that is carried out. This second option is to amend the relevant legislation so that the approach to lifecycle risk assessment is made consistent with REACH.

15.11   Scientific Developments and Challenges

Science is not set in stone and advances are made all the time. Some of the following areas could well force change in the Regulation:

- Increasing analytical sensitivity for chemicals present in the workplace and the environment.
- Test plans informed by genetic or other biochemical assays.
- Improved exposure models.
- More understanding of the combined effects of exposure to many substances simultaneously.
- Improved computational chemistry methods.
- Analysis of the information found in the disseminated dossiers of registered substances.

15.12   Impact on Industry

It is perhaps appropriate to conclude this chapter with what is arguably the most important topic. Chemicals are on the market because society as a whole wishes to have products and services that depend on their availability. Within the present EU economic system, this means that production or import of chemicals must be viable.

15.12.1   Manufacturers and Importers

It is clear that REACH has been a cost on industry. It is to be hoped that the benefits of the dossier development can come from being useful in other geographical/economic zones.

Costs in terms of consultants and testing can be estimated but few companies are totalling the total amount of staff hours committed to it. It may be that the total cost of one 2010 substance has typically been over €100 000 *per registrant*, excluding time spent interacting in the supply chain.

REACH has led to the withdrawal of some substances from the market, with perhaps more to follow when authorisation costs begin to be encountered.

15.12.2   Downstream Users and Consumers

Downstream user associations have been active in helping members, particularly with mixture assessment. Beyond that, they generally lack the technical knowledge to assist suppliers. Some, but by no means all, have helped by development of exposure scenarios. These were generally too late for 2010 but are now proving more useful.

Many downstream user companies are small and medium-sized enterprises and it is clear that the effort to become familiar with REACH has had an impact on internal costs.

15.12.3   Innovation

A study has been made to assess the impact of the application of the REACH regulation on innovation in the industry. Enhancing innovation is one of the primary objectives of the Regulation (expressed in Article 1) (EC, 2013f).

Since REACH was enacted the main impact on industry has been in costs. Companies realise R&D and innovation programmes will need to be adjusted in respect of using phase-in substances. However, the previous new substances regulations have been in place for over 30 years, so there is no major change in respect of discovery of new substances, in respect of cost. However, the need, in REACH, to register at the required Annex level before marketing can be seen as a new additional hurdle.

As a stimulation to innovation, it is clear that one positive force is the aversion of many downstream users to hazardous substances. Authorisation and restriction, while important in specific sectors affected, are affecting a smaller part of the industry. It has not been able to discern overall benefits for consumers, the market and society yet.

The need for small and medium-sized enterprises to have more support cannot easily be understated. Support from industry associations tends to be limited, for reasons of competition. Training courses can help but are costly. National governments have help lines and web sites, but many small and medium-sized enterprises are, sadly, wary of contacting them.

The motivation of companies to innovate and comply fully with REACH will be higher if enforcement is at a high level, so as to ensure a truly fair trading environment. Much suspicion remains of importers getting an easier passage through than producers. The feeling that maintaining a low profile is better than interaction with authorities is prevalent. In such an environment, concerns expressed by non-government organisations about compliance can easily gain more momentum than is justified.

15.13   ECHA Evaluation Report 2012

The ECHA publishes an annual evaluation report, for the purpose of giving general feedback on prominent issues which have occurred during the year, in order to provide learning points for registrants. The main outcomes and key recommendations for industry of the ECHA evaluation report of 2012 are discussed below. The most commonplace shortcomings identified by the ECHA regarding registration dossiers included quality of information on substance identity, exposure assessment and risk characterisation as well as on prenatal and subchronic toxicity studies. The identification of these shortcomings was thereafter used to make key recommendations by the ECHA; firstly, the clear identification of the registered substance, including the provision of analytical data in the registration dossier is required. Substance identity definitions which provide broad descriptions that effectively cover multiple substances within a dossier should be avoided.

Similarly, demonstration of the relevance of the test material to the registered substance and to its uses is essential for dossier completeness. Use and exposure information needs to be reflective of market reality, where descriptions should cover uses, exposure scenarios, operational conditions and risk management measures accurately rather than hypothetically. Finally, the use of alternative approaches as far as possible is recommended, including the use of existing available data and read-across, but it is noted that the data should be scientifically robust and credible. Furthermore, it is emphasised that new testing should only be pursued as a last resort, and where it is proposed the selected test substance needs to be justified clearly, and must reflect the substances of all members of the joint submission.

The registrants should be aware that the ECHA is using IT tools to detect dossiers with 'identified concerns', in addition to a random selections which are assessed against the ECHA's screening criteria. It is also noteworthy that the action is with the registrant to adapt or withdraw existing test proposals in cases where valid new information is received via third party consultations. The Evaluation Report of 2012 also includes a useful summary of the different timelines and stages for interaction between the ECHA and the registrants in the various evaluation processes; a new series of regular webinars with tips for registrants has been made available.

References

EC (European Commission) (2013a) Thematic Studies for Review of REACH 2012. Implementation and Enforcement of Restrictions under REACH. Last updated 2 July 2013. http://ec.europa.eu/environment/chemicals/reach/study4_review_2012.htm (last accessed 29 July 2013).

EC (European Commission) (2013b) Thematic Studies for Review of REACH 2012. Review of the Registration Requirements for 1 to 10 Tonnes Substances and Polymers. Last updated 2 July 2013. http://ec.europa.eu/environment/chemicals/reach/study10_review_2012.htm (last accessed 29 July 2013).

EC (European Commission) (2013c) Review of REACH. Last update 28 June 2013. http://ec.europa.eu/enterprise/sectors/chemicals/documents/reach/review2012/index_en.htm (last accessed 29 July 2013).

EC (European Commission) (2013d) REACH and Nanomaterials. Last update 24 June 2013. http://ec.europa.eu/enterprise/sectors/chemicals/reach/nanomaterials/index_en.htm (last accessed 29 July 2013).

EC (European Commission) (2013e) Thematic Studies for Review of REACH 2012. Technical Assistance Related to the Scope of REACH and Other Relevant EU Legislation to Assess Overlaps. Last updated 2 July 2013. http://ec.europa.eu/environment/chemicals/reach/study8_review_2012.htm (last accessed 29 July 2013).

EC (European Commission) (2013f) Thematic Studies for Review of REACH 2012. Impact of the REACH Regulation on the Innovativeness of EU Chemical Industry. Last updated 2 July 2013. http://ec.europa.eu/environment/chemicals/reach/study5_review_2012.htm (last accessed 29 July 2013).

Eurostat (2012) The REACH Baseline Study 5 Years Update. Summary report, 2012 edition. http://epp.eurostat.ec.europa.eu/cache/ITY_OFFPUB/KS-RA-12-024/EN/KS-RA-12-024-EN.PDF (last accessed 29 July 2013).

Milieu (2012) Technical Assistance Related to the Scope of REACH and Other Relevant EU Legislation to Assess Overlaps. Milieu Environmental Law and Policy 2012-03-12, Milieu Ltd, Brussels, Belgium. www.milieu.be (last accessed 29 July 2013).

Web Reference

EC (European Commission) (2013) General Report on REACH (2013). Report from the Commission to the European Parliament, the Council, the European Economic and Social Committee and the Committee of the Regions. http://ec.europa.eu/enterprise/sectors/chemicals/files/reach/review2012/general-report-draft_en.pdf (last accessed 29 July 2013).

16

Resources, Official Guidance, Further Reading and Centres of Expertise

16.1 Introduction to Resources and Organisations

Various sources and organisations set out useful reference materials for registrants, in the form of guidance, examples and practical details of the rules that must be fulfilled. The specific details in the REACH Regulation itself and its associated legislation are of course essential reading also. This chapter introduces some of the most significantly important and useful sources of practical help to registrants, organised by topic. Regulatory and the ECHA (European Chemicals Agency) guidance, being of authority origin, is listed first throughout the section.

16.1.1 Official Journal

For requirements and criteria set out in legislative instruments it is very convenient to access the *Official Journal* online. The *Official Journal* publishes all European Union (EU) legislation with an intuitive search interface, and contains a number of relevant and useful legal documents which are freely available. The REACH Regulation itself (1907/2006, as amended) is obviously a particularly significant example and sets out in full the details of the scope, time-lines, requirements, exclusions and responsibilities of different parties.

The Classification, Labelling and Packaging (CLP) Regulation 1272/2008 (and the adaptations to technical progress) sets out the criteria by which classification and labelling must be applied (CLP, 2008).

Chemical Risk Assessment: A Manual for REACH, First Edition. Peter Fisk Associates Ltd.
© 2014 John Wiley & Sons, Ltd. Published 2014 by John Wiley & Sons, Ltd.

Regulation 440/2008 (and its corrigenda) presents in full the accepted test methods to follow for each of the required data endpoints set out in Annexes VII–X of the REACH regulation.

Access to EU law can be obtained via the Internet (EUP-Lex, nd).

16.1.2   ECHA and REACH-IT

The ECHA, the centralised EU regulatory authority for REACH and CLP, has made available many pieces of official guidance on a wide range of topics relating to the requirements and approaches needed for registration, to a high level of technical detail. The ECHA guidance documents are free to access through its web site.

The ECHA web site also offers a useful interactive navigator tool to assist users in identifying their obligations under REACH and CLP. The site has many useful features as well as guidance documents.

The ECHA web site also includes the REACH-IT portal, which is the main method for registrants to submit securely their registrations to ECHA online.

16.1.2.1   Guidance Documents

Guidance documents published by the ECHA amount to many thousands of pages of documentation, available as pdf files for download. Guidance documents have been prepared for reference by registrants, SIEF (Substance Information Exchange Forum) members and other involved regulatory authorities; they cover the whole spectrum of topics relating to REACH (requirements, approaches, associated with registration, authorisation, specific substance types and registration types). A full list appears here.

The Guidance on information requirements and chemical safety assessment (IR & CSA) is particularly useful during the process of compiling the technical dossier and CSR (chemical safety report). This comprises a number of documents focussing on many separate aspects of preparation of registration dossiers, approaches, and data needs.

The 'Guidance in a nutshell' series comprises short and more accessible overviews of the corresponding full guidance documents, focussing on the key facts.

Guidance Factsheets are available each of which is associated with a specific full Guidance Document. The factsheets define the scope and key points of the Guidance Document, and what readers of different roles in REACH (e.g. registrants, downstream users) can expect to get from it.

The Practical Guides series are brief manuals with a highly practical focus, intended to provide details and examples on approaches to some aspects of REACH where such guidance has been identified as necessary.

Several ECHA guidance documents have special Appendices relating to application of the approaches to nanomaterials in REACH.

16.1.2.2   REACH-IT Guidance

A series of Data Submission Manuals is available through the ECHA's REACH-IT pages. These guidance documents also have a very practical focus with screen shots and illustrative examples to assist registrants preparing their submission using the recommended IT tools.

The accompanying REACH-IT user manuals tend to be highly detailed and thorough, with explicit guidance oriented around the many different administrative and procedural elements of REACH Registration using REACH-IT. The manuals are supported by useful screen images of key menus and forms which REACH-IT users need to complete.

16.1.2.3   Other Online Tools

The ECHA web site provides free public access to a wealth of reference information useful to notifiers. A visitor to the web site can, for example:

- Find out if a substance of interest has been registered already, or been listed for future registration.
- Identify registrations made by specific registrant companies.
- Access disseminated registration dossiers.
- Find out the current applicable EU harmonised classification and labelling under CLP, and check what self-classification has been applied by notifiers to the CLP Inventory
- Find the current lists of substances that are prioritised for evaluation in the CoRAP (community rolling action plan) programme.
- Check the latest status of test proposal consultations.
- Find information about candidates for authorisation.
- Obtain the latest versions of guidance documents online, and find out about draft amendments to guidance documents during the development process.
- Access minutes from Member State technical committee meetings and other published developments.
- Learn about common problems identified by the ECHA in recent evaluations, and how to avoid them, through ECHA's annual Evaluation Reports.
- Find out about review programs at EC level to assess the impact of REACH, and proposed future changes.

16.1.3   CEFIC and Sector Groups

The European chemicals industry association, CEFIC, has a very important role in supporting companies, at all levels of the supply chain, in REACH by issuing guidance and reference materials. CEFIC's web site includes guidance documents to registrants and to downstream users about their responsibilities and actions.

As well as issuing its own guidance and documents, the CEFIC web site is a repository to check the latest status and links to work done by Sector Groups for specific parts of the chemicals industry.

16.1.4   IUCLID Guidance

IUCLID (International Uniform Chemical Information Database) is a freely-available database program, recommended to registrants as a useful method to compile and prepare to submit REACH registration technical dossiers, among other regulatory and non-regulatory chemical database management purposes (OECD (Organisation for Economic Co-operation and Development) HPV (High Production Volume) program, Biocides Directive etc.). It is well maintained with regular upgrades and various supporting tools to assist the registrant to check their data sets for completeness and see in advance what information will be made public by the ECHA. The program is tailored well for the needs of REACH and prompts the user for most key details.

The latest available version of the IUCLID program, plug-in tools, written guidance, training tutorials, IT details and information about future developments are all available from the web site (IUCLID 5, nd).

16.1.5   ECETOC

This organisation addresses scientific issues of concern to its industry members. It is closely-aligned with the EU chemical industry in all sectors. The 'European Centre for Ecotoxicology and Toxicology of Chemicals' is a very appropriate name. It produces expert reports and methods across a wide range of topics, on the basis of the voluntary input from member companies' scientists. Its work is well-regarded as review material. It also supports CEFIC's Long-Range Research Initiative. It has a very useful web site.

16.1.6   OECD

The Organisation for Economic Co-operation and Development is a large and wide-ranging organisation that includes work on chemicals in its environmental directorate. The intention is to establish a basic understanding of hazard across all member countries. The scope of such studies is not the same as REACH but there is a significant overlap. The OECD then interacts with the United Nations in publication of the agreed and peer-reviewed reports. In reality, the OECD web site is not easy to navigate and the best way to find what has been published is via a standard search engine and CAS (Chemical Abstracts Service) number. The reports are very useful once located!

16.1.7   EU JRC

The EU Joint Research Centre was formerly responsible for general chemicals regulation until the ECHA was set up. Its web site no longer links easily to issues concerning chemicals, although ESIS (the European chemical Substances Information System) is a good source for regulatory status of substances. As indicated by its name, this institute which is sponsored by the Commission still undertakes relevant basic research (Tables 16.1–16.8).

16.2   Facts and Statistics

The ECHA web site, particularly the page http://echa.europa.eu/web/guest/information-on-chemicals/registration-statistics has a large amount of interesting information on registrations made to date, by country and so on.

Table 16.1 *Guidance on general needs and registration of specific chemical types.*

Source	Title	Summary/key features of guidance	Online access
ECHA Guidance on information requirements and chemical safety assessment	Part F – Chemical safety report	Sets out the objectives, content and level of detail to be included in a chemical safety report. Guidance on structure, also templates and format examples for reporting exposure results, are available in separately downloadable appendices	http://echa.europa.eu/web/guest/guidance-documents/guidance-on-information-requirements-and-chemical-safety-assessment
ECHA Guidance Document	Guidance on registration	This fundamentally important guidance sets out clearly the roles of parties in the supply chain, with many worked examples and illustrative figures. It discusses cases outside the scope; gives detailed advice on approach for intermediates; appropriate method for accounting tonnage for the purpose of assessing the tonnage band. Deadlines and details of the registration requirements are also presented clearly. Data sharing and roles of the SIEF and its members are explained. Specific practical explanation of how joint submission works is presented in a useful section. Responsibilities of registrants to maintain the registration with updates, and communicate with customers are set out. Fees, confidentiality and appeal procedures are also explained. This guidance document is among the most important and practical for general information	http://echa.europa.eu/web/guest/guidance-documents/guidance-on-reach
ECHA Guidance Document	Guidance on monomers and polymers	This guidance supports prospective registrants in assessing whether a substance meets the definition of a polymer or monomer and the position of each in the scope of REACH	http://echa.europa.eu/web/guest/guidance-documents/guidance-on-reach

(continued overleaf)

Table 16.1 (continued)

Source	Title	Summary/key features of guidance	Online access
ECHA Guidance Document	Guidance on substance identification	Presents clear definitions and rules for substance nomenclature in REACH, for single-constituent, multi-constituent and UVCB substances, and some advice for analytical requirements. This important guidance assists prospective registrants in understanding the registration substance and defining its identity for registration, with significant implications for approach to the chemical safety assessment and working relationships with other registrants. Supported throughout by numerous useful examples	http://echa.europa.eu/web/guest/guidance-documents/guidance-on-reach
ECHA Guidance Document	Guidance for articles	This guidance assists registrants in identifying whether they have an article or a substance in an article that requires registration	http://echa.europa.eu/web/guest/guidance-documents/guidance-on-reach
ECHA Guidance Document	Guidance on intermediates	Gives detailed guidance on approaches when registering transported or isolated intermediates, and the evidence required by the ECHA	http://echa.europa.eu/web/guest/guidance-documents/guidance-on-reach
ECHA Guidance Document	Guidance on waste and recovered substances	This guidance document discusses various types of wastes/recovered substances and their status and requirements for registration in REACH	http://echa.europa.eu/web/guest/guidance-documents/guidance-on-reach
ECHA Guidance Document	Guidance on Scientific Research and Development (SR&D) and Product and Process Oriented Research and Development (PPORD)	Substances whose use is limited to certain research and development processes and in limited quantities benefit from special requirements which are set out in this guidance	http://echa.europa.eu/web/guest/guidance-documents/guidance-on-reach

ECHA Data Submission Manual	Part 1: How to prepare and submit a PPORD notification	Detailed guide with screen shots showing how prospective registrants of a PPORD substance should complete the technical dossier in IUCLID 5	http://echa.europa.eu/web/guest/support/dossier-submission-tools/reach-it/data-submission-industry-user-manuals
ECHA Data Submission Manual	Part 2: How to prepare and submit an inquiry dossier	Sets out guidance for the inquiry process for prospective registrants of non-phase-in substances	http://echa.europa.eu/web/guest/support/dossier-submission-tools/reach-it/data-submission-industry-user-manuals
ECHA Data Submission Manual	Part 4: How to pass business rule verification ('enforce rules')	Guidance to registrants on how to ensure administrative aspects of the IUCLID dossier (business rules) are fulfilled, without which the dossier is liable to be rejected. This guidance includes useful screenshots demonstrating the required information to set up within IUCLID and online at REACH-IT	http://echa.europa.eu/web/guest/support/dossier-submission-tools/reach-it/data-submission-industry-user-manuals
ECHA Data Submission Manual	Part 14: How to prepare and submit a request for use of an alternative chemical name for a substance in a mixture using IUCLID 5	Guidance on how to apply to use an alternative name for a chemical component of a mixture, in order to protect business interests (e.g. when components are declared in online tools and eSDS, etc.)	http://echa.europa.eu/web/guest/support/dossier-submission-tools/reach-it/data-submission-industry-user-manuals
ECHA Data Submission Manual	Part 16: Confidentiality claims	This guidance discusses the registrant's approach to flagging claims confidentiality in the registration dossier, and presenting justification. This is in the context of the ECHA's obligation to make certain information publicly available. The guidance discusses the possible basis for confidentiality claims and the typical costs associated with doing so. A template claim form and some example forms of words are appended to the document	http://echa.europa.eu/web/guest/support/dossier-submission-tools/reach-it/data-submission-industry-user-manuals

(continued overleaf)

Table 16.1 *(continued)*

Source	Title	Summary/key features of guidance	Online access
ECHA Data Submission Manual	Part 18: How to report the substance identity in IUCLID 5 for registration under REACH	Separate guidance assists registrants to name their substance appropriately in REACH. This guidance exists to support the correct set-up of the technical dossier and REACH-IT entry to document the substance identity for the registration substance	http://echa.europa.eu/web/guest/support/dossier-submission-tools/reach-it/data-submission-industry-user-manuals
ECHA Data Submission Manual	Part 20: How to prepare and submit a substance in articles notification using IUCLID	Guidance to registrants of SVHC substances present in Articles how to go about preparing the required information in IUCLID 5 including how to find and download certain information ready-completed for inclusion in the dossier. The information requirements, approach in IUCLID 5, and submission process are explained with screen shots	http://echa.europa.eu/web/guest/support/dossier-submission-tools/reach-it/data-submission-industry-user-manuals
ECHA Data Submission Manual	Part 22: How to prepare and submit an application for authorisation using IUCLID 5	Guidance to registrants who are applying for authorisation. Includes details on the requirements for the registration dossier in IUCLID 5 (mandatory and optional), supported by screen shots; also guidance on completing the necessary application form, and submitting the dossier via REACH-IT	http://echa.europa.eu/web/guest/support/dossier-submission-tools/reach-it/data-submission-industry-user-manuals

ECHA Guidance Document	Guidance on authorisation application	Very important ECHA guidance for registrants considering applying for authorisation. This is a detailed guidance document covering all aspects of the authorisation application process. A useful flowchart illustrates the steps in considering whether an application for authorisation is needed and appropriate. The approach to conducting the analysis of alternatives, and producing the substitution plan (if indicated) are covered together with guidance on documenting and presenting these in the application dossier. Socio-economic analysis is also covered in this guidance. The role of third parties submitting additional information is also covered	http://echa.europa.eu/web/guest/guidance-documents/guidance-on-reach
ECHA Guidance Document	Guidance on socio-economic analysis – authorisation	Lengthy guidance discussing in detail the methods for conducting socio-economic analysis in the context of an application for authorisation in REACH	http://echa.europa.eu/web/guest/guidance-documents/guidance-on-reach
ECHA Guidance Document	Guidance for Annex V	Straightforward guidance on the various classes of substances exempted from REACH under Annex V of the regulation. This guidance discusses the basis for exemption in each case and gives useful examples for reference	http://echa.europa.eu/web/guest/guidance-documents/guidance-on-reach

Table 16.2 Guidance and resources on property data.

Source	Title	Summary/key features of guidance	Online access
REACH regulation, 1907/2006 as amended	Annexes VII–X – standard information requirements	These annexes specify which properties are required by default in a full registration at a particular tonnage level and where key adaptation criteria apply ('column 2' of the tables in these annexes)	–
Regulation 440/2008 as amended	Test methods pursuant to REACH	Laboratory methods for the majority of data endpoints required in REACH registrations under Annexes VII–X	–
ECHA Guidance on information requirements and chemical safety assessment	Chapter R.2: Information requirements	General advice to registrants on strategy for identifying information requirements, collecting and negotiating access to available existing data, assessing completeness of the data set and generating additional information where required, using testing and non-testing approaches	http://echa.europa.eu/web/guest/guidance-documents/guidance-on-information-requirements-and-chemical-safety-assessment
ECHA Guidance on information requirements and chemical safety assessment	Chapter R.3: Information gathering	Gives advice to registrants on search strategy in the public domain and potentially useful routes to data required or useful for REACH	http://echa.europa.eu/web/guest/guidance-documents/guidance-on-information-requirements-and-chemical-safety-assessment
ECHA Guidance on information requirements and chemical safety assessment	Chapter R.4: Evaluation of available information	Guides registrants in their evaluation of the data set assembled in terms of relevance, reliability and adequacy. Also the use of weight of evidence approaches in the data set	http://echa.europa.eu/web/guest/guidance-documents/guidance-on-information-requirements-and-chemical-safety-assessment

ECHA Guidance on information requirements and chemical safety assessment	Chapter R.5: Adaptation of information requirements	Sets out the methods for registrants to assess if adaptation of the standard data requirements is appropriate on various grounds (exposure-based adaptation; waving on the basis that testing is scientifically unjustified or unnecessary, or technically impossible)	http://echa.europa.eu/web/guest/guidance-documents/guidance-on-information-requirements-and-chemical-safety-assessment
ECHA Guidance on information requirements and chemical safety assessment	Chapter R.6: (Q)SARs and grouping of chemicals	Presents guidance on the appropriate use of (Q)SAR techniques to provide data for the CSA, including development of new algorithms, and the assessment of groups of substances as 'categories'	http://echa.europa.eu/web/guest/guidance-documents/guidance-on-information-requirements-and-chemical-safety-assessment
ECHA Guidance on information requirements and chemical safety assessment	Chapter R7: a, b and c	Lengthy, detailed guidance documents. Describe in detail the data requirements, the role of each property in the chemical safety assessment, waiving criteria, available testing guidelines, established non-testing methods, for example (Q)SAR, guidance to registrants on an integrated testing strategy	http://echa.europa.eu/web/guest/guidance-documents/guidance-on-information-requirements-and-chemical-safety-assessment
ECHA Data submission manual	Part 5: How to complete a technical dossier for registrations and PPORD notifications	Guidance to registrants on compiling the REACH requirements within IUCLID and technical completeness criteria which must be fulfilled for the dossier to be accepted	http://echa.europa.eu/web/guest/support/dossier-submission-tools/reach-it/data-submission-industry-user-manuals
ECHA practical guide	Guide 1: How to report *in vitro* data	Useful guidance on the role of *in vitro* studies in place of *in vivo* toxicity and ecotoxicity data in a substance data set, including when it should be used as part of weight-of-evidence only, and how to document it	http://echa.europa.eu/practical-guides

(continued overleaf)

Table 16.2 *(continued)*

Source	Title	Summary/key features of guidance	Online access
ECHA practical guide	Guide 2: How to report weight of evidence	Discusses the meaning of the concept 'weight of evidence' of the data available in the substance data set, guidance on sources of data to bring together in the evaluation, and how apparently conflicting results can be interpreted; and practical advice on how to document the conclusions in the technical dossier	http://echa.europa.eu/ practical-guides
ECHA practical guide	Guide 3: How to report robust study summaries	Guidance on key points of detail needed for an endpoint study record in IUCLID to be considered robust, for specific properties	http://echa.europa.eu/ practical-guides
ECHA practical guide	Guide 4: How to report data waiving	The guidance sets out the permitted criteria for data waiving in REACH, and includes helpful advice on how registrants can assess if the criteria are met for a particular case. The guidance document includes a clear decision tree and several relevant examples	http://echa.europa.eu/ practical-guides
ECHA practical guide	Guide 5: How to report (Q)SARs	Guidance on how (Q)SAR techniques should be applied to generate data for the REACH data set, including principles of relevance and validity; also covers how to document a (Q)SAR prediction in IUCLID 5	http://echa.europa.eu/ practical-guides
ECHA practical guide	Guide 6: How to report read-across and categories	Guidance on how to develop and use a Category concept in a robust way for the purpose of REACH data assessment, and how to document the Category approach in the dossier and using the Category tools built in within the IUCLID program	http://echa.europa.eu/ practical-guides

ECHA practical guide	Guide 10: How to avoid unnecessary testing on animals	This guidance document explains how to apply various different approaches to filling of data gaps (data waiving, property prediction, (Q)SAR, read-across, *in vitro* methods) with the specific objective of assessing when *in vivo* animal testing is unnecessary and can be avoided in a REACH data set	http://echa.europa.eu/practical-guides
ECHA practical guide	Guide 14: How to prepare toxicological summaries in IUCLID and how to derive DNELs	This guidance explains how registrants should use the updated Endpoint Summary fields correctly in the human health and mammalian toxicology chapter in IUCLID 5.4 to document conclusions for toxicology endpoints	http://echa.europa.eu/practical-guides
IUCLID	IUCLID 5 Guidance and support end user manual	User support and guidance for documenting the property data set in IUCLID 5	http://iuclid.eu/index.php?fuseaction=home.documentation&type=public#usermanual
IUCLID	IUCLID 5 Technical completeness check plug-in user manual	User support and guidance for installing and running the tool to perform technical completeness check within IUCLID 5	http://iuclid.eu/index.php?fuseaction=home.documentation&type=public#usermanual

Table 16.3   Guidance and resources on hazard assessment, PBT and classification and labelling.

Source	Title	Summary/key features of guidance	Online access
ECHA Guidance on information requirements and chemical safety assessment	Part B – Hazard assessment	Provides a useful overview and introduction to the needs under REACH to gather and evaluate property data in a context of the various types of hazard assessment. Classification and labelling, PBT assessment, and PNEC and DNEL derivation are discussed. This document also discusses the approach to evaluate the scope of exposure assessment indicated by the findings of the hazard assessment. The methods are covered in more detail in the relevant Part R guidance documents	http://echa.europa.eu/web/guest/guidance-documents/guidance-on-information-requirements-and-chemical-safety-assessment
ECHA Guidance on information requirements and chemical safety assessment	Part C – PBT assessment	Introductory guidance to the needs under REACH to conduct PBT/vPvB assessment. The methods are covered in more detail in the relevant Part R guidance documents	http://echa.europa.eu/web/guest/guidance-documents/guidance-on-information-requirements-and-chemical-safety-assessment
ECHA Guidance on information requirements and chemical safety assessment	Chapter R.11: PBT assessment	Detailed guidance on approach to assessing PBT/vPvB status for substances using the screening or definitive criteria, and conducting the required risk characterisation and risk management for substances found to meet the criteria	http://echa.europa.eu/web/guest/guidance-documents/guidance-on-information-requirements-and-chemical-safety-assessment

ECHA Guidance on information requirements and chemical safety assessment	Chapter R.8: Characterisation of dose (concentration) – response for human health	Guidance covering the background and approach to determining DNEL (or where more applicable, DMEL) for the various human health effects of registration substances, covering different routes of exposure, different target populations, durations of exposure, and so on. Use of a qualitative approach to risk characterisation in certain cases is also discussed. Covers the use of appropriate assessment factors to reflect the uncertainty associated with extrapolation to humans from the data that exists	http://echa.europa.eu/web/guest/guidance-documents/guidance-on-information-requirements-and-chemical-safety-assessment
ECHA Guidance on information requirements and chemical safety assessment	Chapter R.10: Characterisation of dose (concentration) – response for environment	Guidance covering in detail the approach to deriving PNEC for the various environmental compartments, clarifying the assessment factor approach in terms of the available study data set. Also covers the use of estimation methods (equilibrium partitioning) to produce a usable indicator PNEC in the absence of test data	http://echa.europa.eu/web/guest/guidance-documents/guidance-on-information-requirements-and-chemical-safety-assessment
CLP Regulation 1272/2008 as amended	Classification, labelling and packaging of chemicals	The Regulation defines the categories of hazard applicable in the EU. It includes criteria for classification; labelling and packaging requirements associated with any applicable classifications; harmonised classifications for listed substances (Annex 6); specific concentration limits for deriving CLP classifications for preparations and mixtures	–

(continued overleaf)

Table 16.3 *(continued)*

Source	Title	Summary/key features of guidance	Online access
ECHA CLP inventory	–	Online inventory searchable by CAS or chemical name; gives information on any official CLP classification applying, and also the self-classification combinations used by notifiers to the CLP inventory/REACH registrants	http://echa.europa.eu/web/guest/information-on-chemicals/cl-inventory-database
ECHA practical guide	Guide 7: How to notify substances in the classification and labelling inventory	Guidance to companies on obligations relating to notification to ECHA's CLP Inventory, how to submit and update classification and labelling, in conjunction with other notifiers or submitting alone, with a useful annex on the subject of mixtures	http://echa.europa.eu/practical-guides
ECHA data submission manual	Manual 12: How to prepare and submit a classification and labelling notification using IUCLID	Guidance to support notifiers under CLP to prepare the dossier which is the basis of the notification to the ECHA CLP inventory. This document explains the information which notifiers must complete in IUCLID and the online submission process, supported by screen images	http://echa.europa.eu/web/guest/support/dossier-submission-tools/reach-it/data-submission-industry-user-manuals
ECHA CLP guidance	Guidance on the application of CLP criteria	This useful guidance expands upon the criteria set out in the CLP Regulation to provide practical help and examples to prospective notifiers and registrants. The document sets out guidance in the process of evaluating the classification based on the information available; supported by many examples. A series of annexes discuss special cases in more detail	http://echa.europa.eu/web/guest/guidance-documents/guidance-on-clp

ECHA CLP guidance	Guidance on labelling and packaging in accordance with Regulation 1272/2008	This guidance covers practical aspects of determining suitable labelling and packaging for substances and mixtures once the applicable classifications have been assessed and determined	http://echa.europa.eu/web/guest/guidance-documents/guidance-on-clp
ECHA CLP guidance	Guidance on the preparation of dossiers for harmonised classification and labelling	Guidance supporting manufacturers/importers as well as member state authority users, regarding preparation and submission of a dossier for harmonised classification at EU level. This is mainly aimed at CMR and respiratory sensitising substances though this guidance also discusses the circumstance that a registrant has identified an apparently inappropriate entry in the existing harmonised C&L for a substance	http://echa.europa.eu/web/guest/guidance-documents/guidance-on-clp
ECHA CLP guidance	Introductory guidance on the CLP regulation	This guidance provides useful background on the CLP legislation, in the context of other related legislation and the communication of substance hazards and risks. The approaches for mixtures are also discussed	http://echa.europa.eu/web/guest/guidance-documents/guidance-on-clp
REACH-IT industry user manual	Part 16: How to create and submit a C&L notification using the REACH-IT online tool	This guidance supports notifiers using the online tool at REACH-IT to notify substance classification and labelling. The guidance document explains when this method (rather than preparation of the notification within IUCLID 5) is appropriate and talks through the steps, using screen images	http://echa.europa.eu/web/guest/support/dossier-submission-tools/reach-it/data-submission-industry-user-manuals

Table 16.4 *Guidance and resources on exposure and risk.*

Source	Title	Summary/key features of guidance	Online access
ECHA Guidance on information requirements and chemical safety assessment	Part D – Exposure assessment	Sets out the required approach to assessment of potential exposure throughout the life cycle of the substance. Establishes the requirement to control risks such that RCR < 1 prior to registration. Establishes the use of a tiered modelling system of increasing complexity only if required to refine the scenario	http://echa.europa.eu/web/guest/guidance-documents/guidance-on-information-requirements-and-chemical-safety-assessment
ECHA Guidance on information requirements and chemical safety assessment	Part E – Risk characterisation	For human health, establishes the need to identify the leading health effect and conduct risk characterisation on a quantitative basis where possible using DNEL or DMEL. For the environment, establishes protection targets and sets out the approach for each	http://echa.europa.eu/web/guest/guidance-documents/guidance-on-information-requirements-and-chemical-safety-assessment
ECHA Guidance on information requirements and chemical safety assessment	Chapter R.12: Use descriptor system	Defines the set of standard descriptors by which the sector(s) of use, chemical product category(ies), mode(s) of handling in respect of possible environmental releases, relevant task(s) for human health assessment, are defined in REACH	http://echa.europa.eu/web/guest/guidance-documents/guidance-on-information-requirements-and-chemical-safety-assessment
ECHA Guidance on information requirements and chemical safety assessment	Chapter R.13: Risk management measures and operational conditions	Guidance on iterative approaches to evaluating the effect of possible risk management measures including varying physical aspects of the form of the substance used; duration and context of exposure; PPE and other operational conditions. Guidance is given for workplace exposures, consumer exposures, and the waste processing stage	http://echa.europa.eu/web/guest/guidance-documents/guidance-on-information-requirements-and-chemical-safety-assessment

ECHA Guidance on information requirements and chemical safety assessment	Chapter R.14: Occupational exposure estimation	Guidance on the use of various available tools for REACH exposure assessment for workers, covering the scope of application and the various variables to consider	http://echa.europa.eu/web/guest/guidance-documents/guidance-on-information-requirements-and-chemical-safety-assessment
ECHA Guidance on information requirements and chemical safety assessment	Chapter R.15: Consumer exposure estimation	Guidance on the use of ECETOC TRA and other available tools for REACH exposure assessment for consumers, covering the scope of application and the various variables to consider	http://echa.europa.eu/web/guest/guidance-documents/guidance-on-information-requirements-and-chemical-safety-assessment
ECHA Guidance on information requirements and chemical safety assessment	Chapter R.16: Environmental exposure estimation	Sets out in considerable detail the calculations used in a standard environmental exposure assessment, and default values. Default ERC release fractions and site sizes within the official regulatory guidance	http://echa.europa.eu/web/guest/guidance-documents/guidance-on-information-requirements-and-chemical-safety-assessment
ECHA Guidance on information requirements and chemical safety assessment	Chapter R.17: Estimation of exposure from articles	Guidance on suitable approaches to exposure estimation from substances incorporated in articles	http://echa.europa.eu/web/guest/guidance-documents/guidance-on-information-requirements-and-chemical-safety-assessment
ECHA Guidance on information requirements and chemical safety assessment	Chapter R.18: Estimation of exposure from waste life cycle stage	Guidance on quantitative estimation of exposure from disposal of waste by different methods including landfilling and incineration	http://echa.europa.eu/web/guest/guidance-documents/guidance-on-information-requirements-and-chemical-safety-assessment
ECHA practical guide	Guide 15: How to undertake a qualitative human health assessment and document it in a chemical safety report	This guidance sets out the approach to conducting and documenting risk characterisation in the event that no DNEL or DMEL can be defined, for example irritants/corrosives or some CMRs	http://echa.europa.eu/practical-guides

(continued overleaf)

Table 16.4 *(continued)*

Source	Title	Summary/key features of guidance	Online access
EIPPCB	BAT[a] Reference Documents (BREFs)	Regulatory industry sector-specific reference documents in support of the EU IPPC legislation. These set out the best available techniques and may or may not represent widespread and typical practices	—
OECD	Emission Scenario Documents (ESDs)	Industry sector-specific documents designed to support environmental exposure assessment and covering chemical functions, releases and site tonnages	—
ECHA	CHESAR	Integrated tool designed to interrelate with IUCLID 5 data entries to run exposure estimation and risk characterisation in as automated a way as possible. At the time of writing the tool has some limitations particularly where there are challenging chemistries or compositions or where the properties make a normal standard assessment impossible, but in future when these issues are addressed this should be a generally excellent and inclusive method	http://chesar.echa.europa.eu/
EC JRC ICHP	EUSES	Tool for environmental exposure modelling and risk characterisation for the environment and associated pathways (e.g. secondary poisoning and indirect exposure of humans via the environment). EUSES implements the standard EU modelling algorithms and was the recommended method for performing the necessary assessments under the legislation prior to REACH, and still valid and very useful in REACH, though the CHESAR tool may supersede it in future by undertaking the same actions. Highly adaptable when needed, but sometimes not easy for inexperienced users, and contains superseded default release fractions. Still freely available in a context of use in exposure and risk assessment of biocides	http://ihcp.jrc.ec.europa.eu/our_activities/public-health/risk_assessment_of_Biocides/euses

ECETOC	TRA tool	Model and tools for estimating exposure to humans in the workplace and environment, also applying the algorithms of EUSES in a different format, to create what is intended to be an inclusive tool. In practical terms some aspects can be tricky to work with but the workplace scenarios in particular are robust, if conservative	http://www.ecetoc.org/tra
CEFIC	Advanced REACH Tool (ART)	Online higher-tier tool to estimate workplace inhalatory exposure to vapours, sprays and dusts. Requires a fairly high level of knowledge and understanding of the context of use, though obviously alternative models can be explored for comparison. Using this tool it is possible to explore the effects of implementing additional RMM associated with controls and durations, but limitations on exposure associated with PPE are not implemented within the tool. It does not incorporate REACH use descriptors, and the user will need to justify the selections made in the CSA	http://www.advanced-reachtool.com/
RIVM	CONSEXPO	Downloadable software to estimate higher-tier exposure to chemicals, primarily to members of the general public. A number of specific and well-defined default scenarios are built in to the program and are supported by downloadable fact-sheets available from the RIVM web site. It does not incorporate REACH use descriptors, and the user will need to justify the selections made in the CSA	http://www.rivm.nl/en/Topics/Topics/C/ConsExpo
NL	Stoffenmanager	Stoffenmanager is a model for higher-tier estimation of worker exposure (industrial or professional) through the inhalatory route. Requires a fairly high level of knowledge and understanding of the context of use, though obviously alternative models can be explored for comparison. It can be easily adapted for different conditions, durations and PPE standards but as it does not incorporate REACH use descriptors, the user will need to justify the selections made	https://www.stoffen-manager.nl/

(continued overleaf)

Table 16.4 (continued)

Source	Title	Summary/key features of guidance	Online access
EC JRC	ESR RCRs	Past published EU risk assessments under the preceding legislation. An EU-wide approach was used. These documents were peer reviewed in depth	–
Various via CEFIC	SPERCs and their fact sheets	Detailed guidance from industry groups, including SPERCs. Release rates, site sizes, WWTP details, produced by industry sector-focussed working groups and published through CEFIC. Intended to be sector-specific defaults, but in practice ECHA requires further evidence in individual registrations that exposure levels are realistic	–
ECHA Guidance on information requirements and chemical safety assessment	Chapter R.19: Uncertainty analysis	Explains the reasons why uncertainty analysis is a necessary part of chemical safety assessment and how to undertake such an analysis	http://echa.europa.eu/web/guest/guidance-documents/guidance-on-information-requirements-and-chemical-safety-assessment

^aBest Available Techniques.

Table 16.5 *Guidance and resources on roles and responsibilities of parties in a supply chain, and communication between them.*

Source	Title	Summary	Online access
ECHA Guidance Document	Guidance on registration	This fundamentally important guidance sets out clearly the basis and roles of parties in the supply chain, with many worked examples and illustrative figures. Data sharing and roles of the SIEF and its members are explained. Specific practical explanation of how joint submission works is presented in a useful section. Responsibilities of registrants to maintain the registration with updates, and communicate with customers are set out	http://echa.europa.eu/web/guest/guidance-documents/guidance-on-reach
ECHA Guidance on information requirements and chemical safety assessment	Part G – Extending the SDS	This document provides guidance to registrants on how to integrate exposure scenarios into the extended data sheet and how to adjust substance-based conclusions when preparing the eSDS for a preparation	http://echa.europa.eu/web/guest/guidance-documents/guidance-on-information-requirements-and-chemical-safety-assessment
ECHA Guidance on information requirements and chemical safety assessment	Chapter R.12: Use descriptor system	This important guidance document defines the descriptor codes that may be used in a REACH technical dossier and documentation, with examples and advice on the appropriate use of the codes	http://echa.europa.eu/web/guest/guidance-documents/guidance-on-information-requirements-and-chemical-safety-assessment
ECHA Guidance on information requirements and chemical safety assessment	Chapter R.13: Risk management measures and operational conditions	This guidance to registrants provides reference information on risk management measures from the ECETOC-TRA risk management measures library, which ultimately forms a core of the communication from registrant to user on approaches necessary to manage risks	http://echa.europa.eu/web/guest/guidance-documents/guidance-on-information-requirements-and-chemical-safety-assessment

(continued overleaf)

Table 16.5 *(continued)*

Source	Title	Summary	Online access
ECHA Guidance Document	Guidance on the compilation of safety data sheets	This guidance to registrants covers content and structure of the safety data sheet and where the required data can be found	http://echa.europa.eu/ guidance-documents/ guidance-on-reach
ECHA Guidance Document	Guidance on labelling and packaging	This ECHA guidance document is in support of compliance with CLP rather than REACH, but in view of the close relevance to REACH is mentioned here. This guidance provides practical examples of applying the classification definitions, choice of packaging, and so on	http://echa.europa.eu/ guidance-documents/ guidance-on-reach
ECHA Guidance Document	Guidance for downstream users	This particularly important guidance is aimed at different types of downstream users in the supply chain and gives practical advice on their obligations and responsibilities under REACH, as well as things to be aware of in the context of the new legislation	http://echa.europa.eu/ guidance-documents/ guidance-on-reach
ECHA Guidance Document	Guidance on requirements for substances in articles	This guidance is intended to be useful to companies involved in the supply chain of articles, to determine the requirements and responsibilities in REACH	http://echa.europa.eu/ guidance-documents/ guidance-on-reach
ECHA Guidance Document	Guidance on the communication of information on the risks and safe use of chemicals	This guidance is to support member states communicating the relevant information, for example to interested parties in the public domain	http://echa.europa.eu/ guidance-documents/ guidance-on-reach
ECHA practical guide	Guide 13: How downstream users can handle exposure scenarios	This practical guide sets out the responsibilities and timings of downstream users to check and act on the exposure scenario information they receive, and communicate up and down in the chain of supply. It includes checklists, decision trees and worked examples	http://echa.europa.eu/ practical-guides

ECHA data submission manual	Part 21: How to prepare and submit a downstream user report using IUCLID 5	Guidance to downstream users needing to submit a chemical safety assessment for uses not covered by the manufacturer's registration	http://echa.europa.eu/web/guest/support/dossier-submission-tools/reach-it/data-submission-industry-user-manuals
ECHA Guidance Document	Guidance on authorisation application	Very important ECHA guidance for registrants considering applying for authorisation. Chapter 5 of the guidance covers the input made by third parties (which might include downstream users of the substance)	http://echa.europa.eu/web/guest/guidance-documents/guidance-on-reach
CEFIC sector groups	Use mappings	Sector groups have made significant progress since the launch of REACH, to aid their members and relevant actors in the supply chain to define the use pattern using relevant and appropriate descriptors following ECHA guidance R12. By the time of writing, many sector groups have issued 'use mappings' defining descriptor codes individually for multiple life cycle steps or possible exposure scenarios, for different types of chemicals used in their industry. This is valuable to both registrant companies which may be unfamiliar with details of the mode of use, and to downstream users which require a convenient method to communicate to their suppliers whilst protecting confidential business information	–
CEFIC	Standard phrases	CEFIC has facilitated the publication of a set of standard phrases for use in exposure scenarios and relevant documents	–

(continued overleaf)

Table 16.5 *(continued)*

Source	Title	Summary	Online access
CEFIC	Process in companies after receiving an (extended) SDS	This is guidance to downstream users to help them act on ESDS that are received from suppliers	–
CEFIC	Messages to communicate in the supply chain on extended SDS for substances	This is guidance to suppliers to help them address appropriately the needs of their customers and prepare the required documentation accurately	–
CEFIC	Standard reply to customers on REACH use communication	This is a suggested letter template for suppliers to use when communicating with downstream users about uses and ES to cover in the chemical safety assessment	–
CEFIC	Guidance on ES development and supply chain communication	This is fairly general guidance at a higher level of detail for appropriate documentation of the relevant ES	–

Table 16.6 Guidance and resources on interaction between registrants.

Source	Title	Summary	Online access
ECHA Guidance Document	Guidance on data sharing	Guidance to prospective registrants on how the legal obligations for sharing of information and study data for registration can be met within the context of competition law and other relevant frameworks. Useful and practical guidance covering among other topics data sharing for avoidance of unnecessary new testing; issues associated with individual registrations of different status (e.g. different tonnage bands, and the inquiry process for non-phase-in substances); SIEF management and the status of SIEF participants with different roles (manufacturers, importers, representatives, other parties and the role of Lead Registrant); legal agreements within and between SIEFs; principles for cost sharing and data rights including copyright; and approaches for negotiation and dispute resolution. Includes detailed discussion and practical examples of methods for cost calculations for data sharing	http://echa.europa.eu/guidance-documents/guidance-on-reach
ECHA Guidance Document	Guidance on substance identification	This important guidance assists prospective registrants in understanding the registration substance and defining its identity for registration, with significant implications for working relationships with other registrants in terms of 'sameness'	http://echa.europa.eu/guidance-documents/guidance-on-reach
ECHA data submission manual	Part 19: How to submit a chemical safety report as part of a joint submission	Practical focussed guidance on approach to issues affecting sharing of a CSR by multiple registrants within a joint submission. Covers in particular approach to completing Part A of the CSR, and approach to issues of CBI within Part B	http://echa.europa.eu/web/guest/support/dossier-submission-tools/reach-it/data-submission-industry-user-manuals

(continued overleaf)

Table 16.6 (continued)

Source	Title	Summary	Online access
ECHA data submission manual	Part 17: How to derive a public name for a substance for use under the REACH Regulation	Guidance to registrants on the derivation of public name if necessary, for example to protect confidential information which the chemical name if published in full might reveal. Also includes some examples and the format through which to apply to the ECHA that this step is necessary	http://echa.europa.eu/web/guest/support/dossier-submission-tools/reach-it/data-submission-industry-user-manuals
REACH-IT industry user manual	Part 18: Co-registrants page	Guidance to prospective registrants on the use of the Co-registrants page from which current information on other legal entities with registrations made or planned for the same substance, and their contact details, is accessible	http://echa.europa.eu/web/guest/support/dossier-submission-tools/reach-it/data-submission-industry-user-manuals

Table 16.7 Guidance and resources on submitting the registration within REACH-IT.

Source	Title	Summary	Online access
REACH-IT industry user manual	Part 1: Getting started with REACH-IT	Explains what the REACH-IT web portal is for and how users use it, and the information flow from the registrant's IUCLID 5 system to the dossier submission at REACH-IT. Useful screen capture images help familiarise the interested user with what to expect to see at the site	http://echa.europa.eu/web/guest/support/dossier-submission-tools/reach-it/data-submission-industry-user-manuals
REACH-IT industry user manual	Part 2: Sign-up and account management	This manual guides a registrant or other interested party through the processes involved with creating user accounts at REACH-IT and how the account is populated with certain useful information from the IUCLID installation	http://echa.europa.eu/web/guest/support/dossier-submission-tools/reach-it/data-submission-industry-user-manuals
REACH-IT industry user manual	Part 3: Login and message box	This manual explains the login process for users of REACH-IT and the management of messages received through the system	http://echa.europa.eu/web/guest/support/dossier-submission-tools/reach-it/data-submission-industry-user-manuals
REACH-IT industry user manual	Part 4: Late pre-registration	This manual explains the procedure for online pre-registration, which at the time of writing is open only to 'late' pre-registration, that is to allow first-time manufacturers/importers of phase-in substances to benefit from the phase-in deadlines for registration	http://echa.europa.eu/web/guest/support/dossier-submission-tools/reach-it/data-submission-industry-user-manuals
REACH-IT industry user manual	Part 5: Pre-SIEF	Guidance is given for the various activities for members of a pre-SIEF for a registration substances	http://echa.europa.eu/web/guest/support/dossier-submission-tools/reach-it/data-submission-industry-user-manuals

(continued overleaf)

Table 16.7 (continued)

Source	Title	Summary	Online access
REACH-IT industry user manual	Part 6: Dossier submission	This manual explains and describes (with useful screen images for reference) the process of dossier submission online at REACH-IT	http://echa.europa.eu/web/guest/support/dossier-submission-tools/reach-it/data-submission-industry-user-manuals
REACH-IT industry user manual	Part 7: Joint submission	Sets out the procedures associated with registration as part of a joint submission, whether as lead or member, and the various steps and actions at REACH-IT for such registrants	http://echa.europa.eu/web/guest/support/dossier-submission-tools/reach-it/data-submission-industry-user-manuals
REACH-IT industry user manual	Part 8: Invoices	This manual summarises in brief key rules associated with registration fees, and the features of REACH-IT relating to invoicing, status and payment	http://echa.europa.eu/web/guest/support/dossier-submission-tools/reach-it/data-submission-industry-user-manuals
REACH-IT industry user manual	Part 9: Advanced search	Manual to explain how to perform searches within REACH-IT	http://echa.europa.eu/web/guest/support/dossier-submission-tools/reach-it/data-submission-industry-user-manuals
REACH-IT industry user manual	Part 10: Claim of a registration number for a notified substance	Substances 'notified' under the EU legislation for new substances preceding REACH have a special status in REACH and are considered already registered. This manual sets out the procedures for the registrant of such a substance to claim a registration number using REACH-IT	http://echa.europa.eu/web/guest/support/dossier-submission-tools/reach-it/data-submission-industry-user-manuals

REACH-IT industry user manual	Part 11: Online dossier creation and submission for inquiries	Step-by-step guidance to the creation and submission of an inquiry dossier to the ECHA, which is necessary for prospective registrants of new substances	http://echa.europa.eu/web/guest/support/dossier-submission-tools/reach-it/data-submission-industry-user-manuals
REACH-IT industry user manual	Part 15: Manage your group of manufacturers or importers	This manual covers the submission of classification and labelling notification to the CLP by a group of manufacturers/importers	http://echa.europa.eu/web/guest/support/dossier-submission-tools/reach-it/data-submission-industry-user-manuals
REACH-IT industry user manual	Part 16: How to create and submit a C&L notification using the REACH-IT online tool	This guidance supports notifiers using the online tool at REACH-IT to notify substance classification and labelling. The guidance document explains when this method (rather than preparation of the notification within IUCLID 5) is appropriate and talks through the steps, using screen images	http://echa.europa.eu/web/guest/support/dossier-submission-tools/reach-it/data-submission-industry-user-manuals
REACH-IT industry user manual	Part 17: Legal entity change	Useful guidance to all parties in REACH on the implications of changes in legal entity in various contexts (e.g. merger, acquisition, sale) and the steps that need to be taken	http://echa.europa.eu/web/guest/support/dossier-submission-tools/reach-it/data-submission-industry-user-manuals
REACH-IT industry user manual	Part 18: Co-registrants page	Guidance to prospective registrants on the use of the Co-registrants page from which current information on other legal entities with registrations made or planned for the same substance, and their contact details, is accessible	http://echa.europa.eu/web/guest/support/dossier-submission-tools/reach-it/data-submission-industry-user-manuals

Table 16.8 *Guidance and resources post-registration.*

Source	Title	Summary	Online access
ECHA data submission manual	Part 15: How to determine what will be published on the ECHA web site from the registration dossier	This guidance explains the dissemination process and contains key messages to registrants about confidentiality and data rights associated with the process, and the nature of the information disseminated by the ECHA. It explains the use of the IUCLID plug-in tool which is available to help registrants check in advance how their dossier would be represented in the publicly-disseminated version published on the Internet by the ECHA in due course	http://echa.europa.eu/web/guest/support/dossier-submission-tools/reach-it/data-submission-industry-user-manuals
ECHA practical guide	Practical guide 12: How to communicate with the ECHA in dossier evaluation	This longer than typical practical guide brings together a summary of the background to different types of ECHA evaluation and the communication steps involved	http://echa.europa.eu/practical-guides
ECHA practical guide	Practical guide 8: How to report changes in identity of legal entities	This useful guidance indicates clearly the implications in terms of responsibilities and costs in case of various different types of changes to business circumstances of registrants and only representatives	http://echa.europa.eu/practical-guides

References

CLP (2008) CLP-Regulation (EC) No 1272/2008.
EUR-Lex (nd) Access to European Union law. http://www.advancedreachtool.com/ (last accessed 30 July 2013).
IUCLID 5 (nd) Home page. http://www.iuclid.eu/ (last accessed 30 July 2013).

Web References

ART Advanced Reach Tool 1.5, http://www.advancedreachtool.com/ (last accessed 30 July 2013).
CHESAR (CHEmical Safety Assessment and Reporting Tool) (u.d.) http://chesar.echa. europa.eu/ (last accessed 30 July 2013).
ECETOC http://www.ecetoc.org/tra (last accessed 30 July 2013).
ECHA Registration Statistics. http://echa.europa.eu/web/guest/information-on-chemicals/ registration-statistics (last accessed 30 July 2013).
ECHA (u.d.) Guidance on Information Requirements and Chemical Safety Assessment, http://echa.europa.eu/web/guest/guidance-documents/guidance-on-information-requirements-and-chemical-safety-assessment (last accessed 30 July 2013).
ECHA (u.d.) Guidance on REACH, http://echa.europa.eu/web/guest/guidance-documents/guidance-on-reach (last accessed 30 July 2013).
ECHA (u.d.) Guidance on CLP, http://echa.europa.eu/web/guest/guidance-documents/guidance-on-clp (last accessed 30 July 2013).
ECHA (u.d.) Data Submission and REACH-IT Industry User Manuals, http://echa. europa.eu/web/guest/support/dossier-submission-tools/reach-it/data-submission-industry-user-manuals (last accessed 30 July 2013).
ECHA Practical Guides, http://echa.europa.eu/practical-guides (last accessed 30 July 2013).
ECHA (u.d.) C&L Inventory Database, http://echa.europa.eu/web/guest/information-on-chemicals/cl-inventory-database (last accessed 30 July 2013).
EUR-Lex Access to European Union Law, http://eur-lex.europa.eu (last accessed 30 July 2013).
European Commission (2012) EUSES. Last update 19 January 2012, http://ihcp.jrc.ec. europa.eu/our_activities/public-health/risk_assessment_of_Biocides/euses (last accessed 30 July 2013).
IUCLID 5 (u.d.) Documentation, http://iuclid.eu/index.php?fuseaction=home. documentation&type=public#usermanual (last accessed 30 July 2013).
National Institute for Public Health and the Environment ConsExpo, http://www.rivm.nl/ en/Topics/Topics/C/ConsExpo (last accessed 30 July 2013).
Stoffenmanager 5.0 (u.d.) https://www.stoffenmanager.nl/ (last accessed 30 July 2013).

Appendix A
Substance Classification and Labelling under REACH

Classification and labelling are used to communicate hazards associated with substances through their use on packaging and in safety data sheets. The European Regulation governing classification and labelling is Regulation 1272/2008, known as CLP 2008. (CLP stands for classification, labelling, and packaging). This Regulation replaces the previous legislation (Dangerous Substances Directive, DSD, and Dangerous Preparations Directive, DPD)[1] and brings the European Union classification and labelling in line with the globally harmonised system (GHS) of classification and labelling.

Classification under CLP 2008 has been subject to a long transition period. Substances are no longer in the transition period, so have to be classified according to CLP. Until 2015 they must also be classified according to DSD, but labelled and packaged according to CLP alone. However, the transition for mixtures (previously called preparations) does not end until 1 June 2015. Both DSD and DPD will be fully replaced by CLP 2008 from 1 June 2015.

In summary:

- Substances:
  - Until 1 June 2015: must be classified according to CLP and DSD. Label and package according to CLP.
  - From 1 June 2015: must be classified, labelled, and packaged according to CLP.
- Mixtures (formerly preparations):
  - Until 1 June 2015: must be classified according to DPD, may be classified according to CLP. May be labelled and packaged according to DPD; if classified according to CLP may be labelled and packaged according to CLP.

---

[1] CLP 2008 is a Regulation, so applies directly to Member States. Both DSD and DPD have to be enacted in Member States legislation; in the United Kingdom this is currently CHIP4.
CHIP 4: Chemicals (Hazard Information and Packaging for Supply) Regulations 2009.

Chemical Risk Assessment: A Manual for REACH, First Edition. Peter Fisk Associates Ltd.
© 2014 John Wiley & Sons, Ltd. Published 2014 by John Wiley & Sons, Ltd.

– From 1 June 2015: must be classified, labelled, and packaged according to CLP. For mixtures already in the supply chain, re-labelling, and re-packaging may be postponed until 1 June 2017.

The CLP Regulation sets out standard phrases and pictograms that are used on labels and in safety data sheets to warn consumers of the hazardous properties of a substance. From time to time, as classifications of substances are reviewed, and when there are changes to the GHS, Adaptations to Technical Progress (ATPs) are issued.

REACH registrations have to include classification and labelling requirements. As information is gathered and assessed, and new data generated, the data may demonstrate that there are hazards associated with the substance that are not covered by any existing classification and labelling. If this occurs then self-classification based on the data available is needed.

Classification and labelling requirements are related to the hazard(s) of a substance and may include classification for physico-chemical (e.g. flammability, explosivity), human health (e.g. acute toxicity, corrosive, irritant, sensitising, carcinogenic, mutagenic), and environmental (e.g. aquatic acute and chronic toxicity) hazard classifications. The hazard classes are set out in Annex I of the CLP Regulation.

The CLP Regulation and the subsequent ATPs apply to substances and to mixtures, previously dealt with under separate legislation (DSD and DPD). The legislation sets generic concentration limits (Table A.1) below which a substance does not have to be taken into account when considering the classification of a mixture including that substance. For some hazards and substances, specific concentration cut-off values may be set. In most cases the requirement is to determine impurities to the 1% w/w level unless they are Acute Toxicity Category 1–3 or Hazardous to Aquatic Environment Acute Category 1 and/or Chromic Category 1. In these cases the concentration needs to be stated to 0.1% w/w.

A specific concentration limit will generally be set when reliable scientific information indicates that the hazard of a substance will be evident at levels below the generic concentration limit. In exceptional circumstances a specific concentration limit that is greater than the generic concentration limit may be set where there is reliable and conclusive evidence that the hazard of a classified substance is not evident at a level above the generic concentration limit.

Table A.1 *Generic cut-off values. There are some endpoints with different concentration cut-off values.*

Hazard class	Generic cut-off values (%)
Acute Toxicity	
Category 1–3	0.1
Category 4	1
Skin corrosion/irritation	1
Serious damage to eyes/eye irritation	1
Hazardous to aquatic environment	
Acute Category 1	0.1
Chronic Category 1	0.1
Chronic Category 2–4	1

Generic concentration limits vary between different hazard categories and can be found in the relevant health hazard sections of Annex I of CLP 2008.

Classification for the environment takes into account both long-term and short-term testing; classification for chronic toxicity may be based on short-term or long-term data. Previously, only short-term toxicity was taken into account in aquatic classification. This change was introduced in the second ATP to CLP 2008. In addition, substances that are readily biodegradable may be classified in chronic categories 1–2.

For substances that are very hazardous to the aquatic environment, that is classified as Acute or Chronic Category 1, with acute toxicity below 1 mg/ml or chronic toxicity below 0.1 mg/ml (not readily biodegradable) or 0.01 mg/ml (readily biodegradable), multiplying (M) factors must be set. They are used to give increased weight to the acutely toxic substance when calculating the toxicity of a mixture. M-factors are used instead of specific concentration limits when calculating the classification of a mixture that includes very hazardous properties.

A.1   Important Differences

There are some important differences between the old classification under DSD and DPD and labelling and CLP. Some of these are to do with nomenclature: preparations are now referred to as mixtures; risk (R) phrases are replaced by hazard (H) phrases; safety (S) phrases are now precautionary (P) statements.

There have been important changes in the approach to physical chemical hazards under CLP 2008 compared to the previous legislation. Classification for physico-chemical hazards includes more categories than the previous legislation, allowing greater differentiation between hazards.

A new type of classification for human health has been introduced: specific target organ toxicity (STOT). This may be single exposure (SE) or repeat exposure (RE). For example, from acute toxicity data a substance may have an acute toxicity classification based on concentrations causing death, and a STOT SE classification, where effects to a specific organ are observed, especially where these do not lead to fatality. However, a substance should not receive two classifications from a single effect, so care should be taken when applying the criteria.

STOT RE is based on repeated exposure and like STOT SE has several categories, and again double classification should be avoided: effects that lead to a reproductive toxicity classification would not also lead to classification as STOT RE based on reproductive effects.

Subcategories now exist for sensitisation; both skin and respiratory sensitisers may be classified as either: 1A, strong sensitisers, or 1B, other sensitisers, where there is sufficient information for this to be distinguished. If there is insufficient evidence to decide on a subcategory, but there is evidence of sensitisation, then substances are classified as Category 1. There are concentration limits that apply to these classifications: tables with the criteria can be found in Regulation (EU) No 286/2011 amending CLP 2008.

The second ATP (Regulation EU 2011/286) introduced new classification criteria for chronic aquatic hazards, concentration limits for sensitisation and classification for substances damaging to the ozone layer.

A.1.1   Physico-Chemical Hazards

- H200: Unstable explosive
- H201: Explosive; mass explosion hazard
- H202: Explosive; severe projection hazard
- H203: Explosive; fire, blast or projection hazard
- H204: Fire or projection hazard
- H205: May mass explode in fire
- H220: Extremely flammable gas
- H221: Flammable gas
- H222: Extremely flammable aerosol
- H223: Flammable aerosol
- H224: Extremely flammable liquid and vapour
- H225: Highly flammable liquid and vapour
- H226: Flammable liquid and vapour
- H227: Combustible liquid
- H228: Flammable solid
- H240: Heating may cause an explosion
- H241: Heating may cause a fire or explosion
- H242: Heating may cause a fire
- H250: Catches fire spontaneously if exposed to air
- H251: Self-heating; may catch fire
- H252: Self-heating in large quantities; may catch fire
- H260: In contact with water releases flammable gases which may ignite spontaneously
- H261: In contact with water releases flammable gas
- H270: May cause or intensify fire; oxidizer
- H271: May cause fire or explosion; strong oxidizer
- H272: May intensify fire; oxidizer
- H280: Contains gas under pressure; may explode if heated
- H281: Contains refrigerated gas; may cause cryogenic burns or injury
- H290: May be corrosive to metals

A.1.2   Health Hazards

- H300: Fatal if swallowed
- H301: Toxic if swallowed
- H302: Harmful if swallowed
- H303: May be harmful if swallowed
- H304: May be fatal if swallowed and enters airways
- H305: May be harmful if swallowed and enters airways
- H310: Fatal in contact with skin
- H311: Toxic in contact with skin
- H312: Harmful in contact with skin
- H313: May be harmful in contact with skin
- H314: Causes severe skin burns and eye damage
- H315: Causes skin irritation
- H316: Causes mild skin irritation

- H317: May cause an allergic skin reaction
- H318: Causes serious eye damage
- H319: Causes serious eye irritation
- H320: Causes eye irritation
- H330: Fatal if inhaled
- H331: Toxic if inhaled
- H332: Harmful if inhaled
- H333: May be harmful if inhaled
- H334: May cause allergy or asthma symptoms or breathing difficulties if inhaled
- H335: May cause respiratory irritation
- H336: May cause drowsiness or dizziness
- H340: May cause genetic defects
- H341: Suspected of causing genetic defects
- H350: May cause cancer
- H351: Suspected of causing cancer
- H360: May damage fertility or the unborn child
- H361: Suspected of damaging fertility or the unborn child
- H362: May cause harm to breast-fed children
- H370: Causes damage to organs
- H371: May cause damage to organs
- H372: Causes damage to organs through prolonged or repeated exposure
- H373: May cause damage to organs through prolonged or repeated exposure

A.1.3 Environmental Hazards

- H400: Very toxic to aquatic life
- H401: Toxic to aquatic life
- H402: Harmful to aquatic life
- H410: Very toxic to aquatic life with long lasting effects
- H411: Toxic to aquatic life with long lasting effects
- H412: Harmful to aquatic life with long lasting effects
- H413: May cause long lasting harmful effects to aquatic life
- H420: Harms public health and the environment by destroying ozone in the upper atmosphere

A.1.4 Supplementary Labelling Requirements under the CLP Regulations

Some R-phrases that do not have simple equivalents under the GHS have been retained under the CLP Regulation and the numbering mirrors the number of the previous R-phrase.

A.1.4.1 *Physical Properties*

- EUH001: Explosive when dry
- EUH006: Explosive with or without contact with air
- EUH014: Reacts violently with water
- EUH018: In use may form flammable/explosive vapour-air mixture

- EUH019: May form explosive peroxides
- EUH044: Risk of explosion if heated under confinement.

A.1.4.2   Health Properties

- EUH029: Contact with water liberates toxic gas
- EUH031: Contact with acids liberates toxic gas
- EUH032: Contact with acids liberates very toxic gas
- EUH066: Repeated exposure may cause skin dryness or cracking
- EUH070: Toxic by eye contact
- EUH071: Corrosive to the respiratory tract.

A.1.4.3   Environmental Properties

- EUH059: Hazardous to the ozone layer

A.1.4.4   Other EU Hazard Statements

Some other hazard statements intended for use in very specific circumstances have also been retained under the CLP Regulation.

- EUH201: Contains lead. Should not be used on surfaces liable to be chewed or sucked by children.
- EUH201A: Warning! Contains lead.
- EUH202: Cyanoacrylate. Danger. Bonds skin and eyes in seconds. Keep out of the reach of children.
- EUH203: Contains chromium(VI). May produce an allergic reaction.
- EUH204: Contains isocyanates. May produce an allergic reaction.
- EUH205: Contains epoxy constituents. May produce an allergic reaction.
- EUH206: Warning! Do not use together with other products. May release dangerous gases (chlorine).
- EUH207: Warning! Contains cadmium. Dangerous fumes are formed during use. See information supplied by the manufacturer. Comply with the safety instructions.
- EUH208: Contains *<name of sensitising substance>*. May produce an allergic reaction.
- EUH209: Can become highly flammable in use.
- EUH209A: Can become flammable in use.
- EUH210: Safety data sheet available on request.
- EUH401: To avoid risks to human health and the environment, comply with the instructions for use.

A.2   CLP Symbols

Nine CLP symbols in a diamond with a red border replace the familiar symbols with an orange background. These symbolise the following substances: oxidizers, flammables, explosives, acutely toxic substances, corrosives, gases under pressure, carcinogens, hazards to the environment, and dermal and respiratory irritants (Table A.2).

Table A.2   *CLP symbols.*

	Usage

Physical hazards
Explosive

Unstable explosives
Explosives, divisions 1.1, 1.2, 1.3, 1.4
Self-reactive substances and mixtures, types A, B
Organic peroxides, types A, B

Flammable

Flammable gases, category 1
Flammable aerosols, categories 1, 2
Flammable liquids, categories 1, 2, 3
Flammable solids, categories 1, 2
Self-reactive substances and mixtures, types B, C, D, E, F
Pyrophoric liquids, category 1
Pyrophoric solids, category 1
Self-heating substances and mixtures, categories 1, 2
Substances and mixtures, which in contact with water, emit
    flammable gases, categories 1, 2, 3
Organic peroxides, types B, C, D, E, F

Oxidising

Oxidizing gases, category 1
Oxidizing liquids, categories 1, 2, 3
Oxidizing solids, categories 1, 2, 3

Compressed gas

Compressed gases
Liquefied gases
Refrigerated liquefied gases
Dissolved gases

(*continued overleaf*)

Table A.2   *(continued)*

	Usage

Corrosive

Corrosive to metals, category 1

No pictogram required

Explosives, divisions 1.5, 1.6
Flammable gases, category 2
Self-reactive substances and mixtures, type G
Organic peroxides, type G

Health Hazards
Toxic

Acute toxicity (oral, dermal, inhalation), categories 1, 2, 3

Corrosive

Skin corrosion, categories 1A, 1B, 1C
Serious eye damage, category 1

Irritant

Acute toxicity (oral, dermal, inhalation), category 4
Skin irritation, categories 2, 3
Eye irritation, category 2A
Skin sensitization, category 1
STOT SE category 3
　　Respiratory tract irritation
　　Narcotic effects
Not used with the 'skull and crossbones' pictogram for skin or
　　eye irritation if:
　　　　the 'corrosion' pictogram also appears
　　　　the 'health hazard' pictogram is used to indicate
　　　　　　respiratory sensitization

Table A.2   *(continued)*

	Usage
Health hazard	
	Respiratory sensitization, category 1 Germ cell mutagenicity, categories 1A, 1B, 2 Carcinogenicity, categories 1A, 1B, 2 Reproductive toxicity, categories 1A, 1B, 2 STOT SE categories 1, 2 STOT RE categories 1, 2 Aspiration hazard, categories 1, 2
No pictogram required	Reproductive toxicity – effects on or via lactation
Environmental hazards **Environmentally damaging**	
	Acute hazards to the aquatic environment, category 1 Chronic hazards to the aquatic environment, categories 1, 2
No pictogram required	Chronic hazards to the aquatic environment, categories 3, 4

A.2.1   Comparison of DSP/DPD with CLP 2008

Tables to convert from DSP/DPD classifications to the equivalent under CLP 2008 are given in the Regulation: they are reproduced in the tables shown later in this chapter. A multi-lingual glossary of terminology is available on the ECHA web site (ECHA, 2013). This includes definitions of substance names, technical terms and hazard phrases, and includes terms used in previous legislation.

The translation of old risk phrases into hazard phrases is not necessarily straightforward, as the criteria are not necessarily the same. For example, the cut-offs for acute oral toxicity classification have changed and classification specifies the route of exposure.

Guidance on application of the classification criteria is available from the ECHA web site. The guidance is reviewed and updated periodically, so it is necessary to make sure that the most up-to-date guidance is used. To ascertain existing classifications of substances it must be remembered that changes to CLP in the ATPs include new classification. Existing classifications can be found in the Regulation and the ATPs, and they are available in excel format from the Institute for Health and Consumer Protection (IHCP; Ex-ECB) (IHCP, nd). At the time of writing the Excel tables do not include changes from the third ATP.

A.2.1.1   Explosive Properties

There is no direct translation for:

- E; R2 'Risk of explosion by shock, friction, fire or other sources of ignition'
- E; R3 'Extreme risk of explosion by shock, friction, fire or other sources of ignition'.

   The CLP Hazard Phrases are:

- H200: Unstable explosive
- H201: Explosive; mass explosion hazard
- H202: Explosive; severe projection hazard
- H203: Explosive; fire, blast or projection hazard.

A.2.1.2   Oxidising Properties

Tables A.3–A.6

A.2.1.3   Corrosive to Metals

Table A.7

A.2.1.4   Flammable

Tables A.8–A.10

A.2.1.5   Acute Oral Toxicity

Tables A.11 and A.12

   Dusts are solid particles of a substance or a mixture suspended in a gas (usually air) which are generally formed by mechanical processes. Mists are liquid droplets of a substance or mixture suspended in a gas (usually air), generally formed by condensation of supersaturated vapours or by physical shearing of liquids. Dusts and mists generally have sizes ranging from less than 1 to about 100 μm (Tables A.13–A.15).

A.2.1.6   Aspiration Hazards

Table A.16

A.2.1.7   Skin Corrosion/Irritation

Corrosion is the destruction of skin tissue, namely visible necrosis through the epidermis and into the dermis in at least one of three tested animals after exposure up to four hours. Corrosive reactions are typified by ulcers, bleeding, bloody scabs and, by the end of observation at 14 days, by discoloration due to blanching of the skin, complete areas of alopecia and scars or if the results are based on the results of a validated *in vitro* test or if the results can be predicted, for example from strong alkali or acid reactions indicated by a pH < 2 or > 11.5. Measurement of pH alone may be adequate but assessment of acidic or alkali reserve is preferable; methods are needed to assess buffering capacity (Table A.17).

Table A.3 *Organic peroxides.*

Directive 67/5348/EEC	Hazard class and category	Hazard phrase	Criteria
	Org. Perox. A	H240: Heating may cause an explosion	Any organic peroxide which, as packaged, can detonate or deflagrate rapidly shall be defined as organic peroxide
	Org. Perox. B	H241: Heating may cause a fire or explosion	Any organic peroxide possessing explosive properties and which, as packaged, neither detonates nor deflagrates rapidly, but is liable to undergo a thermal explosion in that package shall be defined as organic peroxide
O; R7	Org. Perox. CD	H242: Heating may cause a fire	Type C: any organic peroxide possessing explosive properties when the substance or mixture as packaged cannot detonate or deflagrate rapidly or undergo a thermal explosion shall be defined as organic peroxide Type D: any organic peroxide in laboratory testing i) detonates partially, does not deflagrate rapidly and shows no violent effect when heated under confinement; or ii) does not detonate at all, deflagrates slowly and shows no violent effect when heated under confinement; or iii) does not detonate or deflagrate at all and shows a medium effect when heated under confinement;
	Org. Perox. EF	H242: Heating may cause a fire	Type E: any organic peroxide which, in laboratory testing, neither detonates nor deflagrates at all and shows low or no effect when heated under confinement shall be defined as organic peroxide Type F: any organic peroxide which, in laboratory testing, neither detonates in the cavitated state nor deflagrates at all and shows only a low or no effect when heated under confinement as well as low or no explosive power shall be defined as organic peroxide

Table A.4 *Oxidising gas.*

Directive 67/5348/EEC	Hazard class and category	Hazard phrase	Criteria
Gas			
O; R8	Ox. Gas 1	H270: May cause or intensify fire; oxidiser	Any gas which may, generally by providing oxygen, cause or contribute to the combustion of other material more than air does

Table A.5 *Oxidising liquid.*

Directive 67/5348/EEC	Hazard class and category	Hazard phrase	Criteria
Liquid			
O; R9	Ox. Liq. 1	H271: May cause fire or explosion; strong oxidiser	Any substance or mixture which, in the 1 : 1 mixture, by mass, of substance (or mixture) and cellulose tested, spontaneously ignites; or the mean pressure rise time of a 1 : 1 mixture, by mass, of substance (or mixture) and cellulose is less than that of a 1 : 1 mixture, by mass, of 50% perchloric acid and cellulose
–	Ox. Liq. 2	H272: May intensify fire; oxidiser	Any substance or mixture which, in the 1 : 1 mixture, by mass, of substance (or mixture) and cellulose tested, exhibits a mean pressure rise time less than or equal to the mean pressure rise time of a 1 : 1 mixture, by mass, of 40% aqueous sodium chlorate solution and cellulose
–	Ox. Liq. 3	H272: May intensify fire; oxidiser	Any substance or mixture which, in the 1 : 1 mixture, by mass, of substance (or mixture) and cellulose tested, exhibits a mean pressure rise time less than or equal to the mean pressure rise time of a 1 : 1 mixture, by mass, of 65% aqueous nitric acid and cellulose

Table A.6   *Oxidising solid.*

Directive 67/5348/EEC	Hazard class and category	Hazard phrase	Criteria
Solid			
O; R9	Ox. Sol. 1	H271: May cause fire or explosion; strong oxidiser	Any substance or mixture which, in the 4 : 1 or 1 : 1 sample-to-cellulose ratio (by mass) tested, exhibits a mean burning time less than the mean burning time of a 3 : 2 mixture, by mass, of potassium bromate and cellulose
–	Ox. Sol. 2	H272: May intensify fire; oxidiser	Any substance or mixture which, in the 4 : 1 or 1 : 1 sample-to-cellulose ratio (by mass) tested, exhibits a mean burning time equal to or less than the mean burning time of a 2 : 3 mixture (by mass) of potassium bromate
–	Ox. Sol. 3	H272: May intensify fire; oxidiser	Any substance or mixture which, in the 4 : 1 or 1 : 1 sample-to-cellulose ratio (by mass) tested, exhibits a mean burning time equal to or less than the mean burning time of a 3 : 7 mixture (by mass) of potassium bromate and cellulose

Table A.7   *Corrosive to metals.*

Directive 67/5348/EEC	Hazard class and category	Hazard statement	Criteria
–	Corrosive to metals	H290: May be corrosive to metals	Corrosion rate on either steel or aluminium surfaces exceeding 6.25 mm/year at a test temperature of 55 °C when tested on both materials

Table A.8   *Flammable gas.*

Directive 67/5348/EEC	Hazard class and category	Hazard statement	Criteria
F+; R12	Flam. Gas 1	H220: Extremely flammable gas	Gases at 20 °C and 101.3 kPa: (a) are ignitable when in a mixture of 13% or less by volume in air; or (b) have a flammable range with air of at least 12% regardless of the lower flammable limit
	Flam. Gas 2	H221: Flammable gas	Gases at 20 °C and 101.3 kPa that have a flammable range while mixed in air

Table A.9 *Flammable liquid.*

Directive 67/5348/EEC	Hazard class and category	Hazard statement	Criteria
R10	Flam. Liq. 1	H224: Extremely flammable liquid and vapour	If flashpoint $< 23\,^\circ$C and initial boiling point $\leq 35\,^\circ$C
	Flam. Liq. 2	H225: Highly flammable liquid and vapour	If flashpoint $< 23\,^\circ$C and initial boiling point $> 35\,^\circ$C
	Flam. Liq. 3	H226: Flammable liquid and vapour	If flashpoint $\geq 23\,^\circ$C
F; R11	Flam. Liq. 1	H224: Extremely flammable liquid and vapour	If initial boiling point $\leq 35\,^\circ$C
	Flam. Liq. 2	H225: Highly flammable liquid and vapour	If initial boiling point $> 35\,^\circ$C
F+; R12	Flam. Liq. 1	H224: Extremely flammable liquid and vapour	–
F+; R12	Self-react. CD	H242: Heating may cause a fire	–
	Self-react. EF	H242: Heating may cause a fire	–
	Self-react. G	None	–
F; R15	–	–	No translation possible
F; R17	Pyr. Liq. 1	H250: Catches fire spontaneously if exposed to air	The liquid ignites within 5 min when added to an inert carrier and exposed to air, or it ignites or chars a filter paper on contact with air within 5 min

Table A.10 *Flammable solid.*

Directive 67/5348/EEC	Hazard class and category	Hazard statement	Criteria
	Flam. Sol. 1	H228: Flammable Solid	Burning rate test (a) wetted zone does not stop fire and (b) burning time $< 45\,$s or burning rate $> 2.2\,$mm/s
			Metal powders burning time $\leq 5\,$min
	Flam. Sol. 2	H228: Flammable solid	Burning rate test (a) wetted zone stops the fire for at least 4 min and (b) burning time $< 45\,$s or burning rate $> 2.2\,$mm/s
			Metal powders burning time $5–10\,$min
F; R11	–	No direct translation possible	–
F; R17	Pyr. Sol. 1	H250: Catches fire spontaneously if exposed to air	The solid ignites within 5 min of coming into contact with air

Table A.11   *Acute oral toxicity.*

Directive 67/5348/EEC	Hazard class and category	Hazard statement	Criteria
T+; R28	Acute Tox. 1	H300: Fatal if swallowed	$LD_{50} \leq 5\,mg/kg$
	Acute Tox. 2	H300: Fatal if swallowed	$LD_{50} = 5-50\,mg/kg$
T; R25	Acute Tox. 3	H301: Toxic if swallowed	$LD_{50} = 50-300\,mg/kg$
Xn; R22	Acute Tox. 4	H302: Harmful if swallowed	$LD_{50} = 300-2000\,mg/kg$

Table A.12   *Acute toxicity dermal.*

Directive 67/5348/EEC	Hazard class and category	Hazard statement	Criteria
T+; R27	Acute Tox. 1	H310: Fatal in contact with skin	$LD_{50} \leq 50\,mg/kg$
	Acute Tox. 2	H310: Fatal in contact with skin	$LD_{50} = 50-200\,mg/kg$
T; R24	Acute Tox. 3	H311: Toxic in contact with skin	$LD_{50} = 200-1000\,mg/kg$
Xn; R21	Acute Tox. 4	H312: Harmful in contact with skin	$LD_{50} = 1000-2000\,mg/kg$

Table A.13   *Acute toxicity by the inhalation of a dust or mist.*

Directive 67/5348/EEC	Hazard class and category	Hazard statement	Criteria
T+; R26	Acute Tox. 1	H330: Fatal if inhaled	$LC_{50} \leq 0.05\,mg/l$
	Acute Tox. 2	H330: Fatal if inhaled	$LC_{50} = 0.05-0.5\,mg/l$
T; R23	Acute Tox. 3	H331: Toxic if inhaled	$LC_{50} = 0.5-1\,mg/l$
Xn; R20	Acute Tox. 4	H332: Harmful if inhaled	$LC_{50} = 1-5\,mg/l$

Table A.14   *Acute toxicity inhalation of a vapour[a].*

Directive 67/5348/EEC	Hazard class and category	Hazard statement	Criteria
T+; R26	Acute Tox. 1	H330: Fatal if inhaled	$LC_{50} \leq 0.5\,mg/l$ per 4 h
	Acute Tox. 2	H330: Fatal if inhaled	$LC_{50} = 0.5-2\,mg/l$ per 4 h
T; R23	Acute Tox. 3	H331: Toxic if inhaled	$LC_{50} = 2-10\,mg/l$ per 4 h
Xn; R20	Acute Tox. 4	H332: Harmful if inhaled	$LC_{50} = 10-20\,mg/l$ per 4 h

[a]A vapour is the gaseous form of a substance or mixture released from its liquid or solid state.

Table A.15   *Acute toxicity inhalation of a gas.*

Directive 67/5348/EEC	Hazard class and category	Hazard statement	Criteria
T+; R26	Acute Tox. 1	H330: Fatal if inhaled	$LC_{50} \leq 100\,ppm\,V$
T; R23	Acute Tox. 2	H330: Fatal if inhaled	$LC_{50} = 100-500\,ppm\,V$
	Acute Tox. 3	H331: Toxic if inhaled	$LC_{50} = 500-2500\,ppm\,V$
Xn; R20	Acute Tox. 4	H332: Harmful if inhaled	$LC_{50} = 2500-5000\,ppm\,V$

Table A.16   *Aspiration hazards.*

Directive 67/5348/EEC	Hazard class and category	Hazard statement	Criteria
Xn; R65	Asp. Tox. 1	H304: May be fatal if swallowed and enters airways	Chemicals known to cause human aspiration toxicity hazards or to be regarded as if they cause human aspiration toxicity hazard
Xn; R65	Asp. Tox. 2	H305: May be harmful if swallowed and enters airways	Chemicals which cause concern owing to the presumption that they cause human aspiration toxicity hazard

Table A.17   *Skin corrosion/irritation.*

Directive 67/5348/EEC	Hazard class and category	Hazard statement	Criteria
C; R35	Skin Corr. 1A	H314: Causes severe skin burns and eye damage	≤ 3 min
C; R34	Skin Corr. 1B	H314: Causes severe skin burns and eye damage	> 3 min ≤ 1 h
	Skin Corr. 1C	H314: Causes severe skin burns and eye damage	> 1 hr ≤ 4 h
Xi; R38	Skin Irrit. 2	H315: Causes skin irritation	1. Mean value of 2.3–4.0 for erythema/eschar or for oedema in at least 2 of 3 tested animals from gradings at 24, 48 and 72 h after patch removal or, if reactions are delayed, from grades on 3 consecutive days after the onset of skin reactions; or 2. Inflammation that persists to the end of the observation period normally 14 days in at least 2 animals, particularly taking into account alopecia (limited area), hyperkeratosis, hyperplasia and scaling; or 3. In some cases where there is pronounced variability of response among animals, with very definite positive effects related to chemical exposure in a single animal but less than the criteria above

A.2.1.8   Eye Irritation

Table A.18

A.2.1.9   Sensitisation

Table A.19 and A.20

A.2.1.10   *Specific Target Systemic Toxicity-Repeated Exposure*

This causes chronic toxicity. The route of exposure could be added to the hazard statement if it is conclusively proven that no other route of exposure causes the hazard.
STOT RE Category 1: Substances that have produced significant toxicity in humans or that, on the basis of evidence from studies in experimental animals, can be presumed to have the potential *to produce significant toxicity in humans* following repeated exposure.

Table A.18   *Eye irritation.*

Directive 67/5348/EEC	Hazard class and category	Hazard statement	Criteria
Xi; R41	Eye Dam. 1	H318: Causes serious eye damage	A substance produces in at least in one animal effects on the cornea, iris or conjunctiva that are not expected to reverse or have not fully reversed within an observation period of normally 21 days
Xi; R36	Eye Irrit. 2	H319: Causes serious eye irritation	When applied to the eye of an animal, a substance produces at least in two of three tested animals a positive response of corneal opacity ≥ 1 and/or iritis ≥ 1, and/or conjunctival redness ≥ 2 and/or conjunctival oedema (chemosis) ≥ 2 calculated as the mean scores following grading at 24, 48 and 72 h after installation of the test material, and fully reverses within 21 days

Table A.19   *Respiratory sensitisation.*

Directive 67/5348/EEC	Hazard class and category	Hazard statement	Criteria
R42	Resp. Sens. 1	H334: May cause allergy or asthma symptoms or breathing difficulties if inhaled	Evidence in humans that the substance can induce respiratory hypersensitivity or positive results from an appropriate animal test

Table A.20 *Skin sensitisation.*

Directive 67/5348/EEC	Hazard class and category	Hazard statement	Criteria
R43	Skin Sens. 1	H317: May cause an allergic skin reaction	Evidence in humans that the substance can induce skin sensitisation by skin contact in a substantial number of persons or positive results from an appropriate animal test

Substance categorised as STOT RE Category 1:

- reliable and good quality evidence from human cases or epidemiological studies;
- observations from appropriate studies in experimental animals in which significant and/or severe toxic effects, of relevance to human health, were produced at generally low exposure concentrations; guidance dose/concentration values are provided below to be used as part of weight-of-evidence evaluation;
- oral, rat < 10 mg/kg (body weight)/day;
- dermal, rat or rabbit < 20 mg/kg (body weight)/day;
- inhalation, rat (gas) < 50 ppm/6 h/day;
- inhalation, rat (vapour) < 0.2 mg/l/6 h/day;
- inhalation, rat (dust/mist/fume) < 0.02 mg/l/6 h/day.

STOT RE Category 2: Substances that, on the basis of evidence from studies in experimental animals, can be presumed to have the potential *to be harmful to human health* following repeated exposure.

Placing a substance in Category 2 is done on the basis of observations from appropriate studies in experimental animals in which significant toxic effects, of relevance to human health, were produced at generally moderate exposure concentrations. Guidance dose/concentration range values are provided below in order to help in classification.

- oral, rat $> 10-< 100$ mg/kg (body weight)/day;
- dermal, rat or rabbit $> 20-< 200$ mg/kg (body weight)/day;
- inhalation, rat (gas) $> 50-< 250$ ppm/6 h/day;
- inhalation, rat (vapour) $> 0.2-< 1$ mg/l/6 h/day;
- inhalation, rat (dust/mist/fume) $> 0.02-< 0.2$ mg/l/6 h/day.

In exceptional cases human evidence can also be used to place a substance in Category 2 (Table A.21).

A.3 Specific Target Organ Systemic Toxicity – Single Exposure

The route of exposure could be added to the hazard statement if it is conclusively proven that no other route of exposure cause the hazard (Table A.22).

Table A.21 *Specific target systemic toxicity – repeated exposure (STOT RE).*

Directive 67/5348/EEC	Hazard class and category	Hazard statement
T; R48/23	STOT RE 1	H372: Causes damage to organs through prolonged or repeated exposure
T; R48/24	STOT RE 1	H372: Causes damage to organs through prolonged or repeated exposure
T; R48/25	STOT RE 1	H372: Causes damage to organs through prolonged or repeated exposure
R33	STOT RE 2	H373: May cause damage to organs through prolonged or repeated exposure
Xn; R48/20	STOT RE 2	H373: May cause damage to organs through prolonged or repeated exposure
Xn; R48/21	STOT RE 2	H373: May cause damage to organs through prolonged or repeated exposure
Xn; R48/22	STOT RE 2	H373: May cause damage to organs through prolonged or repeated exposure

Table A.22 *Specific target organ systemic toxicity – single exposure (STOT SE).*

Directive 67/5348/EEC	Hazard class and category	Hazard statement
T; R39/23	STOT SE 1	H370: Causes damage to organs
T; R39/24	STOT SE 1	H370: Causes damage to organs
T; R39/25	STOT SE 1	H370: Causes damage to organs
T+; R39/26	STOT SE 1	H370: Causes damage to organs
T+; R39/27	STOT SE 1	H370: Causes damage to organs
T+; R39/28	STOT SE 1	H370: Causes damage to organs
Xn; R68/20	STOT SE 2	H371: May cause damage to organs
Xn; R68/21	STOT SE 2	H371: May cause damage to organs
Xn; R68/22	STOT SE 2	H371: May cause damage to organs
Xi; R37	STOT SE 3	H335: May cause respiratory irritation
R67	STOT SE 3	H336: May cause drowsiness or dizziness

A.3.1 Carcinogenic Substances

Classification for CMR (Carcinogenic, Mutagenic, Reprotoxic) properties is complex and should normally be undertaken with appropriate expert input since some judgement is required (Table A.23).

A.3.1.1 Substances Toxic for Reproduction

Hazard statements H360 and H361 indicate a general concern for both the reproductive properties related to fertility and developmental effects (Table A.24).

Table A.23   *Carcinogenic substances.*

Directive 67/5348/EEC	Hazard class and category	Hazard statement	Criteria
Carc. Cat. 1; R45	Carc. 1A	H350: May cause cancer	Chemicals *known* to have carcinogenic potential for humans; the placing of a chemical is largely based on human evidence
Carc. Cat. 1; R49	Carc. 1A	H350i	Chemicals *known* to have carcinogenic potential for humans; the placing of a chemical is largely based on human evidence
Carc. Cat. 2; R45	Carc. 1B	H350: May cause cancer	Chemicals *presumed* to have carcinogenic potential for humans; the placing of a chemical is largely based on animal evidence
Carc. Cat. 2; R49	Carc. 1B	H350i	Chemicals *presumed* to have carcinogenic potential for humans; the placing of a chemical is largely based on animal evidence
Carc. Cat. 3; R40	Carc. 2	H351: Suspected of causing cancer	Suspected human carcinogens

Table A.24   *Substances toxic for reproduction.*

Directive 67/5348/EEC	Hazard class and category	Hazard statement
Repr. Cat. 1; R60	Repr. 1A	H361f: Suspected of damaging fertility
Repr. Cat. 1; R60	Repr. 1A	H360FD: May damage fertility. May damage the unborn child
Repr. Cat. 1; R61	–	–
Repr. Cat. 1; R60	Repr. 1A	H360Df: Peut nuire au foetus. Suspected of damaging fertility
Repr. Cat. 2; R61	–	–
Repr. Cat. 1; R60	Repr. 1A	H360Fd: May damage fertility. Suspected of damaging the unborn child
Repr. Cat. 3; R63	–	–
Repr. Cat. 2; R60	Repr. 1A	H360FD: May damage fertility. May damage the unborn child
Repr. Cat. 1; R61	–	–
Repr. Cat. 1; R61	Repr. 1A	H360D: May damage the unborn child
Repr. Cat. 1; R61	Repr. 1A	H360Df: Peut nuire au foetus. Suspected of damaging fertility
Repr. Cat. 3; R62	–	–
Repr. Cat. 2; R60	Repr. 1B	H361f: Suspected of damaging fertility

Table A.24   *(continued)*

Directive 67/5348/EEC	Hazard class and category	Hazard statement
Repr. Cat. 2; R61	Repr. 1B	H360D: May damage the unborn child
Repr. Cat. 2; R60	Repr. 1B	H360FD: May damage fertility. May damage the unborn child
Repr. Cat. 2; R61	–	–
Repr. Cat. 2; R61	Repr. 1B	H360Df: Peut nuire au foetus. Suspected of damaging fertility
Repr. Cat. 3; R62	–	–
Repr. Cat. 2; R60	Repr. 1B	H360Fd: May damage fertility. Suspected of damaging the unborn child
Repr. Cat. 3; R63	–	–
Repr. Cat. 3; R62	Repr. 2	H361f: Suspected of damaging fertility
Repr. Cat. 3; R63	Repr. 2	H361d: Suspected of damaging the unborn child
Repr. Cat. 3; R62-63	Repr. 2	H361fd: Suspected of damaging fertility. Suspected of damaging the unborn child

A.3.2   Mutagenic Substances

Table A.25

A.3.3   Effect during Lactation

Table A.26

Table A.25   *Mutagenic substances.*

Directive 67/5348/EEC	Hazard class and category	Hazard statement	Criteria
Muta Cat 1: R46	Muta 1A	H340: May cause genetic defects	Chemicals *known* to induce heritable mutations in germ cells of humans
Muta. Cat. 2; R46	Muta. 1B	H340: May cause genetic defects	Chemicals which should be *regarded* as if they induce heritable mutations in germ cells of humans
Muta. Cat. 3; R68	Muta. 2	H341: Suspected of causing genetic defects	Chemicals which *cause concern* for man owing to the possibility that they may induce heritable mutations in germ cells of humans

Table A.26   *Effect during lactation.*

Directive 67/5348/EEC	Hazard class and category	Hazard statement
R64	Lact.	H362: May cause harm to breast-fed children

A.3.4   Aquatic Environment

Table A.27 and A.28

A.3.5   Ozone

Table A.29

Table A.27   *Aquatic environment.*

Directive 67/5348/EEC	Hazard class and category	Hazard statement
N; R50	Aquatic. Acute 1	H400: Very toxic to aquatic life
N; R50-53	Aquatic Acute 1	H400: Very toxic to aquatic life
	Aquatic Chronic 1	H410: Very toxic to aquatic life with long lasting effects
N; R51-53	Aquatic Chronic 2	H411: Toxic to aquatic life with long lasting effects
R52-53	Aquatic Chronic 3	H412: Harmful to aquatic life with long lasting effects
R53	Aquatic Chronic 4	H413: May cause long lasting harmful effects to aquatic life

Table A.28   *Criteria for the aquatic environment chronic category.*

	Chronic 1	Chronic 2	Chronic 3	Chronic 4
96 h LC_{50} (fish)	≤ 1 mg/l	1–10 mg/l	10–100 mg/l	Poorly soluble
48 h EC_{50} (crustacea)	≤ 1 mg/l	1–10 mg/l	10–100 mg/l	substances with no
72 or 96 h ErC_{50} (algae/aquatic plants)	≤ 1 mg/l	1–10 mg/l	10–100 mg/l	acute toxicity recorded and a with a potential to bio-accumulate
Rapidly degradable and/or	No	No	No	No
Potential to bio-accumulate				
Log K_{ow} or	> 4	> 4	> 4	> 4
BCF	≥ 500	≥ 500	≥ 500	≥ 500
Unless chronic toxicity NOEC	–	> 1 mg/l	> 1 mg/l	> 1 mg/l

Table A.29   *Ozone.*

Directive 67/5348/EEC	Hazard class and category	Hazard statement
N; R59	Ozone	EUH059: Hazardous to the ozone layer

A.4   Harmonised Classification and Labelling

For substances, but not for mixtures, harmonised classification and labelling may be proposed. Harmonised classification and labelling may be proposed by anyone in the supply chain (manufacturer, importer or downstream user) as well as by Member States. The process of harmonising classification and labelling is undertaken to protect human health and the environment while at the same time ensuring competitiveness. Usually classification and labelling is harmonised for substances with severe hazards, such as CMRs.

Proposals of harmonised classification and labelling are made on a case-by-case basis by the ECHA Committee for Risk Assessment (RCA). The RCA prepares ECHA opinions on risks to human health and the environment on:

- Harmonised classification and labelling
- Restriction
- Authorisation.

The final decisions on all these are taken by the European Commission.

References

CLP (2008) CLP-Regulation (EC) No 1272/2008.
ECHA (2013) Multilingual Chemical Terminology. echa.cdt.europa.eu/SearchByQuery-Load.do (last accessed 5 August 2013).
ICHP (nd) http://esis.jrc.ec.europa.eu/index.php?PGM=cla.

Appendix B

Further Discussion of Substance Identification and Sameness

B.1 Substance Identifiers

Table B.1 shows the subdivisions of EC Numbers. 'List numbers' are assigned when a substance is pre-registered or inquired about under the REACH Regulation. They are not official because they have not been published in the *Official Journal of the European Union*. The numbering of these categories is open ended, as substances are continually being added.

B.1.1 EC Name

Impurities, additives and minor components are normally not mentioned unless they contribute significantly to the classification of the substance.

B.1.2 CAS Registry Number (CAS# or CAS No.)

In 1965, the CAS Registry Numbers (CAS RNs, CAS Numbers or CAS#) were introduced to provide a unique, unmistakable identifier for chemical substances. The final digit is known as the check digit and serves to verify the number. The REACH process has verified that large numbers of CAS# are obsolete, with some substances having over 20 different CAS# whilst other CAS# cover a number of distinct substances or do not suitably describe substances that they are used for.

B.1.3 SMILES

The Simplified Molecular-Input Line-Entry System (SMILES) consists of a short ASCII string used to describe a structure of chemical molecules in a manner that is readable by a computer. These SMILES strings can be imported by most molecule editing software for conversion back into two-dimensional drawings or three-dimensional models of the

Chemical Risk Assessment: A Manual for REACH, First Edition. Peter Fisk Associates Ltd.
© 2014 John Wiley & Sons, Ltd. Published 2014 by John Wiley & Sons, Ltd.

Table B.1   *The subdivision of EC numbers.*

EC number	
2xx-xxx-x and 3xx-xxx-x	EINECS substances excluding polymers, which were recorded as being commercially available in the EU from 1 January 1971 to 18 September 1981. These were considered registered under Article 8(1) of the Dangerous Substances Directive (67/548/EEC). Under REACH, they are considered phase-in substances.
4xx-xxx-x	ELINCS substances notified under the Dangerous Substances Directive Notification of New Substances (NONS) that became commercially available after 18 September 1981 are considered to be already registered at the tonnage bands.
5xx-xxx-x	No Longer Polymer list (NLP-list) are substances that are no longer considered as polymers by the change in definition made in April 1992.[a] These were commercially available between after 18 September 1981 and 31 October 1993.
List numbers	
6xx-xxx-x	Automatically assigned, for example to pre-registrations of substances with a CAS No.
7xx-xxx-x	Assigned to substances after inquiries by the ECHA Substance ID Team
9xx-xxx-x	Automatically assigned, for example to pre-registrations without a CAS No. or other numerical identifier (including reaction masses of more than one substance)

[a]A polymer compromises of a simple weight majority of molecules containing at least three monomer units which are covalent bound to at least one other monomer unit or other reactant; less than a simple weight majority of molecules of the same molecular weight. In the context of this definition a 'monomer unit' means the reacted form of a monomer substance in a polymer.

molecules. SMILES is generally considered to have the advantage of being easier to read than the International Chemical Identifier (InChI) (Section B.1.4). It also has a wide base of software support with extensive theoretical (e.g. graph theory) backing.

B.1.4   InChI

In July 2006, the IUPAC introduced the International Chemical Identifier (InChI) as a standard for formula representation.

B.2   Substance Analysis

The analytical data should be accompanied by interpretation and there should be enough description given to allow another laboratory to be able to use the same methods. The choice of the analytical methods chosen for the identification and characterisation of the substances depends greatly upon whether the substance is:

1. Organic or inorganic.
2. Covalent or ionic.
3. Solid, liquid or gas at room temperature.
4. Soluble in water and other solvents.
5. Of simple or complex composition.

If one of the usual methods is deemed not useful for identification or quantification, this should be explained in the registration.

As with other technical issues, there are regulatory and commercial considerations behind the methods developed by the authorities and unsupported statements are not acceptable. The ECHA takes scrutiny of information about substance characterisation seriously, for the following reasons:

- Data from different producers may need to be shared, even if the compositions are not identical.
- Impurities or minor constituents may be important or even dominate the regulatory compliance process (i.e. may be substances of very high concern).
- Interpretation of the test data requires insight into the composition.
- Substances which had different European INventory of Existing Commercial chemical Substances (EINECS) or European List of Notified Chemical Substances (ELINCS) numbers may turn out to be the same and it is desirable to rationalise such inconsistencies.

It should be remembered that, at present, polymers[1] are currently outside the scope of REACH, although impurities in them may not be, depending on properties.

Three broad groups of substance are identified here. In each case, methods of production are an important part of the information to be provided in the submission, but without disclosure of confidential business information. A good description of mass balance is vital: the substance composition information should add up to 100% even if some of this is not fully identified due to technical limitations. The broad types are:

- Essentially-pure substances (mono-constituent).
- Multi-constituent substances (MCS).
- Unknown or variable composition, or biological origin substances (UVCB).

A fundamental principle affecting all substances is the ability to identify whether any constituents that might meet the substance of very high concern (SVHC) criteria are present at 0.1% or more, whatever the type of substance. For this reason alone, a realistic description of production methods should be provided, because it will support statements concerning the absence of SVHCs that will not have been analysed for.

B.2.1   Sameness

Experience with the early stages of REACH has shown that the Regulation has helped to rationalise many inconsistencies in choice of CAS/EC number in Europe and, indeed, around the world. Registrants have found that it has been possible (with due respect to confidentiality) to agree a substance identification profile (SIP) that is acceptable to all SIEF (Substance Information Exchange Forum) members. Sameness should best be determined on the highest purity commercial grade. With multi-constituent substances and UVCBs, variability between batches as well as between manufacturers is far more likely and should be taken into consideration.

---

[1] A polymer compromises of a simple weight majority of molecules containing at least three monomer units which are covalent bound to at least one other monomer unit or other reactant; less than a simple weight majority of molecules of the same molecular weight. In the context of this definition a 'monomer unit' means the reacted form of a monomer substance in a polymer.

B.2.2   Impurities

The quantification performed for sameness may not be suitable when quantifying impurities. All components above 1% need to be identified and quantified. Quantification below 1% is needed if the impurity is relevant for classification and labelling (C&L) or persistent bioaccumulative and toxic (PBT) assessment. In some cases this needs to be down to below 0.1% if the substance contains CMR (carcinogenic, mutagenic, and reprotoxic) and PBT Impurities.[2] These substances MUST be quantified, preferably against reference standards or agreed corrected response factors where possible by chemical name [EC, CAS and/or IUPAC Name], EC/CAS Number, molecular weight, and structural information.

The generic cut-off values from the CLP Regulations need to be considered when dealing with the question of whether an impurity is relevant for CLP (Classification, Labelling and Packaging). In most cases the requirement is to determine impurities to the 1% w/w level unless they are Acute Toxicity Category 1–3 or Hazardous to Aquatic Environment Acute Category 1 and/or Chromic Category 1. In these cases the concentration needs to be stated to 0.1% w/w (Table B.2).

It is interesting to note that CLP 2008 state that concentrations of methanol $\geq 3\%$ or dichloromethane $\geq 1\%$ require the use of child-resistant fastenings. It should be noted that high performance liquid chromatography (HPLC) and gas chromatography (GC) quantification are not suitable for determining the amount of volatile organic substances present such as residual solvent. The free methanol and dichloromethane content would be determined by headspace. This gives a good determination of how much methanol is present actually in solution rather than the total potentially present. These methods could also determine the volatile impurities and solvent content.

B.2.3   Departures from the Agreed Norm

Departures from the agreed norm can often be easily accommodated by company-specific additions to the dossier, without requiring the registrant to opt-out of the joint submission. Nevertheless, affected registrants should make a careful assessment case-by-case.

Table B.2   *Generic cut-off values.*

Hazard class	Generic cut-off values (%)
Acute Toxicity	
Category 1–3	0.1
Category 4	1
Skin corrosion/irritation	1
Serious damage to eyes/eye irritation	1
Hazardous to aquatic environment	
Acute category 1	0.1
Chronic category 1	0.1
Chronic category 2–4	1

---

[2] 'Impurities that are relevant for the classification and/or for PBT assessment shall always be specified, irrespective of the concentration'. ECHA (2012) *Identification and Naming of Substances under REACH and CLP*, pg 19.

B.3 Straightforward Organic Substances

B.3.1 Identity

The substance identity is normally determined by qualitative molecular spectroscopic techniques that are sometimes referred to as fingerprinting, due to the pattern matching from the spectra produced from the instruments.

- Ultraviolet-visible spectroscopy (UV-vis) gives an indication on the aromatic content and the presence of chromophores.
- Infrared (IR) spectroscopy by Fourier Transform-Infrared (FT-IR).
- Proton Nuclear Magnetic Resonance (^{1}H-NMR) and Carbon-13 NMR (^{13}C-NMR) spectroscopy.
- Mass spectroscopy (MS) is normally performed after chromatography (either GC or HPLC) gives the fragment pattern information and normally the molecular ion of the substance, hence the molecular weight of the substance can be determined.

Ultraviolet–visible spectroscopy (UV-vis) is very sensitive for aromatic hydrocarbons that absorb between 260 and 285 nm, so the amount of aromatic substances can be determined down to parts per million (mg/kg) levels. Above 5% aromatic content, the quantification should be performed by a suitable HPLC method. The absence of absorbance indicates that the organic substance is aliphatic in nature. This is particularly important when dealing with petroleum and petrochemical derived products. Absorbance in the visible region (400–700 nm) indicates the presence of a chromophore.

Infrared (IR) is used to indicate the presence of saturation, aromatic groups (benzene, naphthalene rings, etc.), hetero-atoms (oxygen, nitrogen, phosphorus, chlorine, bromine, iodine, silica, etc.) and specific functional groups (including hydroxy, nitro, amino, etc. groups). It is rarely used for quantification.

Nuclear magnetic resonance (NMR) should be used for relatively pure substances or well defined MCS. It is probably unsuitable for most UVCB substances because the spectra produced would be far too complex to interpret. ^{1}H-NMR is used for the identification of structures. ^{13}C-NMR is used to give the carbon backbone of a substance. Other forms of NMR are specific for certain elements and these are particularly useful for complex structures. For example, structures containing heteroatoms (nitrogen, oxygen, sulfur, phosphorus) or halogens (fluorine, chlorine, bromine and iodine) or organometallic substances or complexes containing metals (e.g. lithium, sodium, calcium, cadmium, platinum, etc.) or non-metals (e.g. silicon) could be elucidated by the appropriate element NMR (^{14}N-NMR, ^{17}O-NMR, ^{19}F-NMR, ^{35}Cl-NMR, ^{6}Li-NMR, ^{23}Na-NMR, ^{43}Ca-NMR, ^{113}Cd-NMR, ^{195}Pt-NMR, ^{29}Si-NMR etc.). NMR can be used for quantification of organic substances when the usual chromatographic methods are unsuitable. It should be remembered that it would not see components that do not contain that element, for example ^{1}H-NMR and ^{13}C-NMR will not detect inorganic components.

Mass spectroscopy (MS) is normally performed after chromatography as in gas chromatography-mass spectroscopy (GC-MS) or liquid chromatography-mass spectroscopy (LC-MS). Direct introduction into the mass spectrometer can be carried out but this is normally reserved for pure substances.

Differential scanning calorimetry (DSC) or **thermogravimetric analysis (TGA)** can sometimes be useful if the substance decomposes so could not be chromatographed.

B.3.2   Purity and Characterisation

The purity of an organic substance is normally determined by chromatography. This is a separation technique in which a mobile phase (a gas or liquid) containing the substance to be analysed is passed through a stationary phase (usually a solid) in a column. The constituents/impurities have different affinities for each phase; hence some components are retained longer on the stationary phase, resulting in separation of the constituents. A suitable detector is placed at the end of the column to record the components as they emerge.

The most commonly used chromatography methods for the quantification of substances are HPLC for non-volatile compounds and GC or GLC for volatile compounds or gases. Other useful chromatography methods that could be considered for complex substances include chiral chromatography for chiral compounds, ion chromatography (IC) for non-volatile anions/cations and size-exclusion chromatography (SEC) or gel permeation chromatography (GPC) for peptides, proteins or polymers.

The choice of detector can influence the interpretation of the results. This can be a complex and grey area. The normal assumption is that the detector will respond in the same way to each component but this is not strictly correct. It is important to bear in mind that detectors have different sensitivities to different compounds and some are selective so would not see certain substances. Examples of detector blindness include:

- GC with a nitrogen–phosphorus detector (NPD) only sees nitrogen- and phosphorus-containing compounds.
- GC with flame-ionising detector (FID) will only see organic carbon compounds. It will not see carbon tetrachloride, carbon dioxide, inorganics and so on.
- GC-MS or LC-MS using a selective ion detector will only see compounds with that ion.

Most chromatography for the characterisation and sameness is performed by injecting neat substance upon a column. This can overload the detector so may distort the percentage purity (often reducing the apparent purity) and similar components may co-elute, especially geometric isomers. Diluting in a suitable solvent may be a better option before injecting on the GC.

Gas chromatography is a powerful and rapid analytical tool for the identification and quantification of most compounds volatile below about 450 °C. The constituents are separated by their boiling points and interactions with the stationary phase. Most analysis is now performed on 10–30 m capillary columns rather than 1–3 m packed columns, but the latter are still useful for quantification due to their ability to cope with high sample loading. A flame-ionising detector is the most common detector for organic compounds and its response is proportional to the number of carbon atoms in the constituent. It has a bias towards molecules with larger mass. A thermal conductivity detector (TCD) is a non-destructive detector that counts the number of particles passing the detector, so will detect all substances, including inorganic materials. However, it has a bias towards smaller substances that have less mass. The bias in the detectors can be accounted for by calibration or the use of suitable response factors. GC-MS is often essential in confirming the identity of constituents; the order of elution from the column and retention times (the time the component is retained on the column) are important for quantification by GC.

High performance liquid chromatography (HPLC) is useful for non-volatile and non-polar hydrophobic substances using a normal phase column or non-volatile hydrophilic substances using a polar normal phase column. Traditionally, the fixed UV-vis spectrophotometer at 254 nm has been used. Today the diode array UV-vis spectrophotometer allows different wavelengths to be used. These are only useful if the components absorb in the UV-Visible region. Care should be made to determine the response factors. Substances without UV-vis absorbance can be determined by a refractive index detector. LC-MS is rarely used for quantification due to its expense but is often essential in confirming the identifying constituents. The elution order and retention times are important for the quantification by HPLC (Figure B.1).

B.4   Complex Organic Substances

These include organometallics, organic salts and other substances that are not suitable for chromatography. Most substances need to be determined by a series of techniques and a mass balance would be carried out.

NMR quantification is useful on substances that are unsuitable for chromatography because they are reactive, decompose or are non-volatile. Examples include the use ^{1}H-, ^{13}C-, ^{29}Si-NMR and ^{31}P-NMR to give the relative proportions of constituents containing that element. It is not sensitive to components below around 1% and would ignore components that do not contain that element. Therefore, elemental analysis would be recommended to make sure that such components are taken into account (for example inorganic content such as the counter ion and impurities when using ^{1}H-NMR and ^{13}C-NMR).

Elemental analysis requires careful consideration and calculation of a mass balance to determine composition. Some techniques to consider are:

- Carbon, hydrogen, nitrogen, sulfur, and oxygen (CHNSO) could be determined out using a CHN analyser.
- Inductively couple plasma-optical emission spectroscopy (ICP-OES) would be used after the quantitative digestion of the substance in concentrated acids. The related inductively coupled plasma-mass spectroscopy (ICP-MS) is more expensive and is unlikely to be used.
- X-ray fluorescence (XRF) could be used on solid substances.
- Sodium, potassium, and calcium ions could be determined by using a flame photometer.
- Titration could determine the functional group quantity.
- Gravimetric analysis could be used to determine elemental composition.
- Water content could be determined by Karl Fischer titration or other suitable method.
- Volatile organic carbon could be determined by headspace or purge-and-trap GC (e.g. methanol, ethanol etc.).

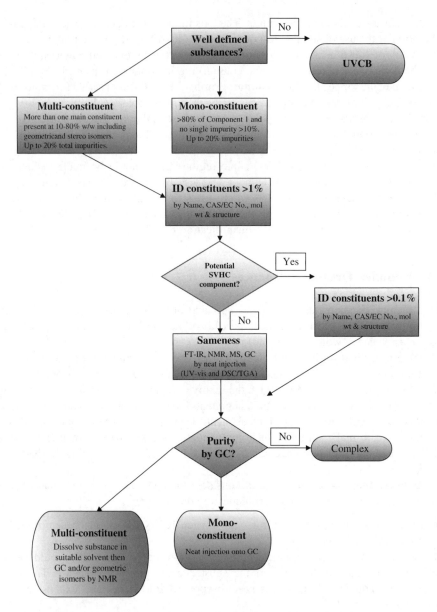

Figure B.1 *Decision Tree for a Straightforward Organic Substance*

B.5   Inorganic Substances

Inorganic substances present more difficulties in terms of compliance and the expertise of producers may be important. This section sets out some example of techniques that could be useful; the actual techniques that are useful for a particular substance will need to be considered on a case-by-case basis. Elemental analysis of some form is essential and can be supported by some of the techniques already listed; ion chromatography may be useful, too. X-ray diffraction (XRD) results may be useful to detect individual constituents in a mixed product.

B.5.1 Structure

The following techniques may be useful for determining structure:

- X-ray diffraction determines the crystal structure of a substance. Amorphous substances show up as an 'amorphous' hump.
- Inorganic NMR, normally in the solid state.
- Thermal methods measure the change in physical/or chemical properties of a substance as a function of temperature. These include thermogravimetric analysis, where the change in weight is monitored, and differential scanning calorimetry, which detects physical transformations such as melting and boiling. This is particularly useful for determining the presence of bound and free water.

B.5.2 Elemental Quantification

The main methods of quantification of inorganic substances are:

- X-ray fluorescence is very useful on solid substances.
- Acid digestion followed normally by ICP-OES. Atomic absorption spectroscopy (AAS) could also be used but this would only be look at each element one at a time. The use of ICP-MS should be reserved for looking at impurities due to its expense.

Flame photometry and ion chromatography is useful for determining ions in solutions.

- Sodium, potassium and calcium ions could be determined by using a flame photometer.
- Anion and cation IC can be used to determine the anion (bromide, chloride, fluoride, nitrate, nitrite, phosphate, sulfate, sulfite etc.) and cation (calcium, magnesium, potassium, sodium etc.) composition of the substance respectively.

Less commonly used traditional methods that could be considered include:

- Acid–base, complexometric, redox, and precipitation titration methods are still useful in elucidating what is happening in solution.
- Gravimetric analysis is the process of converting an element into a definitive compound and weighing the isolated compound.
- Electroanalytical techniques including potentiometry, ion-selective electrodes and voltammetry may also be useful.

B.6 Analysis of UVCBs

The complex nature of UVCBs means that a substance can contain a very large number of components; very little may be known about some of these components. One approach to assessing the physico-chemical and (eco)toxicological properties, and the environmental and human health exposures, is known as the blocking method. Constituents are grouped into blocks on the basis of their composition, physico-chemical, degradation and toxicological properties. A key value for each property (for example water solubility, log K_{OW} and LE_{50}/LC_{50} or no observed effect concentration (NOEC) from (eco)toxicology studies) is selected for each block and these block properties are used as a basis for the chemical safety assessment. Therefore, a relatively small number of blocks are considered instead of the large number of constituents. This has the effect of making the task of assessing a large number of constituents more manageable and reducing the impact of missing data on the assessment.

Blocks are normally created from components with similar retention times from chromatography (GC or HPLC). Less common is the use of mass spectrometry techniques such as field ionisation mass spectrometry (FIMS) or high resolution mass spectrometry (HRMS), which have been used to determine the composition of petrochemicals.

Appendix C

Tools for REACH Compliance: IUCLID, Chesar and In-House Databases

C.1 International Uniform Chemical Information Database (IUCLID)

The International Uniform Chemical Information Database (IUCLID) is a freely-available database program, recommended to registrants as a useful method to compile technical registration dossiers. Its outputs are compatible with the ECHA's systems and although other formats for registrations are also possible the IUCLID is the most widely used.

The IUCLID installation kit can be obtained from the IUCLID web site (www.iuclid.eu). The web site has a 'download' section, where the user is prompted to create a user account before being granted access to the download itself. Various resources in the form of guides and videos are available online to assist with the installation.

There are two different types of IUCLID depending on the requirements of the user: served IUCLID and stand-alone IUCLID. Served IUCLID uses a centralised database that can be shared by a group of users who have access to the same information via an established network. Stand-alone IUCLID uses a localised database that is not shared and will hold data accessible by the individual user only. Both are available for download from the IUCLID web site. The installation of a served IUCLID requires a higher level of technical expertise, whereas getting started with the stand-alone IUCLID is a more straightforward process.

Served IUCLID is useful when a larger group of individuals within a company are working on a substance or a set of substances simultaneously. This allows for editing of the same data set without overwriting data and has the advantage of not having to collate data at the end of the compilation of each of the relevant property sections (e.g. toxicology, ecotoxicology, environmental fate). The stand-alone IUCLID is useful, for example, if a single user is in charge of an entire data set or for the purpose of

Chemical Risk Assessment: A Manual for REACH, First Edition. Peter Fisk Associates Ltd.
© 2014 John Wiley & Sons, Ltd. Published 2014 by John Wiley & Sons, Ltd.

importing received endpoint study records (EPSRs) (e.g. from a contract laboratory) for checking before incorporating returned work onto the server. An example of when stand-alone IUCLID may be the appropriate choice could be if a freelance subcontractor is completing a specific property area. The import and export of data between the served and stand-alone IUCLID is straightforward, the features within the two are identical apart from the aspect of data sharing. Completion of the IUCLID data set for REACH requires in-depth knowledge of a range of different property areas. Other than consultancies which specialise in this area, few companies will possess the skill set to cover all the property areas with sufficient expertise.

Once set up, IUCLID is an essential tool for the capture, storage, submission and exchange of data on a registration substance, presented in the format required by the OECD. A substance data set in IUCLID is parallel to the European Chemicals Agency (ECHA) requirements for the completion of a registration dossier.

A IUCLID data set features the following sections:

1. General information (including information on substance identification and composition).
2. Classification and labelling and persistent, bioaccumulative and toxic (PBT) assessment.
3. Manufacture, use and exposure.
4. Physical and chemical properties.
5. Environmental fate and pathways.
6. Ecotoxicological information.
7. Toxicological information.
8. Analytical methods.
9. Residues in food and feeding stuffs.
10. Effectiveness against target organisms.
11. Guidance on safe use.
12. Literature search.
13. Assessment reports.
14. Information requirements.

Sections 1 to 7 are important for REACH and a completed chemical safety report (CSR) must be attached to section 13. IUCLID is also used for other registrations (e.g. biocides) and not all sections visible in the data set are applicable to REACH work. Furthermore, dependent on the substance use pattern, annual tonnage level and pre-existing information on its safety profile, the minimum data requirements vary dependent on which REACH Annex the registration substance falls under (Chapter 6 of this book).

Most of the property sections will require the selection of a key study, per endpoint, which needs to be entered into IUCLID as a full length EPSR by selecting 'all fields' from the top of the EPSR template. All available studies are assigned a reliability score according to a Klimisch scale (1–4), where reliability 1 and 2 studies are deemed acceptable for fulfilling key criteria (Klimisch *et al.*, 1997). Broadly speaking, a reliable study should be compliant or equivalent in methodology to current OECD test guidelines, which are available on the OECD web site (www.oecd-ilibrary.org). Studies of lower reliability or additional reliable studies should be included in the data set as supporting studies for completeness, which necessitates filling in an abbreviated version of an EPSR ('basic fields' options in the drop down menu at the top of the EPSR). In general, the study selected as key should be the most recent and reliable available study; it should represent the overall conclusion drawn from the data available for the endpoint as well as support the justification for classification or non-classification. Expert skill is required for interpreting sources and evaluating reliability. It is important to assess the data available per endpoint accurately and to enter the information from sources into IUCLID in the appropriate manner, keeping in mind the centrality of IUCLID for the ECHA as the primary screening point of submissions. IUCLID is the first point of the review process for the ECHA which employs automated screening measures to assess dossiers, including comparison of submissions between registrants. Compiling a IUCLID data set of high standard should, therefore, be a priority to all registrants as a starting point for a successful registration.

In cases where data on the registered substance is not available, an endpoint may also be filled with a reliable study from a structurally analogous substance by read-across, or in some cases it may be appropriate to take a weight of evidence-based approach, where supporting information is included to build a case for the overall conclusion of a specific endpoint. More detailed information on read-across and weight of evidence can be found in the ECHA practical guides 2 and 6 (http://echa.europa.eu/practical-guides). In cases where data may be omitted according to Annex requirements, the IUCLID data set must still be populated for completeness. In this instance, a data waiver, or in some cases a test proposal, may be added into the data set instead of a usual EPSR.

Before beginning to work on a substance data set in IUCLID, the user must first register with REACH IT (reach-it.echa.europa.eu) in order to be able to create a 'legal identity' within IUCLID; this contains company-specific information. The IUCLID database has an inbuilt library of limited reference substances that needs to be downloaded and imported during the installation process. If the registration substance is not part of the pre-existing list, a reference substance will need to be created. When a reference substance has been made available, the substance data set itself can be linked to the reference substance. It is worth noting that EC numbers cannot be typed into the Substance identification of a IUCLID substance data set but can only be added by linking through a properly formed reference substance.

The IUCLID interface, which provides the platform for working on the registration substance data set, is provided under 'Substance'. However, wherever transfer of completed dossiers from one legal identity to another is necessary (e.g. between different registrants), the data should be exported as 'inherent templates', which can be copied over to populate the substance specific data set in IUCLID after import to its new legal entity. An option for printing an .html file is also available, in cases where the user wishes to have a record of a data set or an EPSR that is easily accessible and readable in a format outside the IUCLID programme. This function can be useful, for example, when providing an overview of an EPSR to a party without installed IUCLID programme.

Other important features of IUCLID include the Technical Completeness Check (TCC) tool and the CSR tool. The TCC tool is important for checking that the data set meets the requirements for submission at the tonnage level at which the substance is produced per annum. Failure to submit a dossier that passes the completeness checks will result in automatic rejection of the dossier by the ECHA. Finally, the CSR tool is available for the generation of the CSR, which is part of the final dossier that will be submitted to the ECHA for registration.

The different property sections in IUCLID comprise EPSRs and endpoint summaries that summarise the EPSRs; for example, the acute toxicity endpoint consists of acute oral toxicity, acute inhalation toxicity, and acute dermal toxicity endpoints. The purpose of an endpoint summary is to summarise the existing data, highlight the findings of the key studies, and put forward a self-classification proposal for each endpoint. The CSR tool constructs the final CSR based on the endpoint summaries and draws tables out of the result sections of the actual endpoint study records in IUCLID. It is, therefore, important to populate the relevant fields coherently and accurately, to help the data transfer from IUCLID into its final report form.

It is noteworthy that beyond the storage and organization of data within the IUCLID data set, it is important for the registrant to hold records of substance-specific available studies within a manageable filing system. This becomes significant where larger sets of data are maintained and managed, and can have implications to later data and cost sharing. This is covered in more detail, with advice, in the next section.

Finally, a help section is available in IUCLID to guide the user with information requirements when entering data. Problems that may sometimes be encountered with practical use of IUCLID include crashing, which can lead to temporarily locking the user out of being able to edit the last EPSR that has been in use. Also, the use of a virtual private network (VPN) connection may at times lead to issues with connection to served IUCLID.

C.2   IUCLID and PPORDs

The following manuals should be consulted before making a submission for PPORD:

- *Data Submission Manual 1: How to prepare and submit a PPORD notification*
- *Data Submission Manual 4: How to Pass Business Rule Verification ('Enforce Rules')*
- *Data Submission Manual 5: How to complete a Technical Dossier for Registrations and PPORD Notifications.*

C.3   Submission of PPORD to ECHA

Companies that wish to benefit from this exemption must submit a PPORD notification to the ECHA through the latest version of IUCLID 5 or submit it via REACH-IT (ECHA, nd). This must include information on substance identity, its classification, information related to the PPORD programme and the quantity of the substance expected to be manufactured or imported during the five-year period of exemption.

C.3.1   IUCLID Submission

The notifier (manufacturer or importer) must submit an electronic notification providing the ECHA with the following information:

- The identity of the substance (section 2 of Annex VI).
- The classification of the substance as (section 4 of Annex VI).
- The estimated quantity (section 3.1 of Annex V).
- The list of customers with which the PPORD cooperation is carried out, including their names and addresses.

IUCLID sections	Article 9(2)	Annex VI
1.1 Identification	Identity of the manufacturer or importer or producer of articles	Section 1: General registrant information
1.1 General information 1.2 Composition 1.4 Analytical information	Identity of the substance	Section 2: Identification of the substance. The variations in composition of the substance that may be foreseen under the scientific experimentation are taken into consideration in the information to be reported
1.8 Recipients	Listed customers	
1.9 PPORD	Estimated quantity	Section 3.1
2 Classification and labelling	Classification and labelling	Section 4: Classification and labelling

The IUCLID 5 TCC plug-in can be used to detect any missing information and pre-check certain 'Business Rules'.

Apply the IUCLID 5 fee calculation plug-in to estimate the fee associated to your notification.

C.3.2   Notification by REACH-IT

Step by step instructions can be found in the Industry User Manual – Part 6: Dossier Submission.

The ECHA will provide the notification number after verifying the completeness of the information and the payment of the relevant PPORD fee.

The ECHA then performs the assessment on the received information. It may impose conditions to the PPORD exemption that the manufacturer or importer of the substance has to comply with and must inform relevant customers involved in the PPORD. If a substance is subject to restriction or authorisation, the respective decisions will specify how they apply to PPORD. They will also define the maximum quantities of the substance that can benefit from a PPORD exemption.

C.4   Chesar

This section gives a brief overview of Chesar and the relationship with IUCLID; further details are available on the Chesar web site (http://chesar.echa.europa.eu/).

C.4.1   Introduction

Chesar (Chemical Safety Assessment and Reporting Tool) is an application developed by the ECHA. It aims to help companies to successfully perform their CSAs and prepare their CSRs and exposure scenarios for supply chain communication.

Chesar provides an efficient and structured way to perform a standard safety assessment for all identified different uses of a substance. As at the time of going to press, Chesar works well for a mono-constituent (stand assessment) substance, however for multi-constituent and unknown or variable composition, or of biological origin (UVCB) substances, the programme may be limited.

For Chesar to run efficiently and correctly, a user must have sufficient information on the properties, the uses and tonnages and the conditions associated with the safe use of the substance. The programme then calculates exposure estimates compared to the predicted no-effect levels based on these inputs. The comparison is to determine whether safe use of the substance can be established. For substances where hazards have been identified but predicted no-effect levels are available, Chesar supports building the exposure scenario with the use qualitative risk assessment and characterisation.

Workers' and Consumer exposure estimations provided by Chesar are calculated using 'ECETOC TRA worker' (TRA, Targeted Risk Assessment) and 'ECETOC TRA consumer' tools respectively while environmental exposure estimates are based on EUSES 2.1 fate model. All these programmes are embedded into Chesar.

The programme is a web-based application which can either be run as a stand-alone or shared/distributed version (hosted on a server connected to the organisation's computer network), but the database is stored on your computer that is no information about your substance is available on the Internet.

The current version of Chesar works only with IUCLID 5.4 and above.

C.4.2   Chesar Functionalities or Organisation of the Tool

Chesar is divided into seven major groups of functionalities which are called 'boxes'. The boxes are connected and contribute to the generation of the CSR and/or exposure scenario for each substance. The functions of each box are summarised below in Boxes C.1–C.7.

Box C.1   Substance management

In this box, the user or assessor can import pre-generated information (properties and hazard conclusion including PBT) in IUCLID into Chesar. In this box, the scope of the assessment is also listed depending on information on the hazard conclusion completed in IUCLID; the type of risk characterisation required is described by Chesar.

Box C.2   Use management

In this box, the user or assessor describes the uses of the substance in a way to ensure consistency between the description of the uses, exposure scenario building, and exposure assessment. This is achieved by using Chesar's built-in life cycle tree structure, reporting relevant uses of the substance, considering human health, and environmental aspects including tonnage break-down to the different uses.

For each use identified by the assessor, an environmental contributing scenario is automatically generated by Chesar. In addition, the assessor can create other contributing scenarios for both the environment and human health as necessary. The information entered in Box C.2 is used by the program in Box C.3 to manage quantitative exposure assessment for the substance.

Box C.3   Assessment management

In this box, the appropriate exposure assessment method is selected and one or more quantitative exposure is performed for individual contributing scenario. The assessor may use the default conditions in Chesar or refine the assessment by providing information relevant to the safe use of the substance.

Box C.4   ES building and CSR management or generation

In Box C.4, exposure assessments per contributing scenario performed in Box C.3 can be consolidated and final exposure scenarios can be built and reported.

Box C.5   SDS ES management

In this box, an exposure scenario for communication is created for each exposure scenario for the CSR created in Box C.4, which is, in turn, based on information on uses and contributing scenario created in Box C.2.

Box C.6   Library management

In this box, a user can define, create, import and store the following elements:

- SPERC: Developed by sector organisations, it allows the user or assessor to define typical conditions of use from an environmental perspective and related release estimate.
- Determinant types: Are conditions of use reported in the exposure scenario section of the CSR. Determinant type can take several different values depending on the assessment. For each value, one or several standard phrases can be associated so that the condition of use is linked to the extended Safety Data Sheet (eSDS).
- Standard phrases: Are used to communicate the conditions of safe use.

Box C.7   User management

In this box, a user is able to create a legal entity, which is useful to identify and trace back each object created to its author. The management of users allows users access to the same Chesar database. It is also possible for a legal entity to be imported into Chesar; this can be done through the legal entity pain in the program.

C.4.3   Assessment workflow of Chesar

The properties information of a substance is imported into Chesar from IUCLID. There is interaction between Chesar and IUCLID, so that Chesar takes only the information it needs to run the assessment completely from IUCLID.

The properties information from IUCLID used by Chesar is taken from the physico-chemical properties and hazard assessments such as environmental fate, including eco-toxicological information, PBT, and toxicological (human health) data. The information imported from IUCLID to Chesar is read-only information that a user cannot edit.

The assessment workflow of Chesar is summarised in Figure C.1.

C.5   Advice on Storing of Data Outside of the IUCLID

Although it is necessary to complete the IUCLID data set properly by full data entry of existing studies, it is often the case that access to the same information will be required outside of the program. The storage and organization of source documents is an important aspect of the registration process and should be considered carefully. This section aims to introduce some principles regarding data management outside of IUCLID.

A data set for a single substance can be relatively simple but where a company is responsible for registering multiple substances the data set can be of considerable size and can become unmanageable without appropriate measures. Good data management is important in order to avoid potential future issues, which could lead to duplication of work and time wastage. Since registrants are answerable for their submissions for years to follow, for example through emergence of new registrants and associated demand for data sharing, it is imperative to ensure that IUCLID entries are kept up to date and consistent with their original source documents. Additional reasons for good data management are:

- Scientific practices such as prediction (QSARs etc.) rely heavily on quantity of reliable data. Providing this in a readily accessible and usable format saves time and enables an automated workflow.
- Sharing of data between co-registrants requires some form of centralised collation or other practices.
- Sharing of costs between co-registrants (for example within a consortium) can require fairly complex calculations based on nominal values of studies, and can include compensation for intersecting areas. Properly formed records are an important basis to this.
- The registrant is responsible for the information provided and may receive queries on the registration years after submission. Maintenance of a 'data trail' is important regarding this.

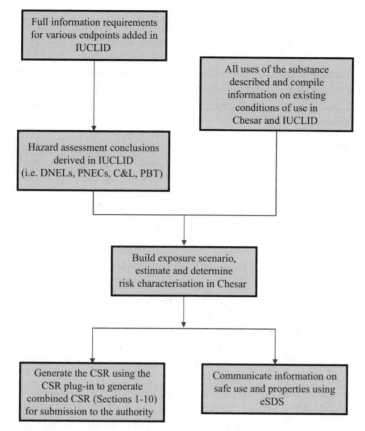

Figure C.1 *Overview of Assessment Workflow of Chesar. (Source: http://chesar.echa .europa.eu/web/chesar/support/manuals-tutorials, last accessed 5 August 2013.)*

Larger companies tend to have pre-existing solutions that have been designed with most such issues in mind, but the majority of registrants is likely to find that data management will be dealt with by non-technical personnel under time constraints. The main principles of good data management are discussed in the following sections with emphasis on common problem areas. Where it is known that a larger data set that requires a higher level understanding of data management systems needs to be handled, recruitment of experienced database designers should be considered.

C.5.1 Structure

It is important to establish a good database structure. For example, normalisation, which is a data management process that allows for a more organised handling of existing data, avoids duplication and inefficient data filing, but it can also make working on the data difficult. It is, therefore, important to decide whether the design and implementation of a particular system is cost effective and to identify database requirements well in advance to avoid later issues.

C.5.2   Identifiers

Every entry of the data set should have a unique identifier, or ID. For the ease of recognition and communication it is recommended that these use a defined alphanumeric format, or delimited format similar to CAS or EC numbers. The format should be clearly distinct from others (particularly dates, registry numbers, study reference numbers etc.) and it should enable sorting with ease. For example, here are some arbitrary IDs with invented meanings:

- S00202 – study # 202. Simple, fairly easy to say.
- 1234-2013-1001 – report # 1001, in the year 2013, of project # 1234. The order of terms here was chosen based on the way it will sort.
- 31.23.14.89.7 – internal substance ID with sign bit. Whilst it is similar to both dates and IP addresses, it is clearly distinct and has the added benefit of splitting digits for ease of human readability, pronunciation and memorisation.

Additionally, it is worth considering how the registered substances will be referred to within the database. A CAS or EC number can be useful but this can lead to inconsistencies when describing substances that do not have a registered number. This may be further complicated by some substances having only one or the other, or neither. It is, therefore, advisable to assign internal IDs and include CAS, EC and any other internal identifiers alongside the ID. It may be beneficial to use a check digit system, like those of CAS, EC and ISBN, to provide a further layer of validation. In essence, a check digit system is a way of verifying that the assigned number is a valid ID within the identifier system, for example the CAS registry.

Furthermore, it is advisable to include ID codes when naming items in the IUCLID (e.g. EPSRs), to facilitate finding the correct entry in IUCLID or the correct row in your data set from the IUCLID. The data these identifiers refer to will need to be persistent into the future, so the IDs must also be persistent. It may be useful to use IUCLID chapter headings as prefixes for IDs within the non-IUCLID data set, to reflect the endpoint of the data.

C.5.3   Mechanistic Issues

It is important to set up the numbering system of your database correctly, in order to avoid later issues once there are more data entries to manage. A common issue becomes apparent when sorting IDs in different programs. Sorting can behave differently if the data and identifier types represented are different; for example, if text is sorted character by character whilst numbers are not.

CAS numbers are particularly susceptible to this. For example, it would be expected that CAS# 7732-18-5 comes after 80-08-8, but since eight is larger than seven it does not. Variability in length is also a cause of this issue, as text is compared character by character, left to right. One solution is to pad the number with zeros up to a maximum length. This approach should not pose problems; however, it will be necessary to omit the additional zeros when interfacing with certain databases and resources.

Additionally, if an identifier format contains information (for example, a date rather than an arbitrary consecutive number), the ordering of components within the format can greatly affect the way it is sorted.

C.5.4 Identification and Expression of Substance Identity

Appropriate identification of substances is vital for REACH submission, and it is therefore important to carefully identify and represent substance identities. As much information should be included as possible, including trade names, so that the information is clear even after a considerable period of time. Failure to present data fully is likely to result in further clarification and, thereby, further costs later. More detail can be found in the analytics chapter (Appendix B of this book).

C.5.5 Result Values

Consideration should be given to maintaining records of all results of all studies in an easy to access format. Data are best presented as separate entries of the value itself (as a number) and its associated units (as text). In this way calculations and conversions can be made directly from the database with ease. Organising data in this way facilitates its preparation for statistical analysis or use in QSAR, or other areas (including IUCLID entry). Other information, such as read-across, categories, and waivers, should also be included in the database in a way that reflects their entry in the IUCLID. Having clear records of the different types of data will also facilitate running queries on the data sets, for example in identifying data gaps.

C.5.6 SMILES and Textual Representations of Structure

When using the Simplified Molecular-Input Line-Entry System (SMILES) or another similar system, the database can be searched by entering the SMILES as a search term that is converted into its canonical form, which allows running the search of the whole database. There are open source and commercial software packages that include the function.

C.5.7 Modifications to the Data Set

It is important to keep track of changes that are introduced into the data set. Each modification needs to be recorded, including when, why and by who the change was introduced. Additionally, it is useful to have records of when data were entered into the IUCLID and when they were updated; this can be entered into additional columns in a data sheet. Fairly simple scripts can be employed to enter these, logging and time-stamping all changes automatically.

C.5.8 Composited Fields and Repeated Processes

Adequately documenting data outside of the IUCLID can also serve as a time saving prerequisite to data entry into the IUCLID. For example, by establishing review sheets that summarise study results and feed into an existing database, the data that are entered into the review sheets can also be used to generate text that can easily be copied and pasted into the IUCLID to fill the required data fields. There are many ways to implement this, the simplest involving Microsoft Excel (or similar) equations. Additionally, complex reports, tables or other ways of representing data can also be handled in a similar way; standard 'queries' or 'smart' (configuration-based) generation of table structure and content, templates or the use of workflow and statistical analysis programmes can be achieved.

C.5.9   Duplicates

Avoid duplication of data unless there is a legitimate reason, for example with shared data (where a study tested two substances, so two rows will exist for the results), and with backups and snapshots of data (where the purpose is persistent traceability and the data is no longer written to). Achieving this requires measures to avoid clashes on work that is being done in parallel. There are many such measures, some are simple and robust, others are not. Additionally, it is important to use defined procedures for processing something, so that fields are filled in correctly and only when necessary. Automated methods, using computers, can be extremely beneficial here – using validation and custom processes.

C.5.10   Validation

In constructing a database, errors are likely to be introduced during data entry due to human error. For this reason it is important to have appropriate validation methods in place for easy detection of inconsistencies. The list below details fields/areas that all require validation to some extent, but it may be useful to define rules for every field in your data set to scan for discrepancies.

The methods for validation vary between programs. A program such as Microsoft Excel can use a simple validation strategy based on types, set values, ranges or through custom functions (using VBA macros and/or forms). Database engines (such as MySQL or Microsoft Access) require more effort to validate input but are more flexible; for example, custom forms can use custom code to validate any piece of information in any way deemed necessary. This is more useful, as the design can facilitate the correct entry of data in the first place, by informing the user about how to properly fill in a field. The following points are examples of items where validation may be required:

- *CAS/EC numbers* – must be valid (check the sign digit) and exist in the data set (if used as an identifier).
- *Internal ID codes* – must be consistent. One potential check is for typographical errors (e.g. O characters instead of 0s). It should also be made sure that links between different locations where an ID code is featured are valid.
- *Source references* – ensure consistent style. Each separate and composite field may need to be validated. Most fields here, for example study report titles, are not suitable for validation, but the year or date and laboratory study numbers are.
- *Chemical names* – maintain accuracy for identity and use standardisation/canonical forms for ease of searching and so on.
- *SMILES or other text-based structural representation* – maintain accuracy for identity, use standardisation/canonical forms for searching, and validate elements, bond valances, aromaticity, charge, matching parenthesis, no illegal characters and so on.
- *Klimisch codes* – must comply with a predefined set of values.
- *Results* – must be expressed as they were stated, with units and any other key information (e.g. temperature). Validation should be used to make sure the types and formats of this information match what is expected in the system.

- *Endpoint* – must comply with a predefined set of values.
- Purpose flag – must be in a predefined set of values, for example key study, supporting study.

C.5.11 Checking

Data entry mistakes may not emerge until long after the occurrence, which can make them difficult to resolve. As pointed out in the section discussing validation, databases and other electronic data storage mechanisms are invaluable in this respect. In addition to validation of inputs and interactions with data, other checks should be made to ensure the consistency, validity, accuracy, and completeness of the data. It may be useful to run queries to check the following areas for consistency and completeness:

- *Gaps in key fields*; such as ID codes, key studies, Klimisch codes, results and references.
- *Inconsistencies in the form and content of data*; such as data entered incorrectly (for example O instead of 0 in an ID, or wrong punctuation, or inconsistencies in list delimiters etc.). These are all validation issues, and would normally get resolved there, but weak input validation (for example in Microsoft Excel) may not always be sufficient.
- *Inconsistencies between versions of data*; such as a mismatch between results at two separate locations. This type of issue should only occur when data are intentionally duplicated, so update procedures should be defined to ensure that copies of data are kept up-to-date. An example of where this can occur is the use of Excel lookups to retrieve relevant information.
- *Broken links between relations (orphaned rows)*; cases where the original source of linked data has been deleted.
- *Duplicates*; for example two documents existing under the same ID (but being unrelated), or a new row being added with an existing ID due to human error.
- *Quality checking*; periodic checks that assess integrity of the existing data are required to ensure that changes which have been introduced over time have not introduced mistakes. These checks can refer to a random manual check or a computer-based automated check.

The types of data checks should be run periodically, to detect any problems early. This may be carried out as an automated part of an import process, as a part of formal quality assurance procedures.

C.5.12 Study Reports and Cataloguing

A central part of the registration process and IUCLID entry is the evaluation and presentation of data that are received in the form of study reports. It is important to have a coherent filing system for the study reports, with assigned document IDs, that can be matched with reviewers' IDs, which summarise the study and link to the database. Both types of IDs need to be presented in a source catalogue that enlists the full library of available data. Maintaining a large data set poses a greater risk of duplication and documents going missing, and it is important to employ a cataloguing system that can detect and simplify the handling of difficulties like these. It is also useful to establish a way of naming EPSRs in the IUCLID that includes a reference to the reviewer's code assigned to the corresponding study report. EPSR names should also include a short reference (e.g. author, year) to facilitate the identification of the correct EPSR for a source.

C.5.13   Snapshots and Backups

Snapshots are created as a part of the workflow to store a copy of a particular data set at a specific point in time. For example, a snapshot may be useful for capturing the state of the data at a point when a report is generated. It can become useful when feedback is received on the report at a later point in time when the live data set may have changed. The difference between a snapshot and a general backup is that snapshots are more specific and usually associated with a defined time point in the course of a project or a task. Backups, in turn, are also important for holding copies of information and they should be made frequently to make sure important data is not lost.

Reference

ECHA (nd) PPORD. http://echa.europa.eu/support/dossier-submission-tools/reach-it /ppord (last accessed 5 August 2013).

Klimisch, H.J., Andreae, M. and Tillmann, U. (1997) A systematic approach for evaluating the quality of experimental toxicological and ecotoxicological data. *Regulatory Toxicology and Pharmacology*, **25**, 1–5.

Appendix D
Glossary

The glossary is subdivided into

- Regulations and background.
- Technical.

Regulations and Background

AC (Article Category)	A REACH descriptor code for substances used or present in articles.
AF (Assessment Factor)	Assessment factors are applied to no-effect levels derived from testing on animals to adjust for differences between the test (species, route of exposure) to produce a no-effect level that can be used in the risk characterisation of a substance.
Authorisation	The process in REACH for allowing use of certain substances of very high concern, under very clear conditions.
Base set	The minimum data requirements for notification of a new substance at > 1 tonne/year, and for carrying out a risk assessment under the New Substance Regulations which preceded REACH.
BREF	Reference document on Best Available Technique for industrial processes.
CA (Competent Authority)	Authority responsible for administrating the regulations.
CSA (Chemical safety assessment)	The CSA is the term used under REACH for the process of obtaining information on hazards and use and identifying measures required to reduce risks to acceptable levels.

Chemical Risk Assessment: A Manual for REACH, First Edition. Peter Fisk Associates Ltd.
© 2014 John Wiley & Sons, Ltd. Published 2014 by John Wiley & Sons, Ltd.

CSR (Chemical safety report) A CSR reports the findings of the CSA; it is required for substances produced at 10 tonnes/year and above.

Downstream use Downstream uses are ways in which a substance is used after it is manufactured or imported. They include formulation, professional use, and industrial use such as incorporation into articles.

Directive Legislative act – in the EU it is legalisation that comes into force by transposition that is it has to become enacted in each member state though national law/s, cf. Regulation.

Effect Used to refer to changes in growth rate, development, morphology, function or lifespan of a test organism under experimental conditions.

ECHA (European Chemicals Agency) The Helsinki-based regulator for Chemicals in the EU and EEA.

EINECS (European Inventory of Existing Chemical Substances) This lists all chemical substances that were supplied to the market prior to 18 September 1981.

ELINCS European List of Notified Chemical Substances.

ERC (Environmental release categories) A REACH use descriptor code that indicates how exposure and releases to the environment can occur during an activity.

eSDS (extended safety data sheet) A safety data sheet that includes exposure scenarios for uses of a substance.

ESR (Existing Substances Regulation, EC 793/93) Under this regulation, data have been collected and published in the ESIS database, and some substances were selected to be subject to risk assessment on a priority basis.

Existing substance A substance which appears on EINECS.

Exposure scenario A set of operating conditions and risk management measures that determine how exposure takes place and the level of exposure.

Generic exposure scenario An exposure scenario developed with assumptions and inferences on how exposure takes place and that aids the exposure assessor in evaluating, estimating or quantifying exposures for a product type or group of substances.

Hazardous waste A type of waste for which special requirements for disposal are set out in law.

HEvE (Human exposure via environment) The indirect exposure of man, via intake of contaminated food, to substances released to the environment.

HH (Human health) The human health properties of a substance are those that indicate potential to cause undesirable effects on the health of workers and consumers exposed to the substance. Examples are acute toxicity, skin irritation or development of diseases such as cancer.

HPVC or HPV (High Production Volume Chemical)	According to the European Chemicals Bureau web site a chemical is defined as a high production volume chemical when it is produced or imported in excess of 1000 tonnes/year by at least one industry.
ICCA	International Council of Chemical Associations.
IPPC (Integrated pollution prevention and control)	A European directive which aims to prevent and control pollution by industrial and agricultural activities with a high potential for pollution.
IUCLID (International Uniform Chemical Information Database)	Used for assembly of property and use data for REACH.
LCS (Life cycle stage)	A major activity in the course of a product's life-span from its manufacture/import, use and service life to its final disposal.
M/I (Manufacturer/Importer)	A person or an organisation that places a chemical substance(s) on the EU market and which has responsibility, under the REACH regulation, to register such a substance(s).
New substance	A substance placed on the market after 18 September 1981 that does not appear on EINECS. After a notification is accepted the substance is published in ELINCS.
NONS	Notification Of New Substances Regulations 1993, which apply to substances *not* listed on EINECS. A system of regulating new substances that applied to the whole EU, assessing the hazards and risks posed. Superseded by REACH.
OECD (Organisation for Economic Co-operation and Development) and OECD Guideline	The OECD Guidelines for the Testing of Chemicals are the approved methods for carrying out tests on substances. OECD has also published guidance on testing difficult substances.
PC (Chemical product category)	A REACH use descriptor code that indicates the type of product or application a substance is used in.
Process category	A REACH use descriptor code that defines an activity and the potential level exposure of workers from the activity.
RAC	Risk Assessment Committee (of the ECHA).
Regulation	Legislative act – in the EU it is legalisation that comes into force in all member states without transposition, that is it becomes law in all member states at the same time and with the same wording, cf. Directive.
Restriction	The process within REACH of imposing limits on certain uses of certain substances. Restrictions are ultimately placed on Annex XVII of the REACH Regulation.

RoHS (Restriction of Hazardous Substances)	An EU Directive, enacted in Member State legislation, to restrict the use of certain hazardous substances in electrical and electronic equipment.
RMM	Risk Management Measure
RMO (Risk Management Option)	The process of appraising different options for control of risks from substances.
SEA (Socio-economic analysis)	The process of comparing the risks of continued use of a substance with the benefit to society of continued use. A process applied in the restriction and authorisation processes of REACH.
SEAC	Socio-economic analysis committee of the ECHA.
SDS (Safety data sheet)	Safety data sheet to be provided to users of hazardous substances.
SU (Sectors of use)	A REACH descriptor code that gives an indication of the lifecycle stage and/or the type of industry a substance/product is used.
SVHC (Substance of very high concern)	Substances meeting specific criteria for hazards to human health (carcinogen, mutagen, reproductive toxicity 'CMR'), and/or to the environment – persistence, bioaccumulative and toxic (PBT), and very persistent and very bioaccumulative (vPvB) or of equivalent concern (to CMR, PBT/vPvB) and listed on the 'Candidate List' by ECHA. These being possible candidates for selection for Annex XIV of REACH (the list of substances that require authorisation).
WEEE	Waste electrical and electronic equipment, and the 2002 Directive setting out requirements for its disposal.

Technical

Acceptable Daily Intake (ADI)	The acceptable daily intake is the amount of a substance which can be ingested every day of an individual's entire lifetime, in the practical certainty, on the basis of all known facts, that no harm will result. The ADI is expressed as milligrammes (mg) of chemical per kilogramme body weight of the consumer. The ADI is derived from the most appropriate No Observed Adverse Effect Level (NOAEL) by applying an assessment factor, normally 100.

Acceptable Operator Exposure Level (AOEL)	The acceptable operator exposure level is the maximum amount of active substance to which the operator may be exposed without any adverse health effects. The AOEL is expressed in milligrammes of the chemical per kilogramme body weight of the operator per day. The AOEL is usually derived in terms of a systemic dose and is based on the most appropriate NOAEL (qv) by applying an assessment factor, normally 100, and any necessary correction for the extent of oral absorption.
Acute toxicity – environment	A toxic effect resulting from a short-term exposure. The effect is often mortality but may also include immobilisation.
Acute toxicity – mammalian	Adverse health effects resulting from single or short-term exposure to a substance. In the context of REACH, acute toxicity refers to systemic effects, and exposure is by oral, dermal, or inhalation routes.
ADME	Absorption, Distribution, Metabolism, and Excretion: the parameters which affect the extent to which a substance is taken up, transferred to other tissues and organs, transformed into other substances by the biochemical mechanisms existing within cells, and finally eliminated from the body.
Adsorption	The uptake of a substance from the water phase onto the solid phase. Such adsorption can typically occur from water onto sediments, suspended sediments and soil. Usually expressed as K_{oc}, the adsorption coefficient normalised for the organic carbon content of the soil or sediment.
AF (Assessment Factor)	See below.
Analyte	The substance itself or a degraded or metabolised product that is subject of analysis.
Aneugen	A substance that causes changes to the number of chromosomes
Assessment factor	A number used to derive a predicted no-effect concentration (PNEC) from the result(s) of a single-species test, multi-species mesocosm study or field trial. A NOEC, EC_{50} or LC_{50} obtained from the test, study or trial is divided by the assessment factor, obtained from the Guidance, to give the PNEC. The size of the assessment factor depends on the uncertainty in the extrapolation.

BCF (Bioconcentration Factor)	The ratio between the concentration in an organism (usually fish) and the concentration in water.
Bioaccumulation	A term to describe transfer of a substance from the environment into an organism. Thus bioaccumulation can occur typically from sediment, soil, water, or via the food chain, or any combination of them. The transfer of a substance from water alone into an organism is called Bioconcentration. See also biomagnification.
Bioavailability	The extent to which a substance is available for uptake into an organism.
Bioconcentration	The uptake of a substance into an organism from water. One component of the total process called bioaccumulation. Bioconcentration factors are usually determined as the ratio between the concentration in the organism and the exposure concentration.
Biodegradation	The actions of biological processes to break down a substance; usually implied to mean bacterial action (in soil, water or WWTP) whereby the organisms use the substance as food. Biodegradation can be complete, resulting in complete breakdown to minerals, or partial, producing particular end products, or may not occur at all. There are many standard tests of biodegradability, the ready and inherent studies leading to substances being described as 'readily biodegradable', 'not inherently biodegradable', 'non-biodegradable' and so on.
Biomagnification	An increase in concentration of a substance up the food chain due to bioaccumulation. Described numerically by trophic magnification factor for a whole food chain, or biomagnification factor (BMF) for one step in the chain.
Carcinogens	The causal agents which induce tumours. They include external factors (chemicals, physical agents, viruses) and internal factors such as hormones. Under the EC Dangerous Substances Directive, Carcinogen Category 3 is the lowest hazard category, that is there is evidence of the agent causing tumours in rats or mice but the strength of evidence is not sufficient to place it in a higher category (category 2 or 1; category 1 being for a substance known to be carcinogenic in humans).
Chronic toxicity	A toxic effect resulting from a longer-term exposure. The effect is often a reduction in growth or reproduction.

Clastogen	A substance that causes changes to chromosome structure (structural aberrations).
CMC (Critical micelle concentration)	The concentration at which surfactant molecules begin to aggregate to form assemblies of many molecules. Above this concentration the monomeric surfactant concentration is equal to the CMC.
Constituent	A single chemical species that is part of a multi-constituent substance or a UVCB. It can be defined by its unique chemical identity.
Continental scale	Represents the size of the EU.
Cytogenicity	Capacity to cause structural or numerical damage to chromosomes
Cytotoxicity	Capacity to cause irreversible damage to cells.
Daphnia	A type of invertebrate commonly used in aquatic toxicity tests. Also sometimes referred to as a 'water flea'.
Default values	Information provided in the Guidance to allow the risk assessment to be carried out in the absence of specific data. There are a wide range of default values in the Guidance, including properties of the environment, degradation rates, emission factors, and dilution factors.
Dermal absorption	The dermal absorption of a pesticide is a measure of the amount in contact with the skin that enters the body and is systemically available. The extent to which pesticide active substances can penetrate the skin varies greatly based on chemical properties and formulation. A value for dermal penetration is included in calculations of operator exposure.
Dilution factor	The dilution of an effluent stream in the receiving water.
DMEL (Derived minimal effect level)	The predicted level of exposure at which it is considered that effect on human health will be minimal. This may be a quantitative or a semi-quantitative value. DMELs are derived for non-threshold effects, and for effects such as irritation and sensitisation where available data do not allow a DNEL to be set.
DNEL (Derived no-effect level)	The predicted level of exposure at which it is considered that there will be no effect on human health. It applies to substances with Threshold effects. It is derived from repeated dose toxicity data, by application of assessment factors.

Downstream User	Any natural or legal person established within the community, other than the manufacturer or the importer, who uses a substance, either on its own or in a preparation, in the course of his industrial or professional activities. A distributor or a consumer is not a downstream user.
EC_{50}	The concentration that causes adverse effects (but not necessarily mortality) in 50% of the exposed population. This is the toxicity measure normally used to express the results of short-term *Daphnia* and algal tests.
$E(L)C_{50}$	A shorthand term to describe both LC_{50} and EC_{50} together.
$E(L)L_{50}$	A short hand term to describe both LL_{50} and EL_{50} together
Effect	Used to refer to changes in growth rate, development, morphology, function, or lifespan of a test organism under experimental conditions.
EL_{50}	The 'loading rate' that causes adverse effects in 50% of the exposed population. This is the toxicity measure normally used to express the results of short-term *Daphnia* and algal tests carried out on poorly water-soluble test substances and complex mixtures.
Emission factor	Relates the emission or release of a substance to the amount of substance used in a given process.
Endocrine disrupter	A substance that adversely affects hormonal processes.
Environmental compartment	The parts of the environment considered in the risk assessment. The main ones considered are surface water, sediment, soil (also known as terrestrial), air, and biota (notionally fish, earthworms and food, for the purposes of modelling).
Environmental release categories	A REACH use descriptor code that indicates how exposure and releases to the environment can occur during an activity.
Equilibrium partitioning method	An approach that allows the toxicity to soil and sediment organisms to be estimated from toxicity data on surface water organisms. Expressed either as a means to determine $PNEC_{soil}$/sediment from $PNEC_{water}$, or to calculate concentration of the substance in the water surrounding the soil/sediment.
Essential	A substance required for normal growth of an organism.

EUSES (European Union System for the Evaluation of Substances)	A computer program that carries out many of the calculations in the Guidance automatically.
Existing substance	A substance which appears on EINECS.
Exposure scenario	The set of conditions, including operational conditions and risk management measures, that describe how a substance is manufactured or used during its life cycle and how the manufacturer or importer controls, or recommends downstream users to control, exposures of humans and the environment. These exposure scenarios may cover one specific process or use or several processes or uses as appropriate.
extended Safety Data Sheet	A safety data sheet that includes exposure scenarios for uses of a substance. [eSDS or extSDS]
Forward mutation	A change from the normal form of a gene to a mutant form.
Generic exposure scenario	An exposure scenario developed with assumptions and inferences on how exposure takes place and that aids the exposure assessor in evaluating, estimating or quantifying exposures for a product type or group of substances.
Genetic toxicity	Synonym for genotoxicity.
Genotoxicity	Refers to changes in the genetic apparatus (DNA and/or chromosomes) that can be passed on to offspring. The term is wider than mutagenicity, referring to changes in structure and segregation as well as information content of DNA.
Germ cell mutagenicity	The term used under REACH for evaluation of genotoxicity.
Good Laboratory Practice (GLP)	The organisational processes and the conditions under which studies are planned, performed, monitored, recorded and reported. GLP ensures that the way the work is done is adequately standardised and of a sufficiently high quality to produce reliable results which can with confidence be compared with those of others carrying out the same work and applying the same general principles.
Half-life	The time taken for 50% of the substance to be degraded or removed.
Hazard	The set of inherent properties of a substance, or mixture of substances, that make it capable of causing adverse effects to humans, other organisms or the environment.

Hazardous waste	A type of waste for which special requirements for disposal are set out in law.
Henry's law constant (H)	In its dimensionless form this is the air–water partition coefficient. It is commonly expressed in a form with the dimensions Pa m^3/mol, obtained from the ratio of vapour pressure to water solubility.
Human exposure via environment	The indirect exposure of man, via intake of contaminated food, to substances released to the environment.
Human health	The human health properties of a substance are those that indicate potential to cause undesirable effects on the health of workers and consumers exposed to the substance. Examples are acute toxicity, skin irritation or development of diseases such as cancer.
Hydrolysis	The action of water to break down a substance. It is almost always partial, giving rise to definite end products. It usually depends strongly on pH.
Impurity	A constituent unintentionally present in a substance. The origin of such a substance may be from the starting materials or an incomplete reaction.
In vitro	Refers to testing carried out on isolated cells or tissues (e.g. bacteria, mammalian hepatocytes and lymphocytes, bovine corneas).
In vivo	Refers to testing carried out on living animals.
Intermittent release	A situation where the process is such that emissions or releases occur only infrequently (e.g. less than once per month) and are of short duration (e.g. less than 24 hours).
IR spectrum	A spectrum showing which wavelengths of infra-red light are absorbed by a substance. Used in substance identification.
Isomer	Two compounds are isomers if they have the same chemical structure (sequence of atoms and bonds) but with different shapes, due to the arrangement of bonds around specific atoms.
LC_{50}	The concentration that is lethal to 50% of the exposed population. This is the toxicity measure normally used to express the results of short-term fish toxicity tests.
LD_{50}	The dose that is lethal to 50% of the exposed population. This is the endpoint measured in some mammalian toxicity studies.
Life cycle	The uses and applications of a substance from when it was manufactured through to disposal or destruction.

Limit of quantification	The limit of quantification is the lowest concentration of a contaminant that can be routinely identified and quantitatively measured with an acceptable degree of certainty by the method of analysis. It is also known as the Limit of Determination.
Lipophilic	Lipid- or fat-loving.
LL_{50}	The 'loading rate' that is lethal to 50% of the exposed population. This is the toxicity measure normally used to express the results of short-term fish toxicity tests carried out on poorly water-soluble test substances and complex mixtures.
Loading rate	The nominal concentration of a poorly water-soluble substance or complex mixture added to aquatic toxicity test medium. The term implies the substance or mixture is not fully dissolved in the test medium.
Local effects	Effects (irritation, corrosion, cytotoxicity) occurring at the site of first contact to a substance (skin, eyes, lungs or alimentary canal). Local effects range from minor irritation to irritation or corrosion leading to classification.
Local scale	The area in the immediate vicinity, both in terms of time and distance, of a source of release.
Log K_{ow}	The log10 value of the octanol–water partition coefficient. Also sometimes known as log P.
Long-term toxicity study	Long-term toxicity studies are usually considered those where more than one life cycle stage of an organism is tested. The duration of the exposure period in a long-term study is normally > 2 weeks. The end point derived from a long-term study is often a NOEC. The commonest examples of long-term studies are the fish early life stage (FELS) test and the *Daphnia* reproduction test. An algal growth inhibition test conducted over only 3–4 days is also considered a long-term study, since it assesses effects over multiple generations.
Lowest Observed Adverse Effect Level (LOAEL)	The lowest dose in a study at which an effect (or effects) is considered to be adverse occurs.
M Factor	A multiplying factor is used when calculating the aquatic toxicity of mixtures. It is used to give extra weight to the toxicity of substances very hazardous to the environment. The multiplying factor should be assigned (by manufacturer, importer or downstream user) for substances classified as aquatic environment Acute Category 1 or Chronic Category 1.

Mass spectrometry	Shows the range of mass:charge ratios characteristic of a substance. Used in substance identification.
Mesocosm	Model ecosystem.
Metabolite	The product formed by the metabolism of a compound
Mono-constituent substance	A well-defined substance in which one constituent is present at a concentration of at least 80% (w/w), and the impurities make up no more than 20% (w/w). Intentionally-added substances (other than stabilisers) are not considered in the mass balance.
Multi-constituent substance	A well-defined substance in which more than one constituent is present at between 10% and 80%
Mutagen	An agent that causes a permanent change in the amount or structure of the genetic material in an organism. Under the EC Dangerous Substances Directive, Mutagen Category 3 is the lowest hazard category, that is there is evidence of the agent causing genetic damage in mammals but the strength of evidence is not sufficient to place it in a higher category (category 2 or 1; category 1 being for a substance known to be mutagenic in humans).
Mutagenicity	This term refers to induction of permanent transmissible changes. When used in this way (as in the heading 'Germ cell mutagenicity' in the REACH Chemical Safety Report), it covers changes in single genes or gene segments, blocks of genes or chromosomes. It is also used more specifically for tests that detect either forward or reverse mutations (*in vitro* or *in vivo*).
NAEC (No adverse effect concentration)	The highest concentration of substance which at which no adverse effects have been observed in test organisms. Now referred to as NOAEC.
NAEL (No adverse effect level)	The highest dose level of substance which at which no adverse effects have been observed in test organisms. Referred to as NOAEL.
New substance	A substance placed on the market after 18 September 1981 that does not appear on EINECS. After a notification is accepted the substance is published in ELINCS.
NMR spectrum	A spectrum showing the nuclear magnetic resonance characteristics of a substance. Used in substance identification.

No observed adverse effect level	NOAEL is the highest exposure level in a toxicity study at which there are no statistically significant and/or biologically significant increases in the frequency of adverse effects between the group of animals exposed to the test substance and its respective control group.
NOAEC (No observed adverse effect concentration)	The highest concentration of substance which at which no adverse effects have been observed in test organisms. Effects which are not considered to be adverse may be observed at this concentration. Usually applies to local effects.
NOAEL (No observed adverse effect level)	The highest dose level of substance which at which no adverse effects have been observed in test organisms. Effects which are not considered to be adverse may be observed at this dose level. Usually applies to systemic effects.
NOEC	No observed effect concentration. This is defined as the highest concentration tested that caused no adverse effects on the test organisms compared to controls. This is usually the endpoint measured on long-term toxicity studies.
NOEL (No observed effect level)	The highest dose level of substance which at which no effects have been observed in test organisms, including effects which are not considered to be adverse.
Non phase-in substance	A new substance, one not covered by the definition of a phase in substance.
Non-threshold effect	Some effects do not have a threshold. These include some genotoxic
Oxidation	The action of an oxidising agent to change the chemical composition of a substance; typical oxidising agents in the environment are oxygen (in air or dissolved) or metals
Parent	This refers to an active substance molecule that has not been changed by metabolism.
Partition coefficient	The ratio between the concentration of a substance in one phase, related to the concentration in water. Examples include suspended sediment–water, sediment–water, soil–water, air–water (also known as Henry's constant), fish–water (better known as BCF), and octanol–water (better known as K_{ow}).
PEC	Predicted Environmental Concentration.

Personal Protective Equipment (PPE)	Any device or appliance designed to be worn or held by an individual for protection against one or more health and safety hazards.
Photodegradation	The action of light on a substance to break it down. It may be direct, or via a mediator (photosensitiser) which traps light energy and then transfers it to the substance.
Photo-oxidation	Action of light to generate oxidising agents in the air, such as the hydroxyl radical, which then can oxidise substances.
PNEC	Predicted No-Effect Concentration.
Preparation	A mixture or solution composed of two or more substances.
Probabilistic modelling	Probabilistic modelling is a technique that considers all the possible combinations of consumption and residue levels. It can be used to provide information on the probability of a particular intake occurring.
Process category	A REACH use descriptor code that defines an activity and the potential level exposure of workers from the activity.
QSAR (Quantitative Structure-Activity Relationship)	A mathematical relationship between properties of a chemical compound.
RAR (Risk Assessment Report)	An abbreviation frequently used in the ESR context.
RCR (Risk Characterisation Ratio)	Derived as the ratio PEC/PNEC for environmental endpoints.
Read-across	Read-across is a method of filling in data gaps for a substance by using surrogate data from another substance. Read-across can be between two substances or through a group or category of chemicals. The groups are selected on the assumption that the properties of a series of chemicals with common structural features will show similar trends in their physico-chemical properties and in their toxicological effects or environmental fate properties.
Realistic worst case (reasonable worst case)	A concept central to the risk assessment process. All assumptions made in the assessment should be both realistic and represent a reasonable worst case. It is a common misconception that PEC values are averages, but they should represent the highest value expected to occur in normal use.
Recycling	For example, reuse, recovery of materials or energy recovery.

Regional scale	Taken to represent a heavily industrialised area that receives above average emissions of a substance from all sources. It has a defined specification in size and composition.
Registrant	The manufacturer or the importer of a substance or the producer or importer of an article submitting a registration for a substance.
Regulation	Legislative act – in the EU it is legalisation that comes into force in all Member States without transposition; that is, it becomes law in all Member States at the same time and with the same wording, cf. Directive.
Relevant metabolite	A compound produced as a result of the metabolism of an active substance that is considered relevant for risk assessment purposes, based on both its toxicity and the levels found.
Restriction	The process within REACH of imposing limits on certain uses of certain substances. Restrictions are ultimately placed on Annex XVII of the REACH regulation.
Reverse mutation	Tests for reverse mutation use organisms that have undergone a mutation in a particular gene back to the normal form.
Risk	The possibility that a harmful event arising from exposure to a substance, or mixture of substances, may occur under specific conditions.
Safety Data Sheet	The safety data sheet provides a mechanism for transmitting appropriate safety information on classified substances and preparations, including information from the relevant chemical safety report down the supply chain to the immediate downstream users. The information provided in the safety data sheet shall be consistent with the information in the chemical safety report, where one is required.
Secondary poisoning	Exposure of an organism via the food chain.
Short-term toxicity study	Short-term studies normally only assess effects on one sensitive life stage of an organism. The duration of the exposure period typically ranges between < 1 and 4 days. The end point derived from a short-term study is often an LC_{50} or EC_{50}. The common short-term toxicity studies are the 96-hour fish toxicity test, the 48-hour *Daphnia* toxicity test. An EC_{50} obtained from a 72–96 hours growth inhibition test with a unicellular alga is also considered to be a short-term result.

SimpleBox	A multicompartment steady state model used to model the regional and continental environments.
SimpleTreat	A multicompartment steady state model used to model the fate and behaviour of a substance in biological waste-water treatment plant.
Speciation	The formation or presence of different inter-convertible forms of a substance in the environment, such as ionised forms, complexes of metals with organic molecules.
Steady state	A system that has reached the state where the inflows of a substance are balanced by the outflows or removal of the substance.
STP	Sewage Treatment Plant.
Substance	A chemical element and its compounds in the natural state or obtained by any manufacturing process, including any additive necessary to preserve its stability and any impurity deriving from the process used, but excluding any solvent which may be separated without affecting the stability of the substance or changing its composition.
Substrate	Material in which a chemical additive is present – for example plastic fabric, paper.
Systemic availability	Refers to how easily a substance reaches parts of the exposed animal or person that are distant from the site of contact (skin, eye, lung, gastrointestinal tract). The systemic availability is determined by the ADME properties of a substance.
Systemic effects	Changes resulting from exposure to a chemical other than those occurring at site of application (Local effects). A substance must be systemically available to produce systemic effects.
Teratogen	A substance which causes congenital abnormalities (deformities) in the baby or offspring in the womb.
The 10% rule	An assumption that 10% of the total EU production and use of a substance occurs at the regional scale.
Threshold effect	Changes resulting from exposure to a substance that are not observed below a certain dose.
Toxicokinetics	The study of the uptake and distribution of a substance subsequent to initial contact. See also ADME.
Trophic levels	The succession of steps through which matter and energy can be transferred through an ecosystem, usually via the food chain.
Use Category (UC)	A classification system used in EUSES related to the function of a substance in a particular use. Not applicable to REACH.

UV spectrum	A spectrum showing which wavelengths of ultra-violet and visible light are absorbed by a substance. Used in substance identification.
UVCB	A substance which is of unknown or variable composition, or a reaction mix, or of biological origin. The composition of such substances varies more than a multi-constituent substance. Such a substance does not have 'impurities' but may be considered as having 'major constituents' and 'minor constituents'. Under REACH it may be possible to register UVCBs as single substances, depending on the hazardous properties. UVCBs cannot be identified by chemical composition alone, rather they are identified by their origin/source and method of production.
WAF (Water accommodated fraction)	An ecotoxicity test medium generated by periods of stirring the substance with water, settling and separation of the aqueous layer (without filtration).
WWTP (Waste-Water Treatment Plant)	Also sometimes known as STP or Sewage Treatment Plant.
Zwitterion	A molecule of zero overall charge but containing a positive centre and a negative centre within it.

Index

Chemical Risk Assessment: A Manual for REACH, First Edition. Peter Fisk Associates Ltd.
© 2014 John Wiley & Sons, Ltd. Published 2014 by John Wiley & Sons, Ltd.